AF735928

Global Environmental Studies

Editor-in-Chief:
Ken-ichi Abe

Series Editors:
Daniel Niles
Hein Mallee

More information about this series at http://www.springer.com/series/10124

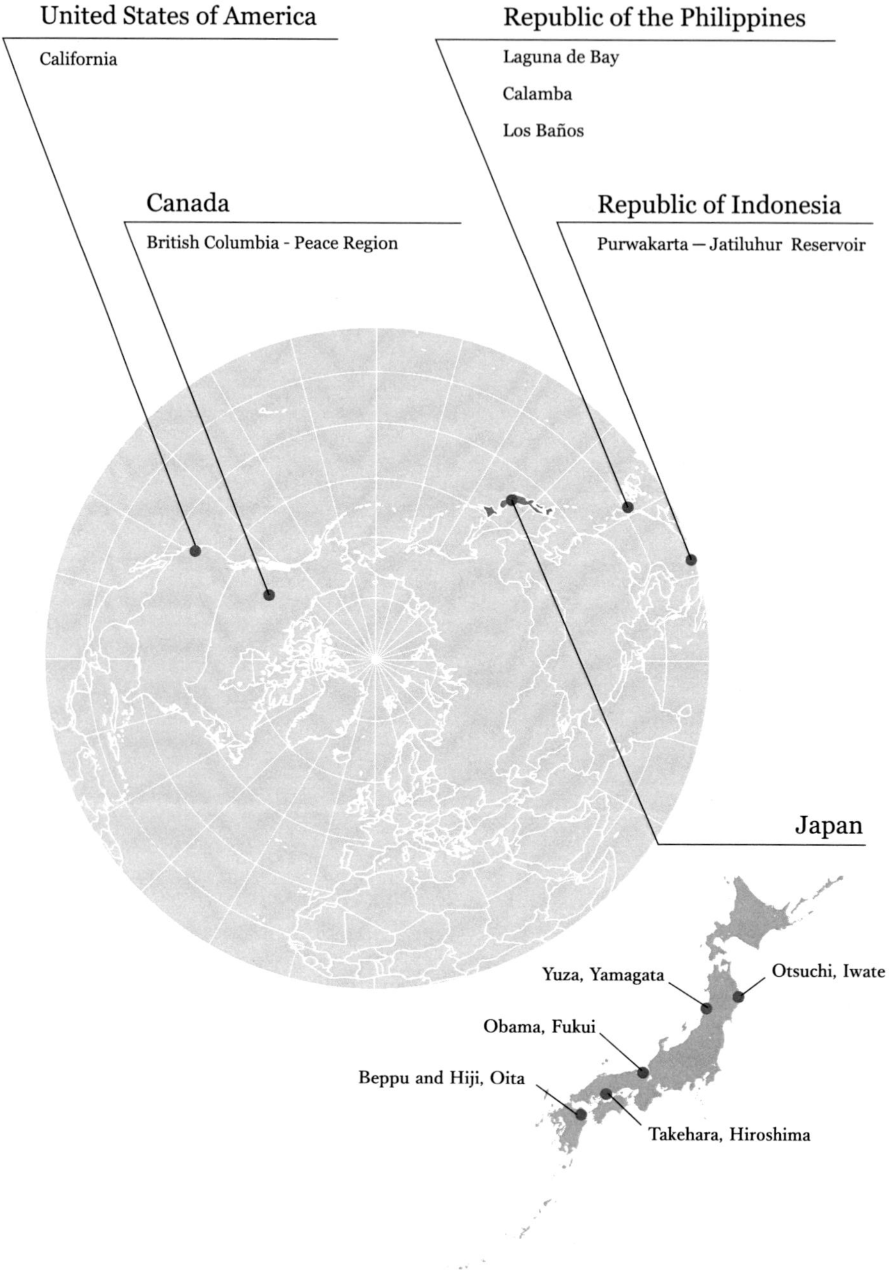
United States of America
California
Republic of the Philippines
Laguna de Bay
Calamba
Los Baños
Canada
British Columbia - Peace Region
Republic of Indonesia
Purwakarta — Jatiluhur Reservoir
Japan
Yuza, Yamagata
Otsuchi, Iwate
Obama, Fukui
Beppu and Hiji, Oita
Takehara, Hiroshima

Aiko Endo • Tomohiro Oh
Editors

The Water-Energy-Food Nexus

Human-Environmental Security in the Asia-Pacific Ring of Fire

Springer

Editors
Aiko Endo
Research Institute for Humanity and Nature
Kyoto, Japan

Tomohiro Oh
Research Institute for Humanity and Nature
Kyoto, Japan

ISSN 2192-6336 ISSN 2192-6344 (electronic)
Global Environmental Studies
ISBN 978-981-10-7382-3 ISBN 978-981-10-7383-0 (eBook)
https://doi.org/10.1007/978-981-10-7383-0

Library of Congress Control Number: 2018934841

© Springer Nature Singapore Pte Ltd. 2018
This work is subject to copyright. All rights are reserved by the Publisher, whether the whole or part of the material is concerned, specifically the rights of translation, reprinting, reuse of illustrations, recitation, broadcasting, reproduction on microfilms or in any other physical way, and transmission or information storage and retrieval, electronic adaptation, computer software, or by similar or dissimilar methodology now known or hereafter developed.
The use of general descriptive names, registered names, trademarks, service marks, etc. in this publication does not imply, even in the absence of a specific statement, that such names are exempt from the relevant protective laws and regulations and therefore free for general use.
The publisher, the authors and the editors are safe to assume that the advice and information in this book are believed to be true and accurate at the date of publication. Neither the publisher nor the authors or the editors give a warranty, express or implied, with respect to the material contained herein or for any errors or omissions that may have been made. The publisher remains neutral with regard to jurisdictional claims in published maps and institutional affiliations.

Printed on acid-free paper

This Springer imprint is published by the registered company Singapore Pte Ltd. part of Springer Nature.
The registered company address is: 152 Beach Road, #21-01/04 Gateway East, Singapore 189721, Singapore

Preface

This book introduced the results and achievements of the "Human-Environmental Security in Asia-Pacific Ring of Fire: Water-Energy Food Nexus" project, funded by the Research Institute for Humanity and Nature (RIHN) (No. 14200097). The book is divided into six parts including the introduction (Part 1), water for energy production (Part 2), water for coastal ecosystem conservation (Part 3), understanding socioeconomic dimension of local resource system (Part 4), governance and management of resource system (Part 5), and inter- and transdisciplinary for approaching nexus issues (Part 6).

We would like to give special thanks to Emeritus Professors Tomoya Akimichi, Masaru Tanaka, and Kazuo Matsushita and Professors Joji Morishita, Jumpei Kubota, and Toru Nakashizuka, who provided their deep insights, critical opinions, and useful comments to improve the project throughout. The support from Professor Makoto Taniguchi, who originally initiated the project and provided continued support for the project, is gratefully acknowledged. Local authorities, NGOs, private entities, government agencies and ministries, and international bodies in and outside of Japan who brought the knowledge and expertise for the practice and management of water, energy, and food are also greatly acknowledged. Finally, we are grateful for the contribution of all the members of the project, who provided their hard work, effort, and thoughtful deliberations to develop the conceptual frameworks, tools, and practices of water-energy food nexus.

We also especially thank our chapter authors for their contributions. Under a very tight schedule, they showed their patience until the end to publish the book. We would like to thank specifically those who generously donated their time to review and provide feedback and critical opinions to improve the quality of the book: Diana Allen, Kenshi Baba, Kimberly Burnett, Robert Delinom, Takahiro Endo, Masahiko Fujii, Jason J Gurdak, Hideki Hamamoto, Deny Hidayati, Karen Ann B. Jago-on, Takahiro Kato, Terukazu Kumazawa, Naoki Masuhara, Shinji Ohsawa, Jun Shoji, Fernando Siringan, Ryo Sugimoto, Osamu Tominaga, Joji Morishita, and

Christopher Wada. Mr. Shun Teramoto and Ms. Okamoto provided appreciated technical editing and administrative support and contributed to the overall efforts of the project. This book would not have been completed without the assistance of all the contributors. We thank them all.

Research Institute for Humanity and Nature
Kyoto, Japan
October 2017

Aiko Endo
Tomohiro Oh

Contents

Global Environmental Studies

The Global Environmental Studies series introduces the research undertaken at, or in association with, the Research Institute for Humanity and Nature (RIHN). Located in Kyoto, Japan, RIHN is a national institute conducting fixed-term, multidisciplinary, international research projects on pressing areas of environmental concern.

RIHN seeks to transcend the common divisions between the humanities and the social and natural sciences, and to develop synthetic and transformative descriptions of humanity in the midst of a dynamic, changeable nature. The works published in the series will reflect the full breadth of RIHN scholarship in this transdisciplinary field of global environmental studies.

Editorial Structure

Editor-in-Chief:
Ken-ichi Abe RIHN Area Studies Environmental Anthropology

Series Editors:
Daniel Niles RIHN Geography
Hein Mallee RIHN China Studies Natural Resource Management EcoHealth

Part I
Introduction

Chapter 1
Introduction: Human-Environmental Security in the Asia-Pacific Ring of Fire: Water-energy-food Nexus

Aiko Endo

Abstract The objectives of this project are to understand the complexity of the water-energy-food (WEF) nexus system and to create scenarios and policy options to reduce trade-offs among resources and to alleviate conflicts of resource users using scientific evidence and under assumptions of uncertainty to maximize human-environmental security. Five different interdisciplinary approaches, scales and clusters will be used in this investigation: (1) the science in/for society; (2) the water-energy nexus; (3) the water-food (e.g., fisheries resources) nexus; (4) the stakeholder analysis; and (5) the interdisciplinary study.

A primary challenge of this nexus study is to analyse the interlinkages between groundwater and fisheries production, regarding the hypothesis that the flow of nutrients from land to ocean affects the coastal ecosystem. This suggests that water use for producing and/or consuming food and/or energy on land might affect fisheries production in coastal areas. To examine this theory, we address two primary objectives; (1) to understand the complexity of the water-energy-food (WEF) nexus system since the relationships of all three resources are interrelated and interdependent; and (2) under scientific evidence and scientific uncertainty to create scenarios and policy options to solve the identified nexus problems, that is, to reduce the number of tradeoffs among three resources and to mitigate potential conflicts among these resource users through transdisciplinary approaches.

Keywords Water-energy-food (WEF) Nexus · Geothermal energy · Hydrothermal enrgy · Hot spring Energy · Hydropower generation · Shale gas · Submarine groundwater discharge (SGD) · Coastal ecosystem · Stakeholder analysis, social network · Scenario planning · Integrated methods · Interdisciplinary · Transdisciplinary

A. Endo (✉)
Research Institute for Humanity and Nature, Kyoto, Japan
e-mail: a.endo@chikyu.ac.jp

© Springer Nature Singapore Pte Ltd. 2018
A. Endo, T. Oh (eds.), *The Water-Energy-Food Nexus*, Global Environmental Studies, https://doi.org/10.1007/978-981-10-7383-0_1

1.1 Background of the Nexus Study

"Nexus" means "a connection or series of connections linking two or more things" or "a connected group or series" according to the Oxford Dictionary of English. This means that the water–energy–food nexus applies to several linkages, such as water–energy, water–food, or water–energy–food groups.

With regard to the historical review of nexus activities, diverse water–energy–food nexus studies and projects have been conducted, based on a variety of actors' interests on different scales around the world since nexus terminology appeared for the first time in 1983, when the United Nations University (UNU) launched a Food-Energy Nexus Programme to acknowledge the important interconnectedness between food and energy issues.

The concept of the water–energy–food nexus was launched in Bonn in 2011, when the German Federal Government organized an international conference titled "The Water Energy and Food Security Nexus: Solutions for the Green Economy" (hereafter the Bonn 2011 Conference) to contribute to the United Nations Conference on Sustainable Development (Rio + 20). The "nexus approach" was introduced in a background paper for the conference in order to show the direction of nexus studies for a sustainable society. This approach aims to increase the efficiency of resource utilization, to reduce trade-offs between resources, to synergize the utilization of different resources, and to improve governance across sectors. This was driven by increasing trade-offs and potential conflicts in the utilization of these resources, which have complex interactions and are interdependent, against the backdrop of climate and social change. The nexus approach focuses on system efficiency rather than on the productivity of isolated sectors (Hoff 2011).

The Global Risks Interconnections Map published by the World Economic Forum (WEF) in 2016 identified food crises, water crises, and energy price shocks as interconnected global risks (WEF 2016). In addition, global water demand for irrigation was the highest in 2000 and will be so again in 2050, as 70% of freshwater withdrawals are used for irrigation (UNESCO 2003). However, water demand for electricity is expected to rise by 2050 (UNESCO 2016). This may lead to serious water resource trade-offs for irrigation, domestic, manufacturing, and electricity in 2050 (OECD 2012).

In summary, the nexus approach emerged within the international community on the premise that (1) social and climate change puts pressure on water, energy, and food resources; (2) demands for water, energy, and food are estimated to increase by 40%, 50%, and 35%, respectively, by 2030 (USNIC 2012); and (3) the number of trade-offs and potential conflicts among these resources with complex interactions will increase. To address these issues, the nexus approach has the potential to enhance water, energy, and food security by increasing efficiency, reducing trade-offs, building synergy, and improving governance across sectors.

1.2 Research Institute for Humanity and the Nature Nexus Project on "Human-Environmental Security in Asia-Pacific Ring of Fire: Water-Energy-Food Nexus"

1.2.1 The Mission of the Research Institute for Humanity and Nature (RIHN)

The mission of the RIHN is to conduct solution-oriented research aimed at exploring how humanity and nature should interact. To achieve this mission, the RIHN has been conducting interdisciplinary research spanning the natural sciences, humanities, and social sciences, and has enhanced transdisciplinary research that involves collaboration with various stakeholders in society. To follow the institute's missions, the RIHN Nexus Project was launched in 2013 as a five-year project and was designed to use interdisciplinary and transdisciplinary approaches.

1.2.2 The Goals of the Project

The goals of this project are to understand the complexities of the WEF nexus system, as these resources have complex interactions, to create policy options to reduce trade-offs among resources, and to solve conflicts among resource users through the use of scientific evidence and uncertainty to maximize human-environmental security.

We target surface water and groundwater for small hydropower plants, geothermal sources, hot springs, and shale gas for the energy nexus. Conversely, we focus on energy for pumping and heating water as water energy. In addition, we address water for fisheries and agricultural production and for the water cycle, which is essential for the ecosystem (Fig. 1.1).

We also suggest that the water used for producing or consuming food or energy on land might affect fisheries production in coastal areas, according to the hypothesis that the flow of nutrients from land to ocean affects the coastal ecosystem. Therefore, we focused on submarine groundwater discharge (SGD). This refers to a trade-off in water resources between land and coastal areas (Fig. 1.2).

With regard to the structure of the project, we worked with approximately 60 researchers in different disciplines, primarily in the seven target research sites located in the following five countries: Indonesia, the Philippines, Canada, Japan, and the USA (Fig. 1.3). The project is designed based on five fundamental groups: (1) the water–energy nexus group, (2) the water–food nexus group, (3) the stakeholder analysis group, (4) the socio-cultural group, and (5) the interdisciplinary group.

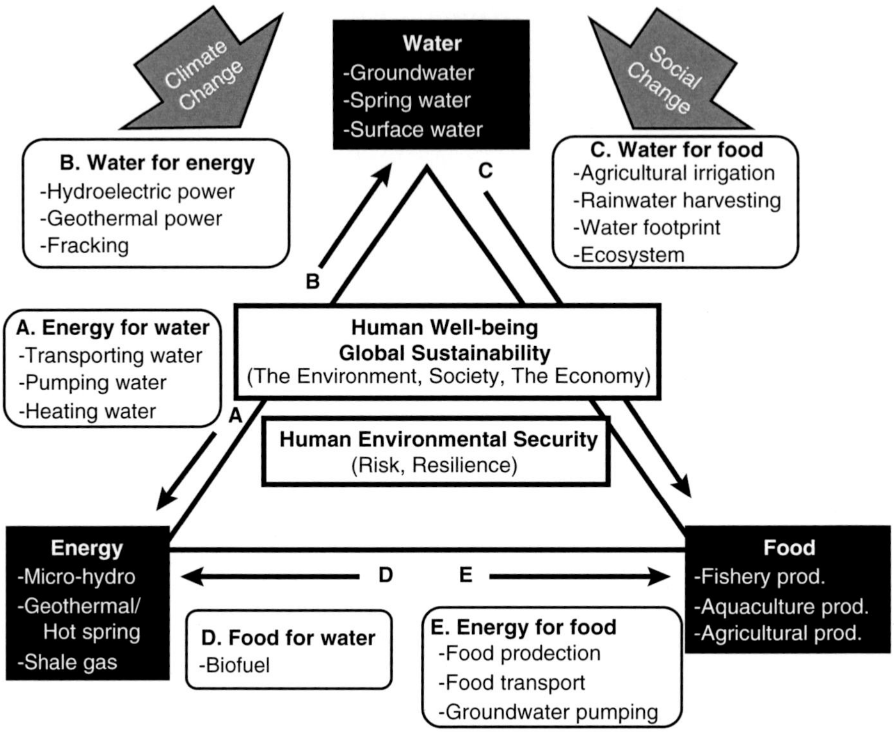

Fig. 1.1 Dynamics of the WEF nexus under the RIHN WEF Nexus project (Source: Authors modified from Endo et al. 2015a)

1.3 Roadmap of the Project

This five-year project was initiated in 2013. During the initial stage, we identified the trade-offs and conflicts among resources and resource users in each research site by working with local experts. Then, we addressed our two primary objectives: A) to understand the complexity of the WEF nexus system, and B) to create policy options to solve the identified nexus problems through the use of scientific evidence and uncertainty. For the first three years, we focused on objective A, and in the last two years, we shifted our focus more toward objective B.

1.3.1 The Water–energy Nexus

The water–energy nexus group in Japan consists of experts on hydrology, hydrogeology, hydrometeorology, geology, geomorphology, hot spring studies, geothermic studies, and limnology. To study the water–energy nexus, they (1) analyzed underground environmental systems, (2) analyzed effective potential energy production

Fig. 1.2 Interlinkages between land and coastal areas through SGD. *SDG* Submarine groundwater discharge

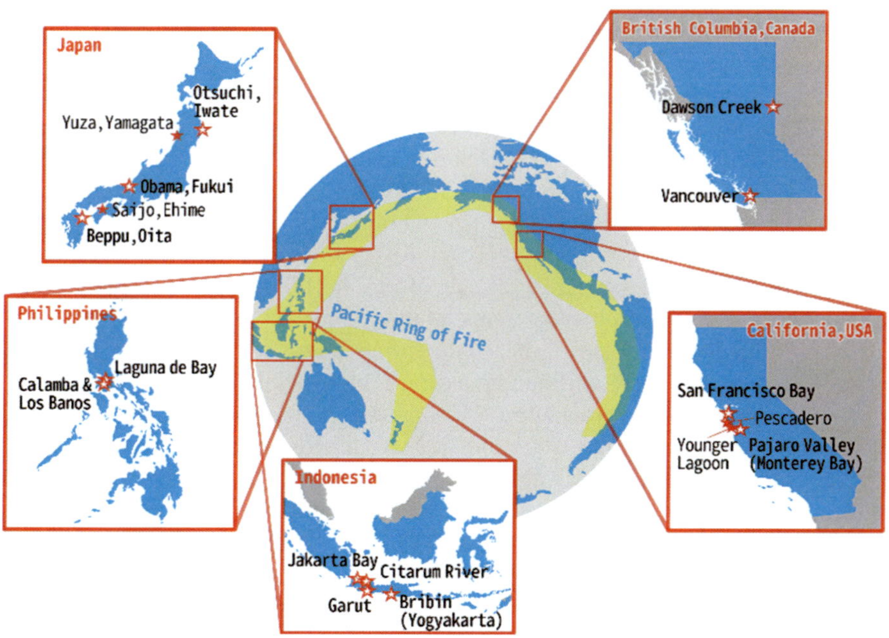

Fig. 1.3 Target research sites

using water from shale gas development, small hydropower generation, and hot spring drainage, (3) examined the changes in river and coastal ecosystems caused by changes in the heat environment, (4) diversified renewable energy sources such as small hydropower generation and hot spring drainage water, and (5) developed a full cascade of uses for hot spring water to produce food and energy.

We analyzed the underground geological structures using microtremor array measurements and explored gravity basement structures to understand groundwater storage and flow direction (Chap. 5). We then conducted a quantitative analysis on how much energy could be produced per kilogram of water from small hydropower plants in Beppu, shale gas production in Canada, and hot spring drainage water in Beppu. We found that shale gas uses water most effectively in the production of energy. However, we should also consider the social and environmental aspects of shale gas development. In Beppu, another finding reveals that changes in the heat environment caused by drainage water from hot spring resorts and hot spring power generation affect river ecosystems. Hot spring drainage creates a more suitable habitat for Nile Tilapia, a foreign species. If new power generation facilities increase the amount of hot spring drainage, then it is possible that other rivers will experience environmental conditions similar to those of the Hirata River (Chap. 10). Furthermore, we examined and found that the thermal energy of SGD affected the coastal environment. We also focused on the heat temperature of SGD. We analyzed the changes in seawater temperature caused by SGD and found that SGD heats seawater during the winter and cools it during the summer. This means that an increase in groundwater use on the land may affect the amount of heat supplied through SGD (Yamada et al. 2016, 2017a, b).

To diversify renewable energy sources, the potential of electricity generated by small hydropower plants was calculated in Otsuchi and Beppu (Fujii et al. 2017). The total was approximately 2000 megawatt-hours, which accounts for 4% of all electric power consumption in Otsuchi. We could also reduce approximately 2000 tons of CO2 emissions (Sawadate 2017) (Chap. 2). Regarding ground heat exchange systems, soil temperature readings in Obama and Otsuchi revealed that the soil temperature in Obama is higher than in Otsuchi. As previous studies on ground warming have shown, further research is needed on how to utilize the energy from ground heat in the application of heat pumps (Chap. 6).

In addition, we developed a full cascade of uses for hot spring water, incorporating the drainage water at different temperatures, from 100° to 35° centigrade, and quality profiles including steam, heat, hot spring water, and hot spring drainage water. The innovation here will be to maximize the benefits from hot spring water for energy and agriculture production. We addressed policies and regulations for the sustainable use of hot spring water and wastewater management, while working with local governments and stakeholders.

1.3.2 The Water–food Nexus

The water–food nexus group includes experts on marine biology, coastal oceanography, and fisheries sciences. They examined the interlinkages between SGD and fisheries resources. Specifically, a change in the SGD rate causes a change in the nutrient flux, which results in a change in primary production that ultimately leads to a change in fisheries resources. The group quantified SGD and found the presence of radon to be one of the potential indicators used to identify SGD, which indicates that spring water may be coming from the seabed, as mentioned in the hypothesis that environmental nutrient flows from land to the ocean may affect coastal ecosystems.

It was revealed that in Obama Bay, there is a positive correlation between the primary production of phytoplankton and radon concentrations as a groundwater tracer of SGD (Sugimoto et al. 2017). Regarding the relationship between SDG and nutrient fluxes, we found that nutrients supplied from SGD contribute significantly to primary production (Sugimoto et al. 2016). It became clear that SGD is the main source of dissolved inorganic phosphorus (DIP) in the coastal area of Hiji and in Obama Bay, and DIP via SGD could play an important role in biological production throughout the year. On the other hand, in Otsuchi Bay, the dissolved inorganic nitrogen (DIN) supply from SGD could make a substantial contribution to primary production during the stratified period (Chap. 8). As a result of addressing SGD and fisheries production, it was found that a greater number of fish were found near SGD (Hata et al. 2016; Utsunomiya et al. 2017) (Chap. 9). The interlinkages between groundwater and fisheries production were demonstrated by this project.

1.3.3 The Stakeholder Analysis Group

To address objective B, experts in public administration, environmental policy studies, consensus building, and social networking in the stakeholder analysis group (1) identified stakeholders and conducted conflict assessments of specific policy issues at different project sites based on individual interviews (Baba et al. 2015a, b), (2) visualized the social networking of stakeholders based on the results of the above-referenced assessments (Kimura et al. 2016) (Chap. 19), (3) held stakeholder meetings and/or workshops to share the results, (4) provided expert knowledge and identified changes in stakeholder perceptions and/or behaviors towards geothermal power and hot springs energy development focusing on the general public of Japan, the Philippines, and Indonesia, through an online survey and deliberation on a regional scale (Baba and Takatsu 2017), (5) established a system of governance for the coexistence of hot spring energy development and conservation, and (6) conducted scenario planning to integrate local and expert knowledge from a transdisciplinary approach (Baba et al. 2016).

More specifically, the experts drafted future scenarios using the Delphi method, held workshops to share and finalize the scenarios while working with relevant stakeholders, and contributed to formulating a local plan to bring the desired scenarios into focus in/for the society based on scientific evidence gathered through the project and scientific uncertainty in the local society (Chap. 22).

1.3.4 The Socio-cultural Group

The socio-cultural group includes experts on jurisprudence, resource studies, ecology, anthropology, and geography. This group (1) studied the management and history of groundwater use (Endo et al. 2015b) (Chap. 12), (2) assessed the ecological and cultural significance of groundwater resources (Mori 2016), (3) shared and developed local and scientific knowledge using a co-production approach (Oh and Tahara 2017), and (4) analyzed the interconnections and interdependencies among WEF resources from the perspective of the urbanization of Asia-Pacific countries.

At the Obama project site, located in Fukui prefecture, we found that changes in groundwater use included changes in users, purposes, and socio-cultural values of groundwater based on an onsite questionnaire survey. We also demonstrated that disparities in the use, interests, and values relating to groundwater in the area are expanding. In addition, we analyzed the changes in the use of deep wells stemming from the modernization and industrialization of a port city and clarified the current status of use of community wells. We also developed transdisciplinary tools, such as mapping, to share the interdisciplinary research results with local stakeholders and contributed to formulating a local plan for groundwater management in Obama.

In Otsuchi, which is a 2011 tsunami-affected area, we cooperated with local stakeholders to conserve the threespine stickleback inhabiting the spring environment (*Gasterosteus aculeatus*) and held a participatory event where stakeholders could research groundwater resources (Chap. 13).

1.3.5 The Interdisciplinary Study Group

The interdisciplinary group includes experts in environmental economics, fisheries economics, computer science, water engineering and modeling, and policy studies. The mission of the group was to determine and develop existing methods, and/or create new discipline-free methods, based on synthesizing and harmonizing team-based production, through the use of information collected from individual scientists in different disciplines from each team for an interdisciplinary research approach (Chaps. 14 and 21). In addition, we designed and visualized a water–energy–food nexus system to identify the interrelationships among WEF resources and to holistically understand the subsequent complexity of WEF nexus systems through an interdisciplinary approach. For more details on the interdisciplinary study, refer to Chap. 20.

1.4 The Water–energy-food System

Beppu in Japan

Figure 1.4 shows the recapitulated terrestrial, especially underground, marine and social systems that we addressed in Beppu. With the background of climate and social change, such as the commencement of hot spring energy development, pumping hot spring water could lead to changes in the underground environment. Changes in underground environment conditions could lead to changes in the coastal environment. On the other hand, the introduction of hot spring energy development leads to changes in the allocation of hot spring water. An increase in hot spring waste water from energy developers could then lead to changes in the river ecosystem.

The water–energy nexus group primarily addressed underground and river systems. The water–food nexus group covers the coastal system, and the interdisciplinary group is currently developing methods to link terrestrial and marine systems. The stakeholder analysis group is in charge of addressing social systems that include stakeholders, such as allocators, distributors, and hot spring inns. The socio-cultural group examined the cultural significance of nexus resources in hot spring resort areas. The goal of our project was to understand these links in these terrestrial and marine systems.

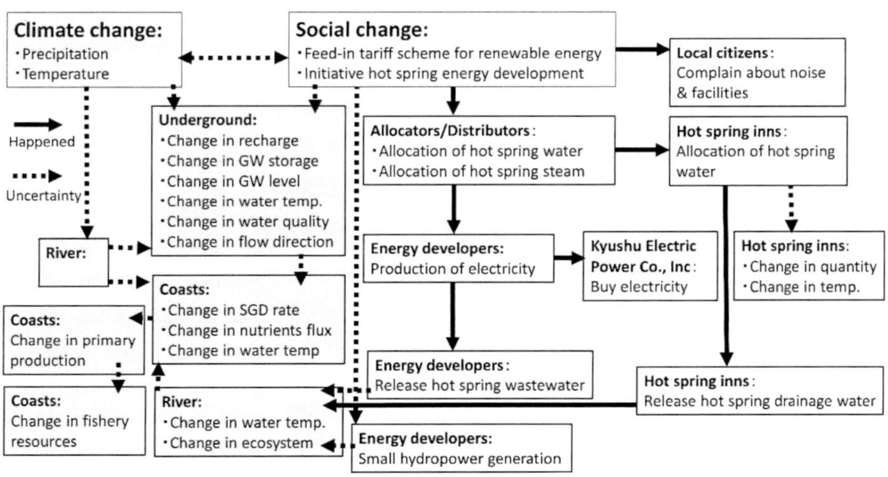

Fig. 1.4 Natural and social events and their linkages in Beppu, Japan. *GW* Groundwater; *SGD* Submarine groundwater discharge

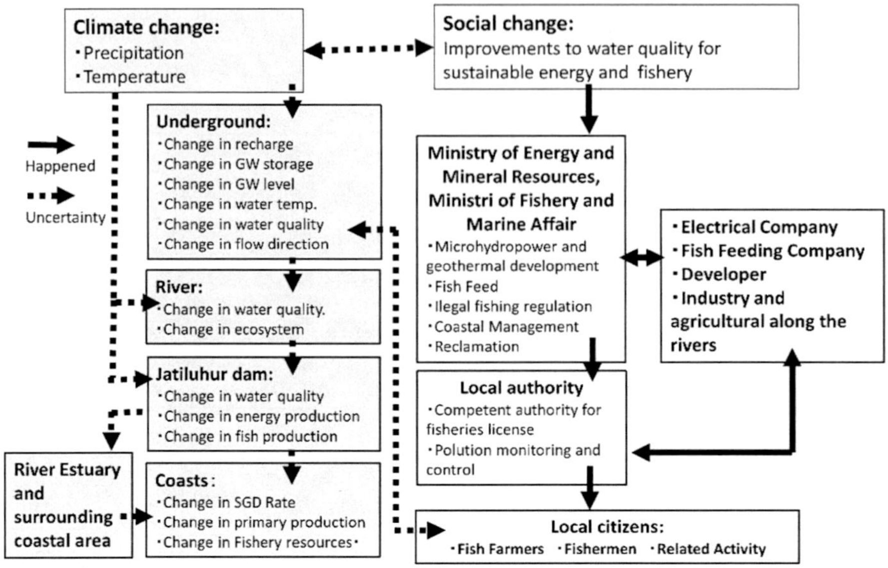

Fig. 1.5 Natural and social events and their linkages in the Citarum River Basin and Jakarta Bay, Indonesia. *GW* Groundwater; *SGD* Submarine groundwater discharge

The Citarum River Basin and Jakarta Bay in Indonesia

According to a recent study, global water quality is projected to deteriorate rapidly in the near future, which would increase risks to health and economic growth. These risks would markedly increase in low-income and lower middle-income countries that are experiencing population and economic growth (Veolia and IFPRI 2015).

The Jatiluhur Dam in Purwakarta, Indonesia, which is one of our project sites, hosts a hydropower plant along with aquaculture activities in the area that contribute to the deterioration of water quality (Fig. 1.5). As a result, the efficiency of energy production using contaminated water has also deteriorated, leading to a trade-off in water resources between energy and food production (Chaps. 16 and 17).

Northeastern British Columbia in Canada

In northeastern British Columbia (BC), surface water and groundwater resources are critical controlling factors for increased shale gas development in the region (Fig. 1.6). An increase in surface water use (and groundwater use) could lead to a change in the hydrologic cycle. This change in the hydrologic cycle (water availability) could then impact future development of shale gas. Use of water for agricultural purposes is currently low (rain-fed), but future growth in this industry could lead to increased water demands. Climate change is also a driver of change in the hydrologic cycle, particularly drought. The quality of surface water and groundwater may also be affected by waste generated by the shale gas industry directly through disposal, leaks, and spills but also indirectly if the surface water or groundwater levels become too low due to use. Shale gas developers are increasingly recycling fracking water to lower their overall use and increasing knowledge of water

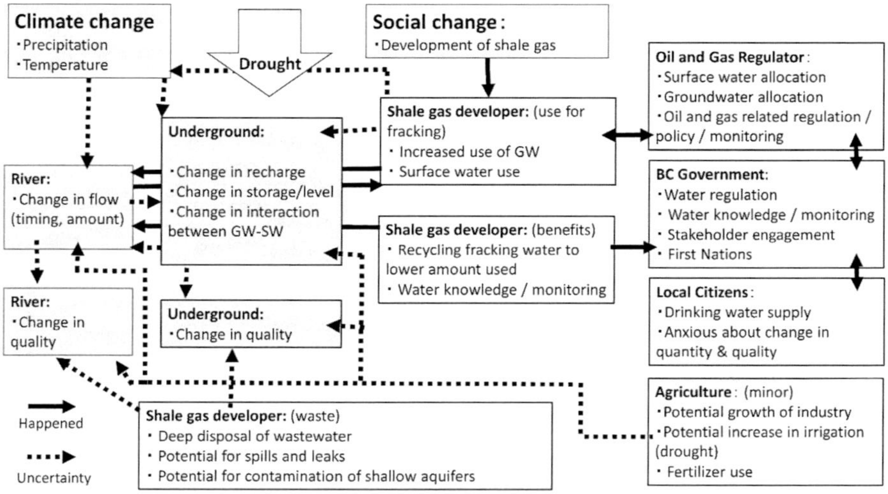

Fig. 1.6 Natural and social events and their linkages in northeastern British Columbia, Canada. *GW* Groundwater; *SW* Surface water; *BC* British Columbia

resources through targeted studies and monitoring. The regulator (The BC Oil and Gas Commission) and various BC ministries are responsible for regulations/policies, water knowledge, monitoring, stakeholder engagement, etc. (Chaps. 3 and 4).

Pajaro Valley in the US

The United States Geological Survey assessed 40 separate aquifers over 109 years from 1900 to 2008 to understand long-term cumulative depletion volumes in the US. Estimated groundwater depletion in the US from 1900 to 2008 was approximately 1000 cubic kilometers (km^3). Furthermore, the rate of groundwater depletion has increased significantly since around 1950, with maximum depletion occurring during the most recent period (2000–2008) when the depletion rate averaged almost 25 km^3 per year, compared to an average of 9.2 km^3 per year over the 1900–2008 timeframe (Konikow 2013).

With a background of groundwater depletion in Pajaro Valley in California, in the US, which is another one of our project sites, groundwater and recycled water from treated household wastewater in Watsonville, California were used for high-value agricultural production, such as in strawberry farming (Fig. 1.7). California is still facing serious water shortage due to the drought of 2012. This extended drought has led to a decrease in groundwater storage and salination, including seawater intrusion in coastal aquifers. Moreover, energy is used for pumping, wastewater treatment, and allocating recycled water. Thus, we can say there is a groundwater resource trade-off between food production and the underground environment (Chaps. 11 and 12).

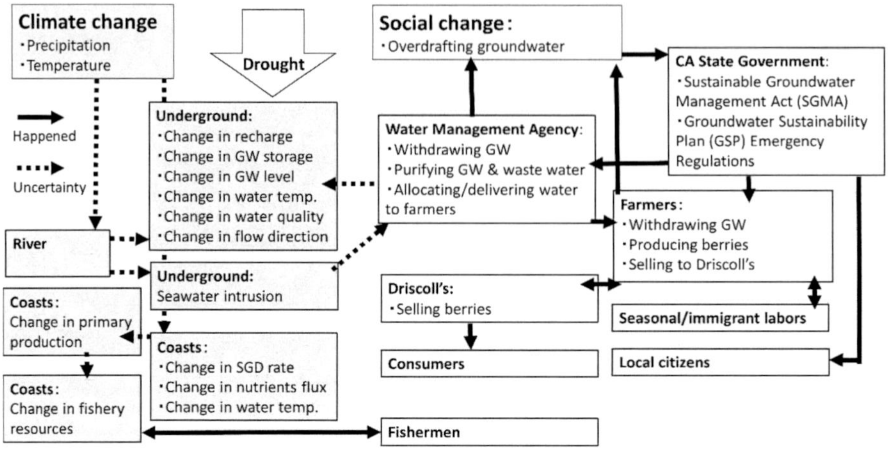

Fig. 1.7 Natural and social events and their linkages in Pajaro Valley, US *CA* California; *GW* Groundwater; *SGD* Submarine groundwater discharge

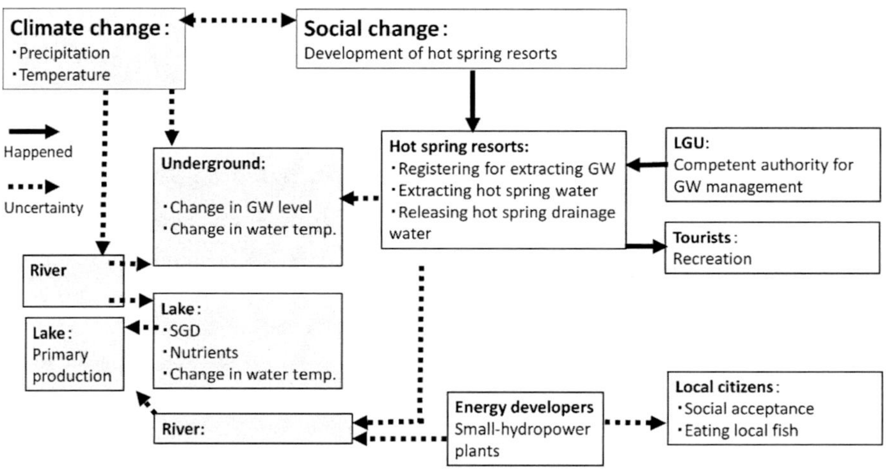

Fig. 1.8 Natural and social events and their linkages in Calamba and Los Banos, the Philippines. *GW* Groundwater; *SGD* Submarine groundwater discharge; *LGU* Local government unit

Calamba and Los Banos in the Philippines

Figure 1.8 presents the natural and social systems addressed by the project in Calamba, where hot spring resorts pump hot spring resources using energy for recreation and release untreated hot spring drainage water into the lake known as Laguna de Bay, where fisheries activities are conducted. Therefore, there is a water resource trade-off between food production from the lake and recreation activities on the land (Chaps. 7 and 18).

1.5 Towards a Transdisciplinary Research Approach

Finally, I would like to introduce our co-production activities from interdisciplinary and transdisciplinary perspectives based on Van der Hel's categorization (2016). He categorized co-production into (1) accountability, which is to enhance scientific accountability to society, (2) impact, which is to ensure the implementation of scientific knowledge in society, and (3) humility, which is to include the knowledge, perspectives, and experiences of extra-scientific actors in scientific knowledge production.

Regarding accountability, we co-organized open, local seminars and events with local governments and publicized our achievements to society through publications. As for impact, we worked with local governments to formulate local plans on sustainable groundwater use and recovery. With regard to humility, we conducted scenario planning with local stakeholders, monitored the temperature of hot spring resources with local people, collaborated with private companies to develop an integrated model, publicized groundwater information, data, and knowledge on the web, as provided by the local people and researchers, and developed a co-playing game to enhance nexus thinking.

Figure 1.9 shows the degree of integration and stakeholder involvement in our project activities. The horizontal axis shows the degree of integration and the vertical axis shows the degree of academic and nonacademic participation. We conducted a Cost-benefit analysis (CBA), identified the linkages between land and ocean, created a map, monitored hot spring resources, developed an integrated physical model and integrated index, and designed and visualized a nexus system for scenario planning. All of the activities are interconnected with and are interdependent on interdisciplinary and transdisciplinary approaches.

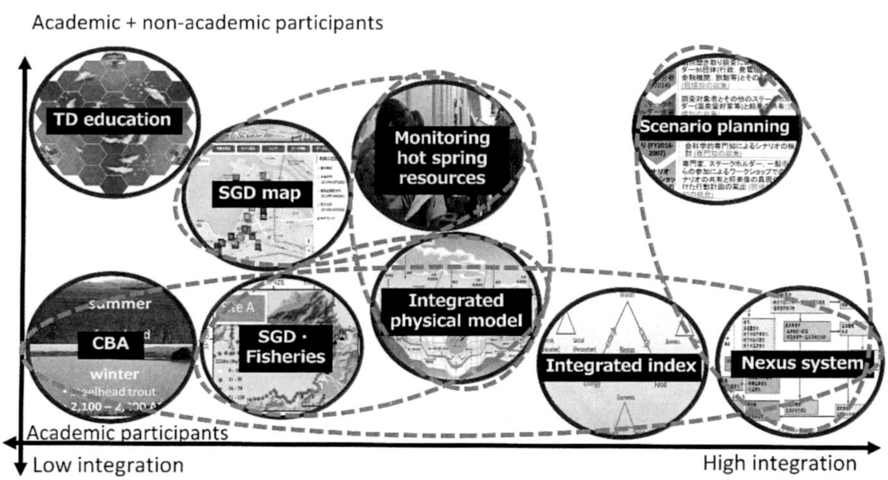

Fig. 1.9 Integration and stakeholder involvement in Beppu, Japan. *CBA* Cost-benefit analysis; *SGD* Submarine groundwater discharge

The academic definition of transdisciplinary is still being developed. The purpose of co-production from the point of humility is to be humble and reflexive about the role of science in society (Van der Hel 2016). Researchers must cooperate with stakeholders to solve global environmental problems at the initial stage when they recognized.

References

Baba K, Takatsu H (2017) Examining the role of facilitator in online deliberation experiments; implications from the examples of social decision making process on use of renewable energy resource. In: Kazuyo Murata (eds) Considering Hanashiai (dialogue) by citizen participation. Hitsuji Shobo, pp 177–198. (in Japanese)

Baba K, Matsuura M, Taniguchi M (2015a) Effectiveness and challenges of stakeholder analysis in co-design and co-product of science and society: implications from a case study of identifying potential issues on utilization of groundwater resource in Obama. Environmental. Science 28(4):304–315. (in Japanese)

Baba K, Takatsu H, Kito M, Kawai Y, Noritake T, Masuhara N, Tanaka M (2015b) Examining harmonious coexistence of geothermal resource between small power generation and hot spring utilization by stakeholder analysis: a case study of Beppu. Environmental. Science 28(4):316–329. (in Japanese)

Baba K, Doi M, Tanaka M (2016) Development of narrative scenario for social implementation of climate change adaptation policy: an integration of community based (bottom-up) approach and Delphi method for experts (top-down) approach in the agricultural sector in Japan. Global Environmental Research 21(2):11–128. (in Japanese)

Endo A, Burnett K, Orencio P, Kumazawa T, Wada C, Ishii A, Tsurita I, Taniguchi M (2015a) Methods of the water-energy-food nexus. Water 7(10):5806–5830. https://doi.org/10.3390/w7105806

Endo T, Ishida K, Yoshitani J (2015b) Karifuoruniashu no chikasuikanri no douko- 2014 nen jizokutekichikasuikanriho wo chusin ni (the trend of groundwater management in California: focusing on the sustainable groundwater management act of 2014). Kasen 71(4):24–28. (in Japanese)

Fujii M, Tanabe S, Yamada M, Mishima T, Sawadate T, Ohsawa S (2017) Assessment of the potential for developing mini/micro hydropower: a case study in Beppu City, Japan. Journal of Hydrology: Regional Studies 11:107–116

Hata M, Sugimoto R, Hori M, Tomiyama T, Shoji J (2016) Occurrence, distribution, and prey items of juvenile marbled sole Pseudopleuronectes Yokohamae around a submarine groundwater seepage on a tidal flat in southwestern Japan. J Sea Res 111:47–53

Hoff H (2011) Understanding the Nexus. Background paper for the Bonn 2011 Conference: the water, energy and food security Nexus. Stockholm environment institute, Stockholm

Kimura M, Baba K and Masuhara N (2016) Visualization of social network that focuses on the common understanding of groundwater resources stakeholders in Obama City, Fukui prefecture, proceedings of 44th annual meeting of environmental systems research 2015, pp 87–92. (in Japanese)

Konikow LF (2013) Groundwater depletion in the United States (1900–2008): U.S. Geological Survey Scientific Investigations Report 2013–5079, p 63.. Retrieved from http://pubs.usgs.gov/sir/2013/5079

Mori S (2016) Kyodozai yusuigyo itoyo kara miru fukkomachizukuri no katachi (evaluation of recovery and town development from the view point of local goods, freshwater three-spined sticklebacks). In: Taniguchi M (ed) Ground design for the future from Otsuchi. Kyoto, Showado, pp 71–95. (in Japanese)

Oh T, Tahara D (2017) Conservation of groundwater for local food in Azumino and Obama. In: Shoji J, Sugimoto R, Tominaga O (eds) Land-ocean interactions through groundwater/submarine groundwater and human society. Tokyo, Koseisyakoseikaku, pp 102–114. (in Japanese)
Organisation for Economic Co-operation and Development (2012) OECD environmental outlook to 2050: the consequences of inaction. OECD Publishing, Paris. Retrieved from https://doi.org/10.1787/9789264122246-en
Sawadate T (2017) Assessment of small hydropower generation potential considering characteristics in the Otsuchi town, Iwate prefecture. (Master's Thesis). Graduate School of Environmental Science, Hokkaido University, p 33. (in Japanese)
Sugimoto R, Honda H, Kobayashi S, Takao Y, Tahara D, Tominaga O, Taniguchi M (2016) Seasonal changes in submarine groundwater discharge and associated nutrient transport into a tideless semi-enclosed embayment (Obama Bay, Japan). Estuar Coasts 39(1):13–26. https://doi.org/10.1007/s12237-015-9986-7
Sugimoto R, Kitagawa K, Nishi S, Honda H, Yamada M, Kobayashi S, Tominaga O (2017) Phytoplankton primary productivity around submarine groundwater discharge in nearshore coasts. Mar Ecol Prog Ser 563:25–33. https://doi.org/10.3354/meps11980
United Nations Educational, Scientific, and Cultural Organization (2003) United Nations World Water Development Report 2003: water for people, water for life. UNESCO, Paris
United States National Intelligence Council (2012) Global trends 2030: alternative worlds. US NIC, Washington, DC, p 137
United Nations Educational, Scientific, and Cultural Organization (2016) United Nations World Water Development Report 2016: water and jobs. UNESCO, Paris
Utsunomiya T, Hata M, Sugimoto R, Honda H, Kobayashi S, Miyata Y et al (2017) Higher species richness and abundance of fish and benthic invertebrates around submarine groundwater discharge in Obama Bay, Japan. J Hydrol 11:139–146
Van der Hel S (2016) New science for global sustainability? The institutionalization of knowledge co-production in future earth. Environ Sci Policy 61:165–175
Veolia and Water North American & IFPRI (International Food Policy Research Institute) (2015) The murky future of global water quality. A white paper. By Veolia & IFPRI.. Retrieved from https://www.veolianorthamerica.com/sites/g/files/dvc596/f/assets/documents/2015/04/IFPRI_Veolia_H2OQual_WP.pdf
World Economic Forum (2016) The global risks interconnections map. The Global Risks Report 2016
Yamada M, Shoji J, Teramoto S, Ohsawa S, Mishima T, Sugimoto R, Honda H, Taniguchi M (2016) Exploration of submarine groundwater discharge using a drone in a coastal area of Hiji town, Oita prefecture, Japan in summer. Journal of Japanese Association of Hydrological Sciences 46(1):29–38
Yamada M, Ohsawa S, Shoji J (2017a) Use of hot spring water and river ecosystem. In: Shoji J (ed) Land-Ocean interactions through groundwater/submarine groundwater: implication for human society. Japan, Kouseisha-kouseikaku, pp 115–126
Yamada M, Shoji J, Ohsawa S, Mishima T, Hata M, Honda H, Fujii M, Taniguchi M (2017b) Hot spring drainage impact on fish communities around temperate estuaries in southwestern Japan. Journal of Hydrology: Regional Studies 11:69–83

Part II
Water-Energy Nexus (Water for Energy Production)

Chapter 2
Assessment of Potential Small hydropower Generation: A Case Study in Otsuchi, Iwate Prefecture, Japan

Masahiko Fujii and Takahiro Sawadate

Abstract Developing renewable energy is crucial not only to reduce greenhouse gas emissions, but also to vitalize local economies by generating new jobs related to their operation and management. Small hydropower (SHP), a baseload renewable energy, has substantial potential in Japan due to its high rainfall and steep geography. However, the SHP sector in Japan is underdeveloped. In this study, the potential power generation by SHP was estimated for the town of Otsuchi, Iwate Prefecture, Japan. The town is famous for its riverine ecosystem, which includes important salmon fisheries and the endangered fish species, three-spined stickleback. Therefore, any future SHP must be installed with careful considerations not to damage the local riverine ecosystem. The total annual potential power generation by SHP in Otsuchi was estimated to be around 2 GWh under the assumption of balancing energy–food and energy–ecosystem nexi to minimize influences on the riverine ecosystem. The potential power generation could meet approximately 4% of the current demand for the entire town and reduce carbon dioxide emissions equivalent to those of around 200 people. The potential power generation is estimated to double by the 2040s because of the greater expected precipitation and changes in land-use patterns in the future; however, such changes are expected to differ among sites in the town. Therefore, attention is necessary when selecting installation locations for new SHP facilities based on estimated potential power generation and anticipated social changes, such as population.

Keywords Small hydropower (SHP) · CO_2 emissions reduction · Climate change · Land-use pattern · Population · Riverine ecosystem

M. Fujii (✉) · T. Sawadate
Graduate School of Environmental Science, Hokkaido University, Sapporo, Japan
e-mail: mfujii@ees.hokudai.ac.jp

© Springer Nature Singapore Pte Ltd. 2018
A. Endo, T. Oh (eds.), *The Water-Energy-Food Nexus*, Global Environmental Studies, https://doi.org/10.1007/978-981-10-7383-0_2

2.1 Introduction

Japan's energy supply has relied on fossil fuels, especially coal and oil, since the 1960s. After the oil crisis of the 1970s, followed by the emergence of global warming awareness after the 1990s, Japan's energy policy gradually shifted towards nuclear power. However, the Great East Japan Earthquake on March 11, 2011, and the ensuing Fukushima Daiichi Nuclear Power Station accident, reduced the country's dependence on nuclear power. Although the dependence on the fossil fuels such as natural gas, coal, and oil has since increased, installation of renewable energy has accelerated recently, especially since a feed-in tariff (FIT), an incentive to develop renewable energy, was introduced in July 2012.

However, the implementation speed of renewable energy is still not satisfactory for several reasons. First, current incentives to introduce renewable energy have predominantly focused on solar power, followed by wind power, both of which depend on short-term weather and do not yield stable energy generation.

To shift from large-scale baseload energy sources such as fossil fuels and nuclear power, stable renewable energy should be introduced. Stable baseload renewable energy sources include geothermal, biomass, and hydropower. Among these, small hydropower (SHP), with a capacity of less than 1000 kW (e.g., Tanabe 2015; Fujii et al. 2017), has a long history in Japan beginning before World War II. This is because Japan has geographic advantages for SHP operations due to its steep geography and high rainfall amounts.

The development of SHP can contribute not only to reducing carbon dioxide (CO_2) emissions, but also to vitalizing local economies by generating new jobs related to the installation and maintenance of SHP facilities. However, the potential power generation depends substantially on location. Therefore, it is important to assess the potential power generation by SHP as well as the possible impacts on the environment from natural and social perspectives.

As a model case of an assessment of potential power generation, this study targeted the town of Otsuchi, Iwate Prefecture, Japan (Fig. 2.1), which has a population of around 12,000 people (Otsuchi Town 2017). The main industry is fisheries, with salmon being the most important target species. The town is also famous as a habitat for three-spined stickleback (*Gasterosteus aculeatus*), an endangered fish species (e.g., Kitano and Mori 2016; Sawadate 2017). The town was heavily damaged by the tsunami caused by the Great East Japan Earthquake on March 11, 2011, and reconstruction following the disaster and comprehensive ground design of the town are urgently required. Designing an energy policy for this town is important, and SHP is considered promising for Otsuchi because of its steep geography. This study aims to contribute to this energy policy, especially to support the development of renewable energy such as SHP.

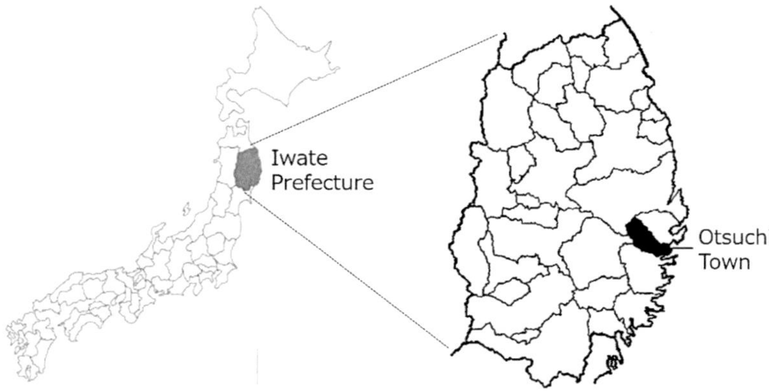

Fig. 2.1 Location of Otsuchi, Iwate Prefecture, Japan

2.2 Methodology

In this study, sites for SHP power generation were identified by assessing their suitability based on several parameters. First, sites had to offer sufficient potential power generation, with a natural head of more than 10 m (Fig. 2.2A). Then, sites that did not meet the following three conditions were excluded: (1) located downstream of salmon hatcheries and habitat, and masu salmon (*Oncorhynchus masou*) habitat (Otsuchi, per. Comm.) to avoid damage to salmon fisheries and conservation efforts; (2) located in three-spined stickleback habitats to protect this endangered species (Kitano and Mori 2016; Sawadate 2017); and (3) located in forests or far from roads, which could impede installation, regular operation, and maintenance (e.g., Sawadate 2017).

The potential power generation by SHP in Otsuchi was estimated based on our previous studies (Tanabe 2015; Fujii et al. 2017; Sawadate 2017). The potential power generation by SHP (P, kWh) can be calculated as:

$$P = g \times H \times Q \times E \times t, \tag{2.1}$$

where g is the gravitational constant (9.8 m s^{-2}), H is the natural head (m), Q is the river discharge (m^3 s^{-1}), E is the efficiency of SHP, for which the typical value of 0.7 (e.g., Fujii et al. 2017) was used in this study, and t is the operation time (h).

The natural head (H) was estimated using a 10-m-meshed grid digital elevation model (DEM) from the Geospatial Information Authority of Japan (2017).

The river discharge (Q) was calculated as the product of precipitation (R, mm km^{-2}), area of the drainage basin (A, km^2), and the discharge coefficient (L):

$$Q = R \times A \times L. \tag{2.2}$$

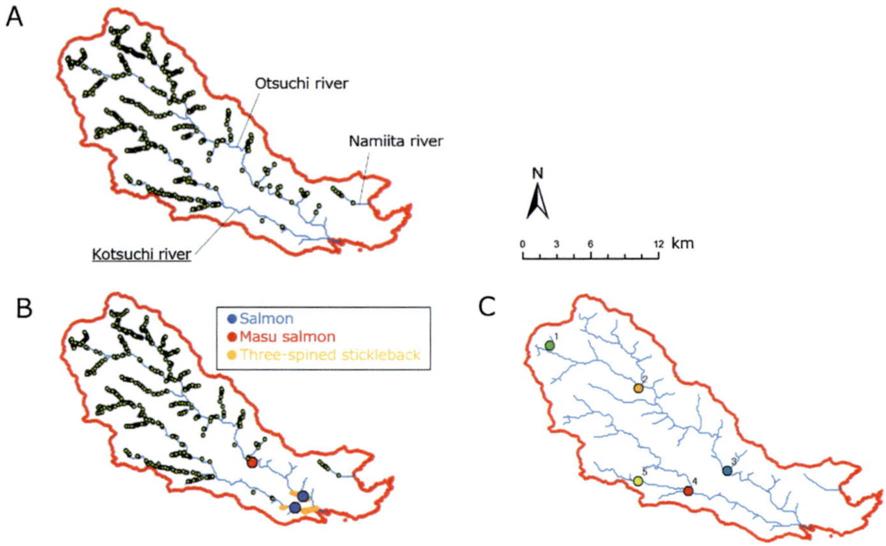

Fig. 2.2 Potential SHP sites in Otsuchi. (A) Potential sites originally evaluated as suitable for power generation by SHP with regard to the natural head (> 10 m) along the Otsuchi, Kotsuchi, and Namiita rivers; (B) potential sites after excluding those unsuitable in terms of riverine ecosystem conservation, i.e., downstream of salmon hatcheries or habitat (blue), the upstream limit of masu salmon habitats (red), and three-spine stickleback habitats (orange); and (C) potential sites after excluding forest areas and sites far from roads where it would be difficult to install and manage SHP facilities. The numbers represent the names of the suitable sites referred to in this study

Table 2.1 Discharge coefficient for each land-use type (Source: Ministry of Land, Infrastructure, Transport and Tourism (MLIT) 2004)

Land-use pattern	Discharge coefficient
Paddy field	0.2
Other agricultural land	0.2
Forest	0.3
Wasteland	0.5
Land for building	0.9
Road	0.9
Railway	0.9
Other land	0.8
Rivers and lakes	1.0
Beach	1.0
Sea waters	1.0
Golf course	0.5

The fixed discharge coefficient (L) was determined according to the land-use patterns (Table 2.1; The Ministry of Land, Infrastructure, Transport and Tourism (MLIT), 2004).

Considering that the typical payback and lifetime for SHP facilities are both on the order of tens of years (e.g., Sawadate 2017), we estimated both current and future potential power generation by SHP by considering possible climate and social changes, such as precipitation, land-use patterns, and local population.

The current precipitation was estimated using 1-km-meshed grid monthly climatology data (the National Land Information Division, National Spatial Planning and Regional Policy Bureau, MLIT 2012). For the future precipitation estimation, projected precipitation data for the 2040s from the Japan Meteorological Agency's Global Warming Projection Vol. 6 (2005) were used.

Current land-use patterns were obtained from 100-m-meshed grid data from the National Land Numerical Information (the National Land Information Division, National Spatial Planning and Regional Policy Bureau, MLIT 2006). Future land-use patterns were estimated with a logistic regression analysis under the assumption that recent changes in land-use patterns, especially transitions from forest and wastelands in the early 1990s into other land-use types in the late 2000s, would continue in the future (e.g., Sawadate 2017).

Population is an important factor for identifying suitable sites for SHP with regard to maintenance by local residents. Current and future population information was obtained from Otsuchi Town (2017) and the National Land Information Division, National Spatial Planning and Regional Policy Bureau, MLIT (2014), respectively.

2.3 Results

2.3.1 *Current Potential Power Generation by Small Hydropower*

Initially, 462 locations were evaluated as potential sites suitable for power generation by SHP with regard to their natural head (> 10 m) (Fig. 2.2B). However, after considering the other parameters, the number of potential sites decreased dramatically (Fig. 2.2C, D). Therefore, only five sites met all conditions; Sites 1–3 along the Otsuchi River, and Sites 4–5 along the Kotsuchi River. There are notable seasonal fluctuations in the potential power generation among these sites, generally consistent with precipitation, although spatial differences are prominent among the sites (Fig. 2.3). The annual potential power generation by SHP at Sites 1–5 is estimated to be 122, 822, 527, 502, and 199 MWh, respectively.

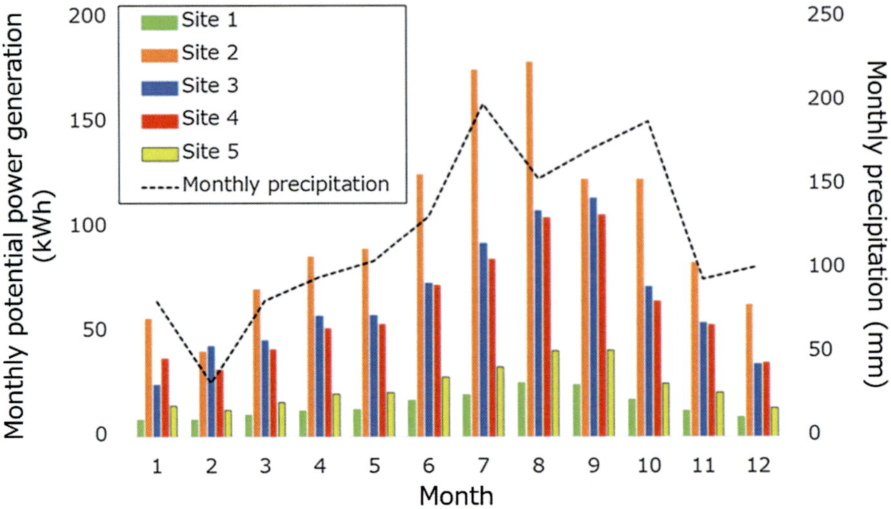

Fig. 2.3 Average monthly precipitation (mm; data from the Japan Meteorological Agency) and estimated monthly potential generated power by SHP (kWh) at each site in Otsuchi

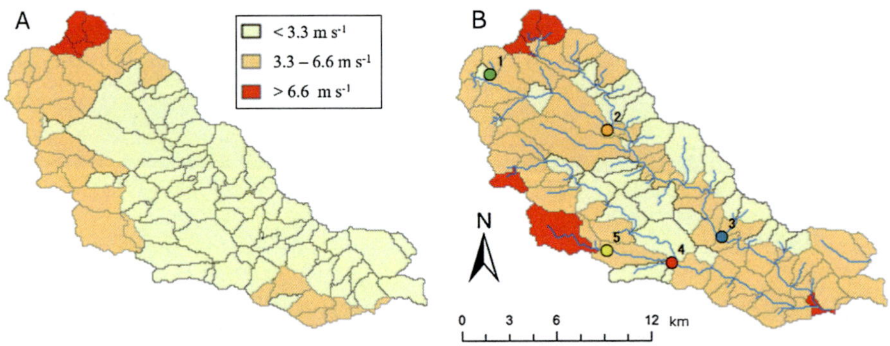

Fig. 2.4 Estimated river discharge ($m^3\ s^{-1}$) in Otsuchi (**A**) at present and (**B**) in the 2040s. The locations of the five suitable sites for SHP (Sites 1–5) are also shown in (**B**)

2.3.2 Future Potential Power Generation by Small Hydropower

River discharge is projected to increase by the 2040s (Fig. 2.4) due to increased precipitation (Fig. 2.5) and higher discharge coefficients caused by changes in land-use patterns, especially due to changes from forest (0.3) to other land (0.8) in the upstream areas of the Otsuchi and Kotsuchi rivers (Table 2.1, Figs. 2.2A and 2.6). Accordingly, the average annual potential power generation in the 2040s is projected to be 1.1, 1.1, 0.9, 1.0, and 0.4 GWh at Sites 1–5, respectively (Fig. 2.5). The degree of increase differs among the sites. Compared with the current potential, the future potential is projected to increase by 1.4–2 times at Sites 2–5, and by 8.6 times at Site 1.

The population is projected to dramatically decrease by 2050, although the degree of decrease differed greatly among regions (Fig. 2.7). The population is

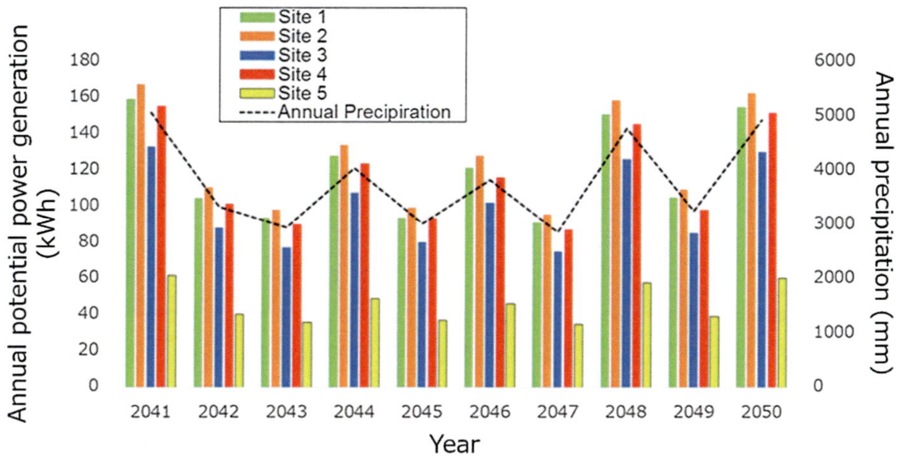

Fig. 2.5 Projected annual precipitation (mm) and estimated annual potential generated power by SHP (kWh) in Otsuchi

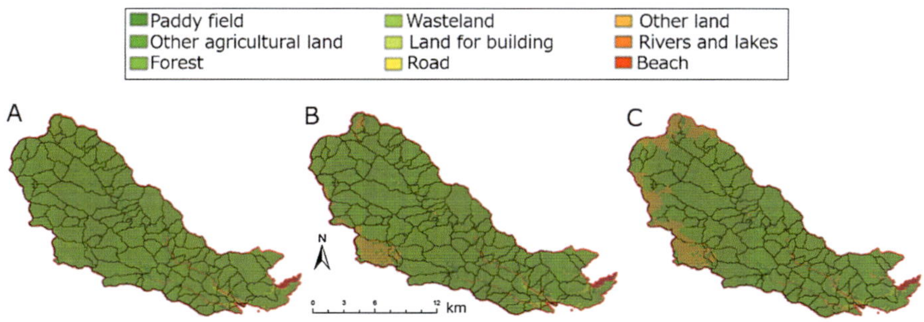

Fig. 2.6 Land-use patterns in (**A**) 1971, (**B**) 2006, and (**C**) 2050 (projected) in Otsuchi

projected to decrease by more than 80% around most of the suitable sites (Sites 1, 2, and 3). The only exception is Site 4, where the projected degree of decrease is lower (60–80%).

2.4 Discussion

2.4.1 Possible Effects of Developing Small Hydropower in Local Communities

The total annual potential power generation of 2.2 GWh is equivalent to around 4% of the total annual electricity demand in the town (~55 GWh in 2012; e.g., Sawadate 2017). One of the advantages of developing SHP is that greenhouse gases such as CO_2 are not emitted during SHP operation. By applying previous life cycle

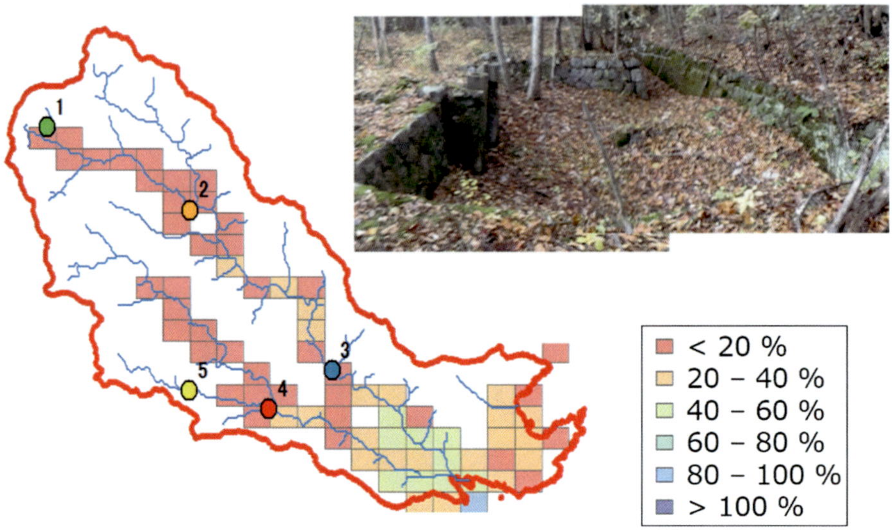

Fig. 2.7 Projected change in population in 2050 (%; relative to population in 2010). The locations of the five suitable sites for SHP (Sites 1–5) are also shown. The image shows the ruins of an SHP facility constructed before World War II at Site 4

assessment results (Imamura and Nagano 2010) and assuming an energy shift from coal-fired power to SHP, the total annual CO_2 reduction related to the operation of SHP at the five suitable sites was calculated to be around 2026 t CO_2, roughly equivalent to the annual CO_2 emissions of 200 Japanese people (National Institute for Environmental Studies 2017).

Conversely, the annual potential power generation of 122, 822, 527, 502, and 199 MWh by each SHP facility at Sites 1–5 corresponds to typical electricity demands for 34, 228, 146, 139, and 55 households, respectively, based on a typical household electric demand of 300 kWh per month in Japan (Federation of Electric Power Companies of Japan 2012). Therefore, the electricity generated by SHP is expected to provide most, or even more than 100%, of the demand in neighboring communities near the suitable sites where electricity is primarily consumed for household or agricultural uses.

The FIT was introduced in July 2012 in Japan. The purchase price of electricity generated by SHP was 29–34 yen/kWh in 2017, depending on the capacity and other conditions (Agency for Natural Resources and Energy 2017). By applying the FIT, the estimated annual revenue obtained by selling the electricity generated to an electric company is estimated to be around 4 million yen (36,364 USD), 24–28 million yen (218,182–254,545 USD), 15–18 million yen (136,364–163,636 USD), 15–17 million yen (136,364–154,545 USD), and 6–7 million yen (54,545–63,636 USD) at Sites 1–5, respectively. These represent relatively large sums of money for local communities, and would be sufficient for generating new jobs for several people to engage in maintaining the SHP facilities, which could contribute to vitalizing the local communities.

2.4.2 Challenges in Developing Small Hydropower in Local Communities

The number of suitable sites for SHP decreased greatly when location requirements were considered. The primary reason was that most potentially suitable sites are located in forests and are far from roads. The only exception is Site 1, which is adjacent to a main road. Considering that around 90% of the area in Otsuchi is categorized as forest (Fig. 2.6) and that the area generally has relatively steep geography and high potential power generation for SHP, further strategies are necessary to enhance the local production and consumption of energy generated by SHP in the future. Because many potential areas are in mountainous regions where depopulation is a major problem, measures for SHP development could potentially also address this issue, for example, by generating new local jobs related to the development and operation of SHP.

Several factors have caused changes in land-use patterns in Otsuchi. Change was especially notable in the upstream areas of the Otsuchi and Kotsuchi rivers. The mountainous area of the district used to be famous for stock farms of native shorthorn cattle ("*akabeko*"). However, Japanese black cattle ("*kuroge wagyu*") have recently become more popular, and have replaced the native shorthorn cattle. As a result, stock farms of native shorthorn cattle have declined. In 2004, 14 wind power plants were installed for operation in the district as a windfarm project to supply municipalities adjacent to the Shinyama Heights. This caused a shift in land use from forest and wasteland to other land (Fig. 2.4). Future increases in the discharge coefficients according to changes in land-use patterns could be greater than estimated in this study if the development of wind power plants accelerates with the promotion of renewable energy policies. This is an example where the development of one type of renewable energy can affect other renewable energy sources.

Among the five suitable sites for SHP, Sites 1, 2, and 3 are not suitable for operating and maintaining SHP facilities locally, considering the expected dramatic decrease in population (Fig. 2.7). Site 4 is an exception to this trend, where the degree of decrease is estimated to be lower (60–80%). Therefore, considering the projected change in population, Site 4 is the most promising location for SHP. The suitability of the potential power generation of SHP at Site 4 is partly supported by the fact there are ruins of an SHP facility that was constructed before World War II, which was abandoned after the facility was destroyed by Typhoon Ione in 1948 (Fig. 2.7). Thereafter, in the 1960s, the national energy policy shifted to fossil fuels and nuclear power.

Environmental factors were also considered in this study. However, the degree of effects differed among fish species. For example, salmon hatcheries and salmon and three-spined stickleback habitats are located in the estuaries of the Otsuchi and Kotsuchi rivers (Fig. 2.2C), and would not substantially affect the development of SHP facilities in the midstream or upstream of the rivers. Conversely, masu salmon travel to the midstream of the rivers, and SHP facilities should not be developed in the downstream area of the rivers to minimize possible conflicts with masu salmon conservation. In this study, all sites where the three species were found were excluded as potential sites for SHP development. However, the number of potential sites for SHP could be increased without any major damage to these species if fish passes are constructed along the river to bypass SHP facilities.

2.5 Conclusions

This study assessed the potential power generation by small hydropower (SHP) in Otsuchi, Iwate Prefecture, Japan. The number of suitable sites for installing SHP facilities decreases when local characteristics from natural and social perspectives are considered, especially with regard to minimizing impacts on the riverine species, such as salmon, an important fisheries target, and three-spined stickleback, an endangered species.

Ultimately, only five sites were identified as optimal for SHP installation, of which total potential power generation is estimated to correspond to 4% of the current total demand in the town with possible CO_2 reductions equivalent to the emissions of 200 Japanese people. However, if the power generated by SHP facilities at each site is consumed primarily within the local community, the power would cover all or most of the local demand for household or agricultural uses, which could vitalize the local communities. The local benefits could be further enhanced if the local community were to sell surplus SHP electricity with fixed prices under the FIT.

Because the operational period of SHP facilities is usually on the order of tens of years, potential power generation in the future must be considered, accounting for the effects of climate change, such as precipitation, and social changes, such as population and land-use patterns. The results of this study suggest that the potential power generation could double current levels by the 2040s because of increased precipitation and changes in land-use patterns. However, the degree of increase differs substantially among the sites. Meanwhile, the population is predicted to decrease more around the optimal sites, raising potential concerns over the local operation and maintenance of SHP facilities. Therefore, the selection of installation locations of new SHP facilities should consider the estimated potential power generation as well as possible future social changes.

Recently, low-head SHP generation has been a focus of development efforts, which can enable generation at sites with natural heads less than 2 m. This would increase the number of suitable sites, especially downstream of these rivers. In many cases, such areas are in downtown districts where the natural head is relatively low, but the social conditions are more suitable, with good access to roads and a substantial number of residents who could maintain SHP facilities daily. Such sites could be promising if energy–food and energy–ecosystem nexi are considered to minimize conflicts with the riverine ecosystem.

Acknowledgements We appreciate the useful comments from Drs. Seiichi Mori, Naoki Masuhara, Hajime Araki, Takenori Takada, and two anonymous reviewers as well as Mrs. Soichiro Tanabe and Yoichiro Fukuda. We would like to thank the officers and citizens of the town of Otsuchi for providing local information. This study was financially supported by the R-08-Init Project, entitled "Human-Environmental Security in Asia-Pacific Ring of Fire: Water-Energy-Food Nexus," Research Institute of Humanity and Nature, Kyoto, Japan.

References

Agency for Natural Resources and Energy (2017). http://www.enecho.meti.go.jp/category/saving_and_new/saiene/kaitori/fit_kakaku.html. Accessed 2 Oct 2017

Federation of Electric Power Companies of Japan website (2012). http://www.fepc.or.jp/enterprise/jigyou/japan/sw_index_04/. Accessed 2 Oct 2017

Fujii M, Tanabe S, Yamada M, Mishima T, Sawadate T, Ohsawa S (2017) Assessment of the potential for developing mini/micro hydropower: a case study in Beppu City, Japan. Journal of Hydrology: Regional Studies 11:107–116

Geospatial Information Authority of Japan (2017) A 10-m-meshed grid digital elevation model (DEM). http://www.gsi.go.jp/ENGLISH/index.html. Accessed 2 Oct 2017

Imamura E, Nagano K (2010) Evaluation of life cycle CO_2 emissions of power generation technologies: update for state-of-the-art plants. In: Central Research Institute of Electric Power Industry (CRIEPI) research report, Rep. No.Y09027, Socio-economic Research Center, 96 pp

Japan Meteorological Agency (2005) Global Warming Projection, vol 6. http://ds.data.jma.go.jp/tcc/tcc/products/gwp/gwp6/index.html. Accessed 2 Oct 2017

Kitano J, Mori S (2016) Toward conservation of genetic and phenotypic diversity in Japanese sticklebacks. Genes Genet Syst 91(2):77–84

Ministry of Land, Infrastructure, Transport and Tourism (MLIT) (2004). http://www.mlit.go.jp/river/pamphlet_jirei/kasen/gaiyou/panf/tokutei/. Accessed 2 Oct 2017

National Institute for Environmental Studies (2017) National Greenhouse Gas Inventory Report of JAPAN, p 752

National Land Information Division, National Spatial Planning and Regional Policy Bureau, MLIT (2006). http://nlftp.mlit.go.jp/ksj/gml/datalist/KsjTmplt-L03-b.html. Accessed 2 Oct 2017

National Land Information Division, National Spatial Planning and Regional Policy Bureau, MLIT (2012). http://nlftp.mlit.go.jp/ksj/gml/datalist/KsjTmplt-G02.html. Accessed 2 Oct 2017

National Land Information Division, National Spatial Planning and Regional Policy Bureau, MLIT (2014). http://nlftp.mlit.go.jp/ksj/gmlold/meta/suikei.html. Accessed 2 Oct 2017

Otsuchi Town (2017). http://www.town.otsuchi.iwate.jp/gyosei/. Accessed 2 Oct 2017

Sawadate T (2017) Assessment of small hydropower generation potential considering characteristics in the Otsuchi town, Iwate Prefecture. Master's Thesis, Graduate School of Environmental Science, Hokkaido University, pp 33. (in Japanese)

Tanabe S (2015) Assessment of the potential of developing mini/micro hydropower. Master's Thesis, Graduate School of Environmental Science, Hokkaido University, p 94 (in Japanese)

Chapter 3
Hazard-specific Vulnerability Mapping for Water Security in a Shale Gas Context

Diana M. Allen, Shannon Holding, and Zachary McKoen

Abstract Northeast British Columbia (NEBC) is estimated to hold large reserves of unconventional natural gas and has experienced rapid growth in shale gas development over recent decades. This industrial development has the potential to impact the quality and quantity of surface water and groundwater. In this study, hazard-specific vulnerability mapping was conducted across NEBC to identify areas most vulnerable to water quality and quantity deterioration due to shale gas development activities. Vulnerability represents the combination of a specific hazard threat and the susceptibility of the water system to that threat. Hazard threats (i.e. potential contamination sources and water abstraction) were mapped spatially across the region. The shallow aquifer susceptibility to contamination was assessed using the DRASTIC approach, while the aquifer susceptibility to abstraction was assessed according to aquifer productivity. Surface water susceptibility to contamination was assessed on a watershed basis to describe the propensity for overland flow (i.e. contaminant transport), while surface water susceptibility to water abstractions was assessed using watershed runoff estimates. The spatial distribution of hazard threats and susceptibility were combined to form hazard-specific vulnerability maps for groundwater quality, groundwater quantity, surface water quality and surface water quantity. The vulnerability maps identify priority areas for further research, monitoring and policy development.

Keywords Shale gas · Hazards · Water security · Vulnerability · Northeast British Columbia

---

D. M. Allen (✉) · S. Holding · Z. McKoen
Department of Earth Sciences, Simon Fraser University, Burnaby, BC, Canada
e-mail: dallen@sfu.ca

© Springer Nature Singapore Pte Ltd. 2018
A. Endo, T. Oh (eds.), *The Water-Energy-Food Nexus*, Global Environmental Studies, https://doi.org/10.1007/978-981-10-7383-0_3

3.1 Introduction

Northeast British Columbia (NEBC), Canada, is estimated to hold significant unconventional natural gas reserves. Shale gas development in NEBC has occurred very rapidly following technological advancements in hydraulic fracturing and directional drilling. In the past 16 years, there has been an 82% increase in the number of shale gas development applications (BC Oil and Gas Commission (BCOGC) 2014). However, the rapid development of shale gas in this region has not been matched by advances in the scientific understanding of the environmental impacts (Council of Canadian Academies 2014). This lack of understanding poses challenges for effective regulation of shale gas activities alongside effective management of the environment, specifically water. Meanwhile, shale gas development is poised to continue to grow significantly in the coming years. The province of BC has a vision of becoming a global leader in secure and sustainable natural gas investment, development, and export and has set a goal of having three liquefied natural gas (LNG) facilities in operation by 2020 (Province of British Columbia 2015).

The growth of shale gas activities in NEBC creates a water-energy nexus that is characterized by growing conflict surrounding water use and protection (Council of Canadian Academies 2014). Shale gas development has the potential to significantly impact water security in the region, both through water consumption and potential contamination (Vengosh et al. 2014). Although the region is sparsely populated, water security for both human and environmental needs may be impacted. The water resources of NEBC require sound management to protect water quality and quantity in relation to the risks to water security by shale gas development.

In this study, hazard-specific vulnerability mapping was conducted across the region to characterize areas most vulnerable to water quality and quantity deterioration due to shale gas development.

3.2 Water Demand and Contamination Hazards in the Peace Region

The Peace Region of NEBC extends from the Rocky Mountains to the west, through the foothills, to the low-lying plains to the east (Fig. 3.1). The majority of the population resides near Fort St. John and Dawson Creek, where shale gas development is focused. Of the four major shale gas plays in NEBC, the Montney Play exhibits the highest levels of development (Fig. 3.1).

Across the region, the climate varies from cold continental in the south to cold subarctic in the north, and is characterized by sustained cold winters and warm summers. Accordingly, the hydrologic regime is typically snowmelt dominated, with a sustained cold winter period characterized by low river discharge and competent river ice. The spring freshet, during which most of the snow melts, extends from approximately mid-April to late June and is characterized by high

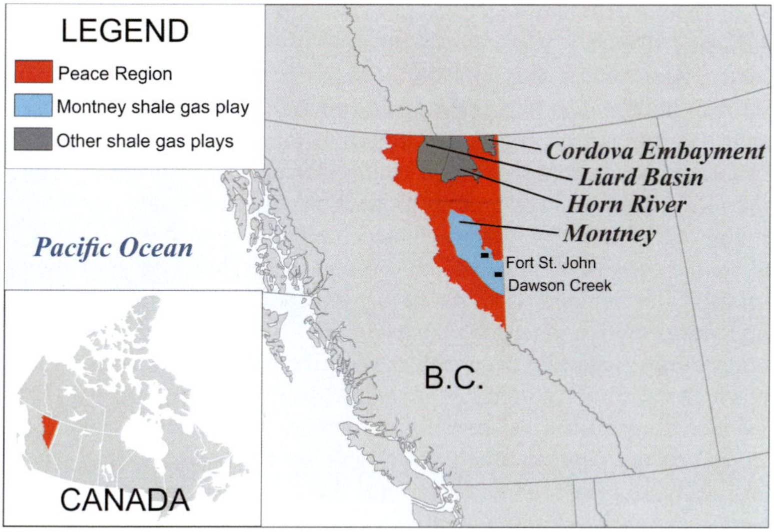

Fig. 3.1 The Peace Region in Northeast British Columbia, Canada, showing the location of four shale gas plays. The Montney Play has the highest levels of development

river discharge. After the spring freshet, river levels generally recede slowly through the summer and autumn until the winter freeze-up. During summer and fall, frontal or convective storm systems bring varying amounts of rain, often resulting in increases in river levels and discharge, and occasionally producing flooding.

Water management in NEBC must therefore adapt to changing hydrological conditions, not only in relation to seasonal variability, but also considering inter-annual variability, climate change, and a growing water demand, particularly by industry. Large quantities of water are required for shale gas development, particularly for hydraulic fracturing - a typical well in the Montney requires between ~10,000–25,000 m^3 water for hydraulic fracturing (Johnson and Johnson 2012). Currently, most of the water used for hydraulic fracturing in NEBC derives from surface water sources, although there is increasing demand for groundwater. At the present time, there is little available information on groundwater resources in NEBC. The unconsolidated aquifers, comprised of glacial or pre-glacial origin, are generally of limited extent, but provide local water supplies. Groundwater is also sourced from bedrock aquifers (Berardinucci and Ronneseth 2002).

Shale gas withdrawals account for less than 1% of total surface water runoff estimates (BC Oil and Gas Commission (BCOGC) 2013), but while water is seemingly abundant across the region, the withdrawals occur over a short timeframe and are concentrated in specific geographic locations. Mountain front watersheds deliver large quantities of water to rivers, particularly during the spring freshet; however, a large portion of the region is semi-arid and water withdrawals are increasing in these areas. The localized nature of withdrawals may lead to conflict with other water users and environmental needs, particularly during seasonal lows or drought periods (Council of Canadian Academies 2014). The BC Oil and Gas Commission (BCOGC) uses the Northeast Water Tool (NEWT) for water management decision-making in NEBC (Chapman et al. 2012). NEWT combines modeled hydrometric data (e.g. monthly and annual averaged surface water flows) with water license and permitting records. It is intended to provide guidance on water availability and support decision-making for new water licensing approvals.

Hydraulic fracturing and oil and gas production generate wastewater. The amount and chemical composition of wastewater depends on the type of fracturing activities, original source of water (fresh, saline, or recycled), subsurface geology, and the phase of well development (i.e. fracturing or production). Although wastewater varies in its composition, it is generally a solution with high concentrations of salts, metals, metalloids, naturally occurring radioactive materials, as well as numerous proprietary chemical constituents. Recognized hazards associated with shale gas activities include spills and leakages resulting from handling, transport or disposal of the chemicals used in hydraulic fracturing or of the wastewater that is produced. Surface spills have a high risk of occurrence due to the large volumes handled and number of trucks used to transport wastewater to (when recycled wastewater is used for fracking) and from well pads. The potential contamination from wastewater poses a threat to drinking water supplies and healthy aquatic ecosystems (Council of Canadian Academies 2014).

The hazard-specific vulnerability mapping conducted in the Peace Region aims to characterize areas most vulnerable to water quality and quantity deterioration due to shale gas development. Vulnerability represents the combination of a specific hazard threat and the susceptibility of the water system to that threat (Eq. 3.1).

$$\text{Vulnerability} = \text{Susceptibility x Hazard Threatt} \tag{3.1}$$

Mapping of vulnerability, therefore, requires spatial datasets that can be integrated in such a fashion to assess susceptibility and the range of hazard threats. For this study, spatial physical data for characterizing the aquifer system and watersheds were acquired from Data BC (https://data.gov.bc.ca/). Spatial hazard data were gathered from the BCOGC public zone GIS data (http://data-bcogc.opendata.arcgis.com/).

The following sections describe the approaches used first to assess the susceptibility of groundwater and surface water to specific hazard threats that may impact the quality and quantity of water. The mapping was carried out in ArcGIS (v. 10).

3.3 Susceptibility and Hazard Threat Mapping

Susceptibility, in the context of water security, refers to the physical characteristics of the aquifer system or watershed that make it more or less susceptible to threats related to contamination or high demand (here termed hazard threats). By aquifer system, we mean the full range of geological materials that form aquifers (permeable units) and confining units (less permeable units).

Table 3.1 Overview of the approach used to assess groundwater and surface water quality and quantity vulnerability

Component		Quality	Quantity
Groundwater	Susceptibility	DRASTIC shallow groundwater susceptibility, where: D is depth to water table;	(Supply): Inverse of aquifer productivity (based on aquifer media from DRASTIC, where high density of domestic wells decreases potential productivity).
		R is recharge;	
		A is aquifer media;	
		S is soil media;	
		T is topography;	
		I is impact of vadose zone; and	
		C is conductivity.	
	Hazard threat	(1) Oil and gas wells (density) related to chemical handling at surface and spills/leaks (all inactive/ abandoned /active wells included).	(Demand): Source well groundwater abstractions (density characterised based on magnitude of abstraction). Includes all source wells (inactive and active) with active wells having a higher magnitude but inactive wells still having some baseline magnitude representing the potential.
		(2) Transportation (density) (pipelines and all oil and gas developed roads).	
		(3) Oil and gas Infrastructure (density) (water hubs, facilities).	
		Each hazard is ranked 1–10. Total hazards weighted: H = (4 x wells) + (3 x roads) + (2 x pipelines) + (1 x facilities)	
	Vulnerability	Total hazards x susceptibility	Demand x supply
Surface water	Susceptibility	Overland flow susceptibility from DRASTIC components S, T and I.	(Supply): Runoff from Northeast Water Tool (NEWT) per watershed. Points of surface water diversion subtracted for existing demand.
	Hazard threat	Density of hazards (from industry groundwater abstraction)	(Demand): Surface water licenses (long and short-term) from NEWT (per watershed).
	Vulnerability	Hazards x susceptibility	Demand x supply

Different approaches were used to map (1) the susceptibility of groundwater and surface water to (2) hazard threats that may result in deterioration of water quality or water quantity. The following sections describe the approaches used. Table 3.1 gives an overview of the approaches used.

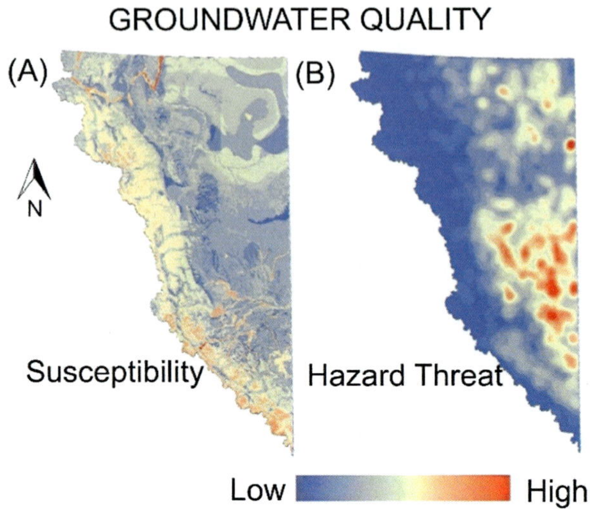

Fig. 3.2 Groundwater Quality (**a**) Susceptibility: DRASTIC shallow aquifer susceptibility assessment. (**b**) Hazard Threat: associated with activities related to shale gas development (facilities, wells, roads, and pipelines). Hazard threat is represented as weighted cumulative density

3.3.1 Groundwater Quality

Groundwater Quality Susceptibility was based on the DRASTIC method (Aller et al. 1987), where D is Depth to water table; R is recharge; A is aquifer media; S is soil media; T is topography; I is impact of vadose zone; and C is conductivity. DRASTIC is internationally known and has been applied to numerous hydrogeological settings. DRASTIC assumes that contamination occurs from ground surface sources; therefore, the method focuses on shallow geological materials and the groundwater contained in these materials within approximately 30 metres (m) of ground surface. The method does not assess the susceptibility of deeper groundwater that may be impacted from contamination originating at greater depth, but provides some indication of the relative susceptibility of shallow groundwater to sources at or just below the ground surface. Holding and Allen (2015) describe the DRASTIC mapping carried out in the Peace Region in detail.

The resulting groundwater susceptibility map is shown in Fig. 3.2a. Areas of higher susceptibility are shown in red with areas of lower susceptibility in blue. Areas of high susceptibility occur predominantly along the mountainous western edge of the region where there is high elevation bedrock. High susceptibility is the result of shallow water tables combined with high recharge rates, relatively high permeability, and limited soil cover. Other high susceptibility areas include river valleys where the vadose zone and aquifer materials have large proportions of sand and gravel.

Hazard Threat to Groundwater Quality was assessed by mapping the potential for contamination of groundwater due to spills and leaks of industrial waste. Areas of high potential are associated with oil and gas wells, the location of transportation routes and pipelines, and the location of oil and gas related infrastructure. First, the spatial density of oil and gas wells was mapped. At the time of mapping, there were 30,711 recorded oil and gas wells of all operation types (active, disposal, abandoned, etc.). All well operation types were given the same hazard weight. Next, the spatial density of roads, pipelines (both permanent and currently in development), and oil and gas related facilities was mapped. At the time of mapping, there were 21 different facility types; all were assigned the same hazard weight.

The spatial density of all features was determined using a search radius of 15,000 km^2 and a 500 m output cell size. The total groundwater hazards layer (Fig. 3.2b) was generated using a weighted sum (i.e. multiplying the individual density maps for facilities, wells, roads and pipelines by weights of 4, 3, 2 and 1, respectively). These weights were applied to reflect the relative likelihood of each being a potential source of contamination. This produced a combined density map that was reclassified to a 1–10 scale (Fig. 3.2b).

3.3.2 Surface Water Quality

Surface Water Quality Susceptibility was assessed based on the potential for overland flow (i.e. the likelihood of surface water, which could potentially be contaminated, remaining on the surface and flowing into lakes, rivers and streams). This was done by combining select component maps from DRASTIC; specifically, soil media, vadose zone impact and topography (slope). For soil media, a reverse ranking to that used in DRASTIC was employed to allow for the least permeable soils to correspond to greater overland flow potential. The soil, vadose zone, and topography maps were assigned weights of 2, 5 and 1, respectively, and then added and reclassified to a 1–10 scale (Fig. 3.3a).

Hazard Threat to Surface Water Quality was the same as for groundwater (described above) because it is based on the likelihood of contaminants being spilled on the surface or shallow sub-surface (Fig. 3.3b).

3.3.3 Groundwater Quantity

Groundwater Quantity Susceptibility was assessed by estimating the aquifer supply available for shale gas development needs and was based on geological materials and existing domestic water use. First, a groundwater productivity map was generated using the aquifer media rating from DRASTIC. Areas of high groundwater productivity have highly permeable aquifer materials. The density of domestic wells was then subtracted from the groundwater productivity map to

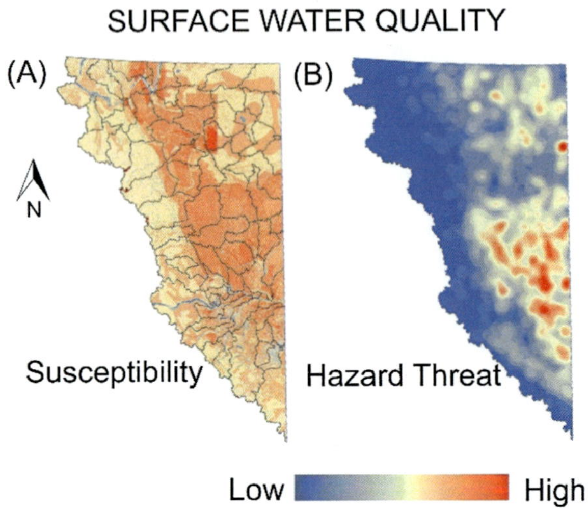

Fig. 3.3 Surface Water Quality (**a**) Susceptibility: Potential for overland flow, representing increased likelihood of migration of surface spills into surface water bodies. (**b**) Hazard Threat: associated with activities related to shale gas development (facilities, wells, roads, and pipelines). Hazard threat is represented as weighted cumulative density (same hazards used for groundwater)

represent reduced groundwater quantity available in these areas. Thus, high groundwater productivity corresponds to low susceptibility to groundwater abstraction impacts (Fig. 3.4a).

Hazard Threat to Groundwater Quantity was assessed by mapping the density of oil and gas groundwater source wells. Density was assessed based on the abstraction rates for each well reported as by the BCOGC (Fig. 3.4b).

3.3.4 Surface Water Quantity

Surface Water Quantity Susceptibility was based on data extracted from NEWT for each watershed within Peace Region. The surface water supply per-watershed was assessed as the average annual runoff (m^3/year) minus the points of surface water diversion for uses other than oil and gas (i.e. domestic, agricultural, municipal) (Fig. 3.5a).

Hazard Threat to Surface Water quantity was the total approved surface water withdrawal volumes per watershed. These include short term water use approvals and long term water licenses (both oil and gas related) (Fig. 3.5b).

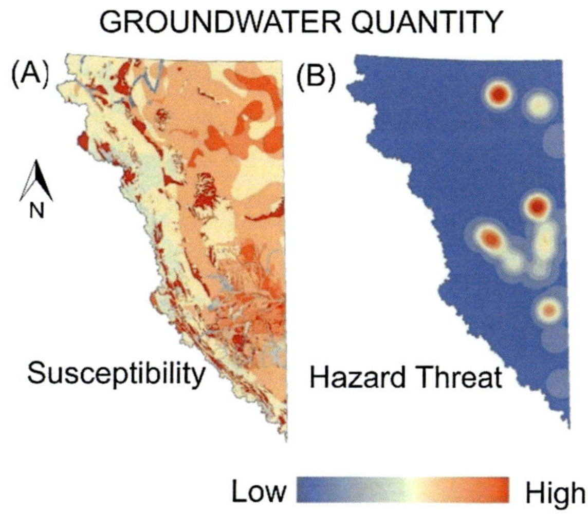

Fig. 3.4 Groundwater Quantity (**a**) Susceptibility: Estimated aquifer supply available for shale gas development needs. (**b**) Hazard Threat: Water withdrawal for shale gas development. Represented as density map based on abstraction rates of source water wells used by the oil and gas industry

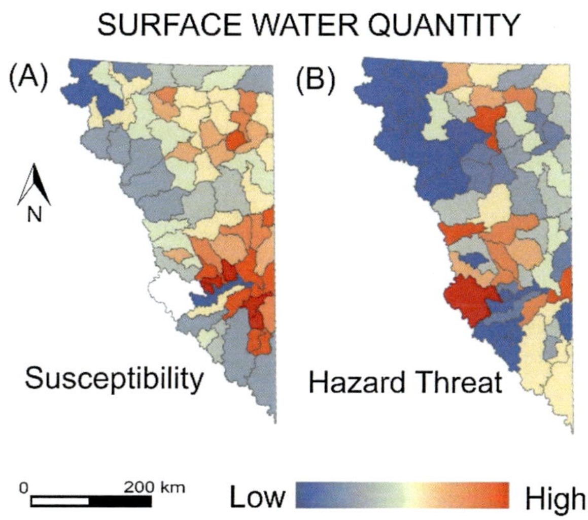

Fig. 3.5 Surface Water Quantity (**a**) Susceptibility: Estimated volume of surface water discharge available for shale gas development needs. (**b**) Hazard Threat: Total approved surface water withdrawal volumes per watershed

3.4 Vulnerability Mapping

Total vulnerability for each of component, groundwater quality, surface water quality, groundwater quantity, and surface water quantity, was calculated by multiplying the susceptibility map by the total hazard map. The final maps were reclassified to a 1–10 scale (Fig. 3.6).

The groundwater quality vulnerability map (Fig. 3.6a) and the surface water quality vulnerability map (Fig. 3.6c) identify broad areas surrounding Fort St. John and Dawson Creek where water quality vulnerability is high.

Fig. 3.6 Vulnerability. (**a**) Groundwater Quality, (**b**) Groundwater Quantity, (**c**) Surface water quality, and (**d**) Surface water quantity

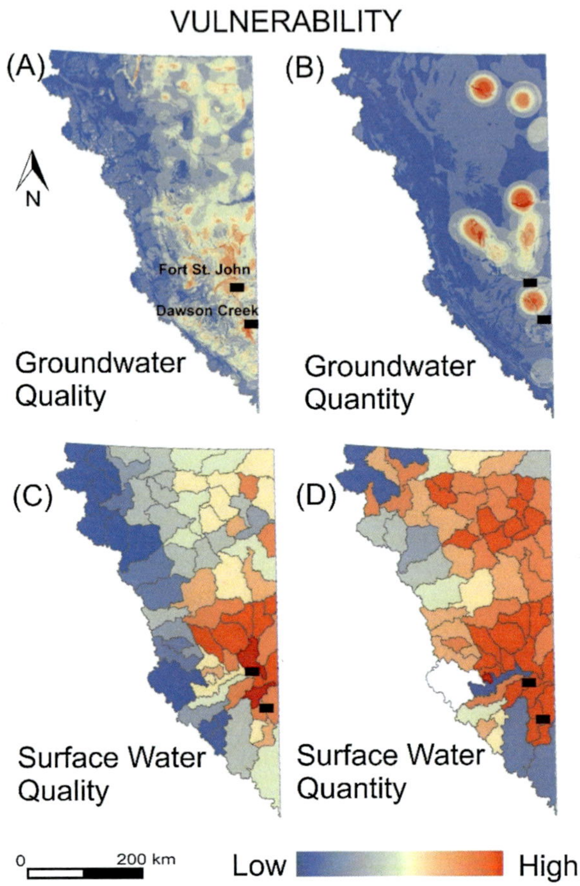

These areas occur where high hazard threat (contamination potential) coincides with high aquifer susceptibility or high overland flow potential. As noted earlier, the Montney Play is situated in this area, and the associated infrastructure presents a risk to both surface water and groundwater quality due to potential spills and leaks of wastewater. While DRASTIC was used for assessing aquifer susceptibility in this study, other methods could be employed.

Areas of high vulnerability for water quantity occur where demand is estimated to represent a significant proportion of estimated supply. The groundwater quantity vulnerability map (Fig. 3.6b) only shows isolated areas of high vulnerability related to industry groundwater source wells, but currently groundwater use by industry is low. If the demand for groundwater grows, this map will look considerably different. The surface water quantity vulnerability map (Fig. 3.6d) shows many watersheds throughout the Peace Region with high vulnerability. These are concentrated in the low-lying, semi-arid regions that experience seasonal summer low flows.

3.5 Conclusions

The maps characterize water security in the Peace Region form a tool to improve resilience to water security risks within a rapidly growing shale gas sector. Identification of vulnerable areas may support water management by:

- Informing policy and regulation decisions;
- Highlighting priority areas for further research, data collection and monitoring;
- Identifying areas in which to focus limited enforcement resources;

This vulnerability mapping approach, using the hazard threat and susceptibility indicators, can be applied to other shale gas areas to assess vulnerability. The approach can also be tailored to different settings as necessary (i.e. excluding surface water or groundwater if not affected).

Acknowledgments This research was supported by the Pacific Institute for Climate Solutions, Victoria, British Columbia, Canada and the Research Institute for Humanity and Nature, Kyoto, Japan.

References

Aller L, T Bennett, J Lehr, R Petty and G Hackett (1987) DRASTIC: a standardized system for evaluating ground water pollution potential using Hydrogeologic settings. EPA-600/2-87-035, National Water Well Association, Dublin, Ohio / EPA Ada. Oklahoma

BC Oil and Gas Commission (BCOGC) (2013) Water use for oil and gas activity: annual report

BC Oil and Gas Commission (BCOGC) (2014) 2013/2014 Annual service plan report

Berardinucci J, Ronneseth K (2002) Guide to using the BC aquifer classification maps for the protection and Management of Groundwater. In: British Columbia Ministry of Water, Land and Air Protection, Victoria, 54 pages

Chapman A, Kerr B, Wilford D (2012) Hydrological modelling and decision-support tool development for water allocation, northeastern British Columbia. In: Geoscience BC summary of activities 2011, geoscience BC, report 2012–1, pp 81–86

Council of Canadian Academies (2014) Environmental impacts of shale gas extraction in Canada. The expert panel on harnessing science and technology to understand the environmental impacts of shale gas extraction. Council of Canadian Academies, Ottawa, p 266

Holding S, Allen DM (2015) Final report: shallow groundwater intrinsic vulnerability mapping in Northeast British Columbia. Prepared for Pacific Institute for Climate Solutions and BC Ministry of forests. Department of Earth Sciences, Simon Fraser University, Lands and Natural Resource Operations, p 41

Johnson EG and LA Johnson (2012) Hydraulic fracture water usage in Northeast British Columbia: locations, volumes and trends. Geoscience reports 2012, British Columbia Ministry of energy and mines, pp 41–63

Province of British Columbia (2015) Liquefied natural gas in British Columbia. https://lnginbc.gov.bc.ca/. Accessed July 2017

Vengosh A, Jackson RB, Warner N, Darrah TH, Kondash A (2014) A critical review of the risks to water resources from unconventional shale gas development and hydraulic fracturing in the United States. Environ Sci Technol 48:8334–8348

Chapter 4
Exploring Future Water Demand and Climate Change Impacts on Water Availability in the Peace Region of British Columbia, Canada

Jennifer R. Dierauer, Diana M. Allen, and Paul H. Whitfield

Abstract In recent decades, the Peace River watershed in Northeast British Columbia (NEBC) has experienced rapid growth in shale gas development activities, resulting in significant increases in surface water and groundwater use and a growing conflict over the use and protection of these water resources. Under a high development scenario, industrial water demand in the Peace River watershed is projected to increase by over 350% by 2030, and future water security in the context of the water-energy nexus is unknown, especially with continued climate warming. In this study, hydrological models are used to simulate the current and future water balance for two headwater catchments of the Peace River watershed, one in the foothills and one in the plains. Both catchments have been impacted by the recent shale gas development, and both contain oil and gas industry water use permits. Climate variables output from three Global Climate Models were used as inputs for the hydrologic models. Water quantity projections for future decades (2020s, 2030s, 2040s, and 2050s) were then compared to the projected water use for low, medium, and high shale gas development scenarios and used to estimate the potential for water scarcity in the region. Results from this study show that areas with high levels of oil and gas industry development may experience water scarcity if rapid industrial growth continues, and improved water management policies will be needed to mitigate the high industrial water demand.

Keywords Shale gas · Climate change · Water security · Water allocation · Northeast British Columbia

J. R. Dierauer (✉) · D. M. Allen
Department of Earth Sciences, Simon Fraser University, Burnaby, BC, Canada
e-mail: jbrand@sfu.ca; dallen@sfu.ca

P. H. Whitfield
Centre for Hydrology, University of Saskatchewan, Saskatoon, SK, Canada

© Springer Nature Singapore Pte Ltd. 2018

A. Endo, T. Oh (eds.), *The Water-Energy-Food Nexus*, Global Environmental Studies, https://doi.org/10.1007/978-981-10-7383-0_4

4.1 Introduction

Shale gas development in Northeast British Columbia (NEBC) has occurred rapidly in recent decades due to technological advancements made in hydraulic fracturing and directional drilling. Hydraulic fracturing requires large volumes of water – on the order of 2000–100,000 m^3 per hydraulic fracturing event for well pads in NEBC (Rivard et al. 2014). While the region has abundant water resources, these multi-stage hydraulic fracturing operations put high demands on local watersheds, requiring large volumes of water in concentrated areas over short time periods. With such sporadic high-volume water demand, water availability is a key issue in NEBC.

The short-term, high volume water demands by the oil and gas industry are managed by the British Columbia Oil and Gas Commission (BCOGC) through the use of short-term water use approvals (ST-approvals). These ST-approvals have a maximum duration of 2 years. In 2015, there were 294 active ST-approvals with 1027 approved withdrawal locations totaling 19 million m^3 of water – approximately 0.015% of the region's average annual runoff volume (BCOGC 2016).

ST-approvals create a fast-changing system of water withdrawals in NEBC. To aid in the management of the region's water resources, the BCOGC developed the Northeast Water Tool (NEWT). NEWT is a Geographic Information System (GIS)-based hydrology decision support tool that combines hydrometric data (estimates of monthly and annual runoff volumes) with water license and permitting records (Chapman et al. 2012). NEWT is used by the BCOGC to manage water allocations – with the goal of balancing environmental flow needs (EFNs) with other industrial, municipal, and agricultural water demands. The EFNs of a stream are defined as the volume and timing of water flow required for proper functioning of the aquatic ecosystem.

NEWT represents an important step forward in water resource management for this data-scarce region; however, it has several limitations. The primary limitation is that the estimated runoff volumes represent long-term averages (Chapman et al. 2012). The hydrologic regime in NEBC is snowmelt-dominated, and streamflow is highly variable. Thus, in any one year, the observed conditions may differ significantly from the long-term average runoff volumes in NEWT. This highly variable hydrologic regime makes it difficult to identify reliable surface water sources for the high-volume fracking operations in NEBC – especially during drought conditions when industrial water abstractions from rivers and lakes are often limited or suspended. Due to the high variability of streamflow, other surface water sources are required to meet the shale gas industry water demands. In NEBC, many oil and gas companies currently rely on water source dugouts, i.e. ponds or pits where water from snowmelt, rainfall, or groundwater inflow accumulates (BCOGC 2016). The water source dugouts used by the oil and gas industry store excess precipitation and runoff from wet periods (spring snowmelt) for use in dry periods (late summer) when streamflow is insufficient. Unlike abstractions from streams and lakes, oil and gas industry dugout water use is not suspended during drought conditions.

With continued climate warming and a shift in the snow-to-rain ratio, the reliability of dugout water sources and summer streamflow in NEBC is uncertain. Continued shale gas industry development may lead to increased freshwater demand, and in the context of this water-energy nexus, future water security in the region is unknown. To address this knowledge gap, this study aims to estimate the potential for water scarcity in the Peace River watershed in the context of future climate change and growing water demand.

4.2 Study Area

The Peace River region of NEBC extends from the Rocky Mountains to the west, through the foothills, to the low-lying plains to the east (Fig. 4.1a). Most of the shale gas development is focused in the low-lying plateau area, although some activity extends into the foothills. Of the four major shale gas plays in NEBC, the Montney Play exhibits the highest levels of development (Fig. 4.1a).

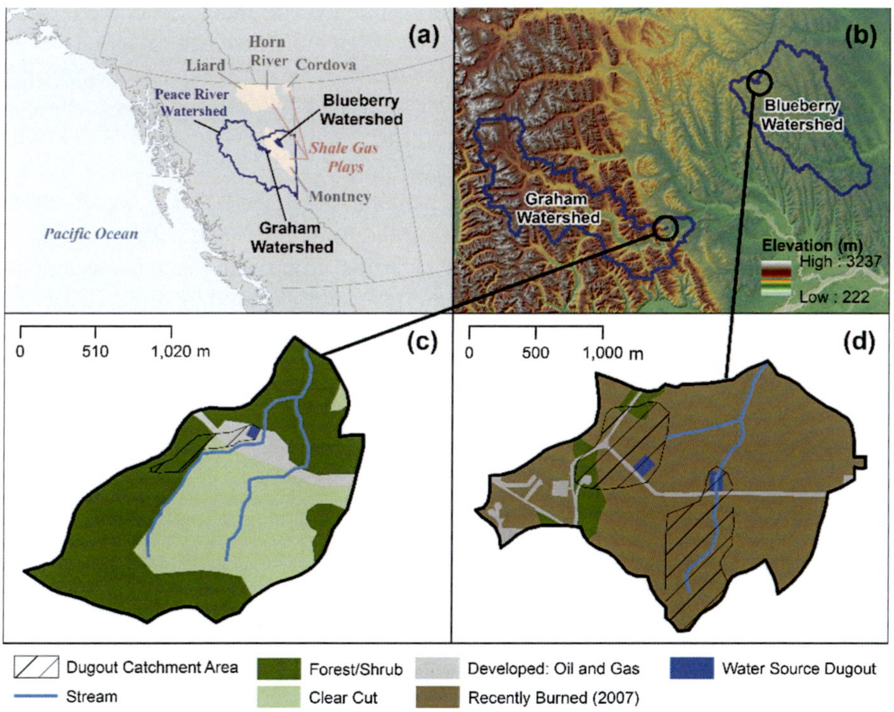

Fig. 4.1 Study area including (**a**) Peace River watershed and shale gas plays, (**b**) Graham and Blueberry watersheds, and current land use and oil and gas industry water source dugout locations in the (**c**) Graham headwater catchment and (**d**) Blueberry headwater catchment

For this study, two small headwater catchments within the Montney Play area were chosen, the 2.3 km^2 Graham River headwater catchment in the foothills and the 3.2 km^2 Blueberry River headwater catchment in the plains (Fig. 4.1b). Both catchments are located upstream of active gauging stations, the watershed boundaries of which are shown in Fig. 4.1b. Land use in each catchment was digitized from satellite imagery (Google Earth 7.1.8.3036 (32-bit) 2014, 2015) and is shown in Fig. 4.1c and d. At the time of this study, both watersheds contained water source dugouts that were in use by the oil and gas industry. The Blueberry catchment contained 2 dugouts, with a combined total allocated (i.e. approved abstraction) water volume of 26,136 m^3/year. The Graham catchment contained 1 dugout with a total allocated water volume of 1500 m^3/year.

4.3 Future Water Demand

The shale gas industry development scenarios used in this study were developed by the Pembina Institute and correspond to scenario options 1 through 6 in Table 1 of Kniewasser and Horne (2015). These scenarios include current and improved water management policies, and low, medium, and high liquefied natural gas (LNG) development. The improved policy scenarios incorporate 25% water recycling and 25% deep aquifer saline (non-potable) water use. The LNG development scenarios also assume 1, 3, and 5 LNG terminals (i.e. storage and distribution facilities) for the low, medium, and high development, respectively.

Future shale gas industry freshwater demand is projected to peak in 2030 (Fig. 4.2), with a high development scenario exhibiting more than a 350% increase in freshwater demand compared to the 2015 levels. Improved water management policies have the potential to decrease future freshwater demand by almost 50%.

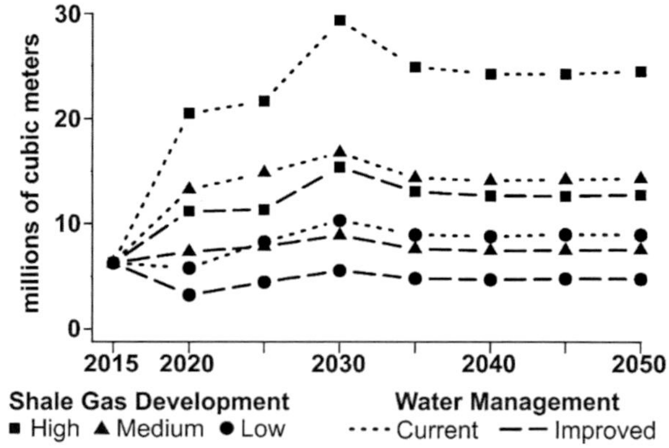

Fig. 4.2 Projected shale gas industry freshwater demand from Kniewasser and Horne (2015). High, medium, and low development correspond to 5, 3, and 1 Liquid Natural Gas (LNG) plants, respectively. The improved water management scenario incorporates 25% water recycling and 25% saline water use

4.4 Climate Change Projections

Statistically downscaled forcing datasets based on three models from Phase 5 of the Coupled Model Intercomparison Project (CMIP5) under representative concentration pathways (RCPs) 4.5 and 8.5 were used for the climate change scenarios in this study. The three models from the CMIP5 ensemble (CNRM-CM5–1, CanESM2-r1, ACCESS1–0-r1) were selected to capture the widest spread in projected future climate while using a small subset of the full ensemble, following Cannon (2015). Daily climate time series downscaled with the bias-correction/constructed analogues with quantile mapping reordering (BCCAQ) method were obtained from the Pacific Climate Impacts Consortium (PCIC) data portal (Pacific Climate Impacts Consortium 2014) covering the period of 1950 to 2100.

The Global Climate Model (GCM) ensemble projects increases in temperature and precipitation for both the Blueberry (plains) and Graham (foothills) catchments. For both catchments, the mean annual temperature is projected to be 2 °C warmer in the near future (2020–2050) as compared to the historical period (1970–2000). The largest increases in temperature are for the coldest (January) and warmest (July) months. Total annual precipitation is projected to increase by approximately 60 mm/year relative to the historical period. Fig. 4.3 shows the historical and near future climate for the Blueberry headwater catchment. Historical and near future climate in the Graham catchment (not shown) exhibits similar seasonal climate patterns and projected changes.

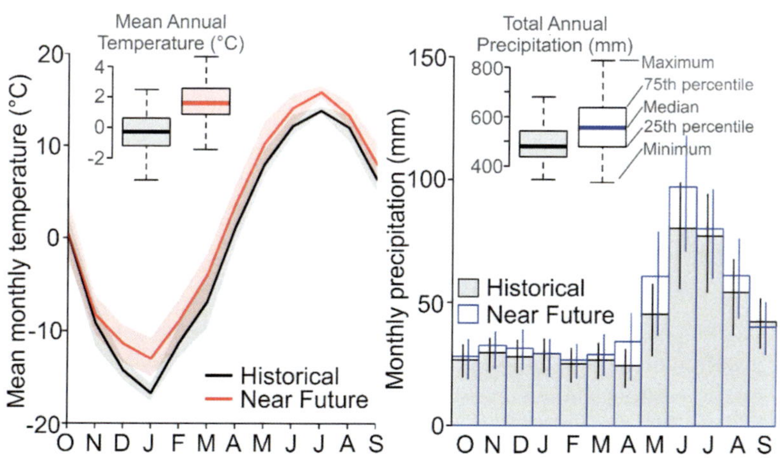

Fig. 4.3 Historical (1970–2000) versus near future (2020–2050) climate from model ensemble for the Blueberry headwater catchment. Shading (temperature) and vertical lines (precipitation) indicate the 30-year inter-quartile range. Inset boxplots show corresponding annual values

Table 4.1 Graham and blueberry headwater catchment hydrological response units (HRUs)

	Blueberry Headwater catchment		Graham headwater catchment	
HRU / land use	Area (km^2)	% in dugout catchment	Area (km^2)	% in dugout catchment
Water source dugout	0.030	100%	0.005	100%
Developed: Oil & gas	0.169	21%	0.132	13%
Clear cut / recent burn	2.854	23%	0.861	1%
Forest	–	–	1.315	3%
Shrub	0.164	30%	–	–
Total	3.217	24%	2.314	3%

4.5 Hydrologic Modelling

Frozen soils, snow accumulation, and snowmelt are all important components of the hydrologic cycle in the Peace River region, and much of the region is data scarce. The physically based cold regions hydrological modelling (CRHM) platform was chosen as the modelling code for this study because of its proven ability to simulate snow processes in diverse settings such as prairie (Fang and Pomeroy 2007; Pomeroy et al. 2007) and alpine basins (Pomeroy et al. 2012). The CRHM models were structured as a set of four Hydrological Response Units (HRUs) corresponding to the major land cover features as listed in Table 4.1. Within each HRU, physically based modules were sequentially linked to simulate the dominant hydrological processes.

Within the CRHM modelling framework, dugout water storage was simulated as detention storage. In both watersheds, the following assumptions regarding dugout water storage and withdrawal were made: (1) each dugout has a maximum storage volume equal to the total annual allocated water, (2) the allocated water volume is withdrawn at a constant rate over the licensed withdrawal period (May 1st to August 31st), and (3) the dugouts are clay-lined and above the groundwater table.

4.6 Near Future (2020–2050) Versus Historical (1970–2000) Hydrology

In both catchments, higher temperatures in the near future (relative to the historical period) result in decreased late-season (April–May) snowfall (Figs. 4.4a and 4.5a), a shift towards earlier snowmelt as evidenced by reduced snow water equivalent (SWE) in March and April (Figs. 4.4b and 4.5b), and increased March runoff (Figs. 4.4d and 4.5d). Despite an overall increase in annual precipitation, hydrologic modelling results show no significant changes in total annual runoff in the Blueberry catchment between the historical and near future periods (Fig. 4.4d). Increases in total annual precipitation in the near future are offset by increased actual

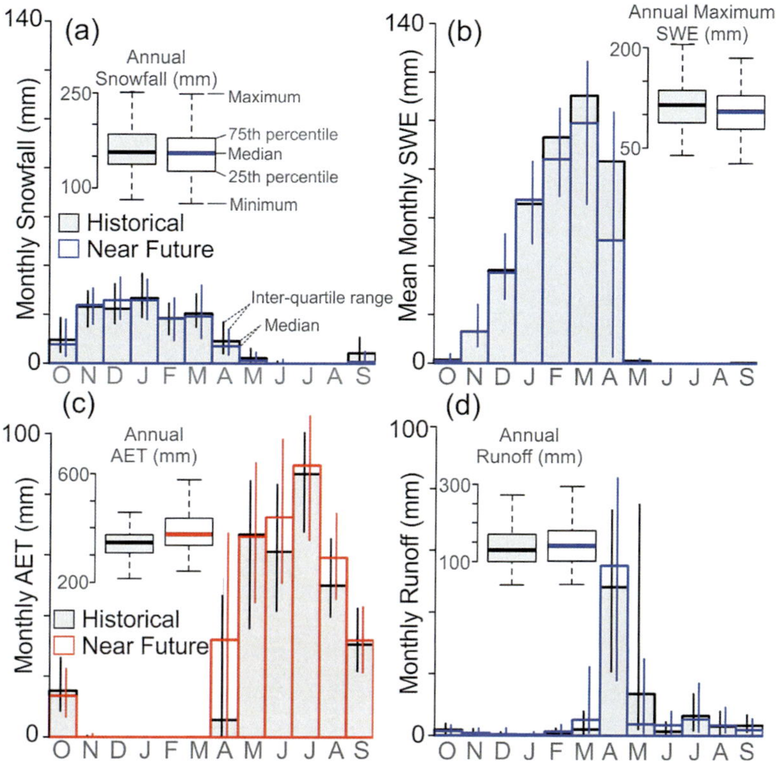

Fig. 4.4 Blueberry headwater catchment (plains) modelled historical (1970–2000) versus near future (2020–2050) (**a**) monthly snowfall (represented as mm of snow water equivalent - SWE), (**b**) mean monthly SWE, (**c**) monthly actual evapotranspiration (AET), and (**d**) monthly runoff. Inset boxplots show corresponding annual values

evapotranspiration (AET; Fig. 4.4c), resulting in no net change in total annual runoff at the annual time scale. In the Graham catchment, however, precipitation increases are larger than AET increases, resulting in an increase in total annual runoff in the near future compared to the historical period (Figs. 4.5c and d).

4.7 Demand Versus Supply

To determine the potential for future water scarcity, the range of projected mean annual runoff (MAR) from the GCM ensemble was compared against future water demand for each time period under the six shale gas development scenarios. Currently, the BCOGC determines the maximum amount of water available for allocation based on 15% of the MAR with lower allocation percentages for winter low flow months. Future water demand volumes were calculated by multiplying the

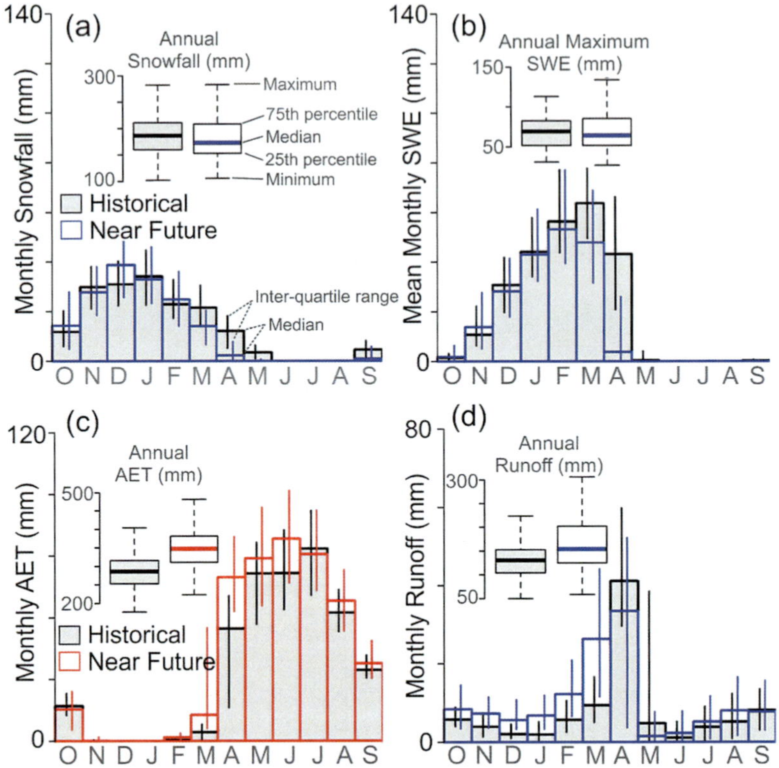

Fig. 4.5 Graham headwater catchment (foothills) modelled historical (1970–2000) versus near future (2020–2050) (**a**) monthly snowfall (represented as mm of snow water equivalent - SWE), (**b**) mean monthly SWE (**c**) monthly actual evapotranspiration (AET), and (**d**) monthly runoff. Inset boxplots show corresponding annual values

2015 allocation volumes by the projected percent change in development relative to 2015. Allocation levels in the Blueberry headwater catchment exceeded 15% of MAR (over-allocation) for all time periods under the high development scenario (Fig. 4.6).

The hydrological modelling results show no significant change in the total annual runoff for the Blueberry headwater catchment (Fig. 4.4d boxplot inset). Therefore, the observed annual runoff from the Blueberry watershed gauge (ID: 07FC003) was used to estimate future water supply at the annual scale. Assuming mean total annual runoff will remain constant through the near future, allocation levels within the 1770 km^2 Blueberry watershed do not exceed 3% under any of shale development scenario combinations. While the Blueberry water supply will likely be sufficient to meet demand at the annual time scale, water scarcity may occur during the warmer months following the spring freshet (May–August) when water demand is high. Modelling results show a decrease in summer (May–August) runoff (relative to historical) in the near future (Fig. 4.4d) for the Blueberry catchment. Median

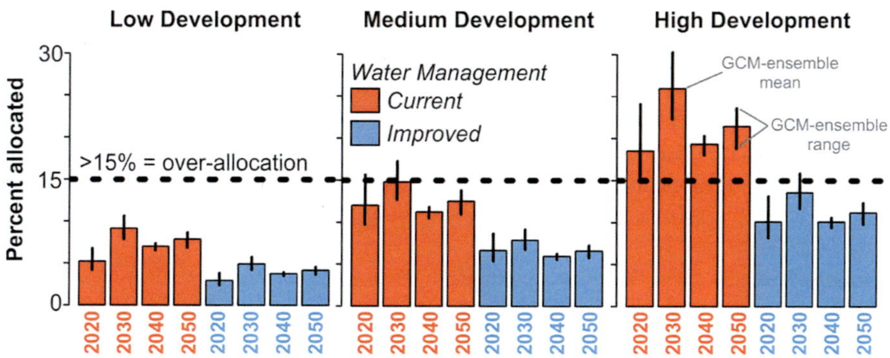

Fig. 4.6 Blueberry headwater catchment percent of mean annual runoff required to meet shale gas industry freshwater demands. Results are presented as centered 10-year averages (2020 = 2015–2024, 2030 = 2025–2034, etc.)

summer runoff in the historical and near future periods are 64.6 mm and 55.0 mm (a reduction of 14.8%) in the Blueberry headwater catchment and 40.6 mm and 35.5 mm (a reduction of 12.6%) in the Graham headwater catchment, respectively.

During drought periods, the BCOGC suspends shale gas industry water use from streams and lakes, but water stored in dugouts remains available for industrial use. With the projected decrease in summer streamflow volumes, dependency on water source dugouts may increase. Therefore, the future reliability of water source dugouts was analyzed by tabulating the CRHM-simulated monthly minimum dugout water levels (results not shown). During the 2020–2050 period, the modelled water source dugout maintained a minimum water level of at least 20% volume in the Graham catchment; however, minimum August dugout water levels decreased by up to 41% relative to the historic period. During late summer (August) in drought years, the dugout water supplies were exhausted in the Blueberry catchment in both the historical and near future periods. This result suggests that, in the plains region of the Peace River region, water source dugouts may not provide a reliable summer water source for the oil and gas industry during drought years. One of the assumptions in this study is that the total dugout storage volume is equal to the total annual ST-approval allocation volume. The reliability of water source dugouts could be improved by increasing the total storage volume to 1.5 or 2 times the total annual allocated volume.

4.8 Conclusions

Future increases in water demand may be met by water source dugouts or stream diversion points distributed throughout the shale gas development area and not focused only where development already exists. Thus, water quantity in larger watersheds in the Peace Region of NEBC will likely be adequate to meet the demands of the shale gas industry. Areas with high levels of oil and gas industry development may experience water scarcity, especially during drought conditions and under high development scenarios. If the shale gas industry in NEBC continues to expand, improved water management policies, e.g., water recycling and saline water use, will be needed to meet water demands, especially when competing interests (e.g. agricultural, domestic) are accounted for – which was not done in this study.

Acknowledgments This research was supported by the Pacific Institute for Climate Solutions, a Simon Fraser University graduate entrance scholarship to Jennifer Dierauer, and the Research Institute for Humanity and Nature, Kyoto, Japan.

References

BCOGC (2016) Water management for oil and gas activity 2015 Annual report'. Downloaded from https://www.bcogc.ca/node/13261/download on 20 March 2017

Cannon AJ (2015) Selecting GCM scenarios that span the range of changes in a multimodel ensemble: application to CMIP5 climate extremes indices. J Clim 28(3):1260–1267

Chapman, A., Kerr, B., Wilford, D (2012) Hydrological modelling and decision-support tool development for water allocation, Northeastern British Columbia; In: Geoscience BC summary of activities 2011, geoscience BC, Report 2012–1, 81–86

Fang X, Pomeroy JW (2007) Snowmelt runoff sensitivity analysis to drought on the Canadian prairies. Hydrol Process 21:2594–2609

Google Earth 7.1.8.3036 (32-bit). July 11, (2014) Blueberry river watershed, British Columbia, Canada. 56.96°N, 121.89°W, http://www.google.com/earth/index.html, Accessed 13 Mar 2017

Google Earth 7.1.8.3036 (32-bit). December 30, (2015) Graham river watershed, British Columbia, Canada. 56.42°N, 122.51°W, http://www.google.com/earth/index.html, Accessed 13 March 2017

Kniewasser, M. and M. Horne (2015) 'B.C. shale scenario tool technical report', Pembina institute, pp. 30. Downloaded from http://www.pembina.org/reports/pembina-bc-shale-scenario-tool-technical-report-20150618.pdf January 2017

Pacific Climate Impacts Consortium (2014) Statistically downscaled climate scenarios. Downloaded from http://tools.pacificclimate.org/dataportal/downscaled_gcms/map/ September 2016

Pomeroy JW, Gray DM, Brown T, Hedstrom NR, Quinton WL, Granger RJ, Carey SK (2007) The cold regions hydrological model: a platform for basin process representation and model structure on physical evidence. Hydrol Process 21:2650–2667

Pomeroy J, Fang X, Ellis C (2012) Sensitivity of snowmelt hydrology in Marmot Creek, Alberta, to forest cover disturbance. Hydrol Process 26:1891–1904

Rivard C, Lavoie D, Lefebvre R, Séjourné S, Lamontagne C, Duchesne M (2014) An overview of Canadian shale gas production and environmental concerns. Int J Coal Geol 126:64–76

Chapter 5
Monitoring Hot Spring Aquifer Using Repeat Hybrid Micro-gravity Measurements in Beppu Geothermal Field, Japan

Jun Nishijima, Kento Naritomi, Yayan Sofyan, Shinji Ohsawa, and Yasuhiro Fujimitsu

Abstract Repeat hybrid micro-gravity measurements were conducted to detect the gravity change caused by hot spring water production around Beppu in eastern Kyushu, Japan. An A10 #017 absolute gravimeter (Micro-g LaCoste) and a CG-5 #549 gravimeter (Scintrex) were used for this study in intervals of three to four months at eight gravity stations. According to the results obtained with the absolute gravimetry, a gravity change of up to 33 μgal was detected at the Beppu Geothermal Research Laboratory (BGRL) reference station. The observed absolute gravity was compared with the groundwater level, and there was a good correlation between the gravity changes and the groundwater level changes. Based on the precipitation, groundwater level, and soil character, the effect of the water content changes in the unsaturated zone was estimated precisely by using a Gwater-1D. This calculation can explain that the gravity seasonal changes were caused by the groundwater level changes. After removal of noise effects (e.g., tidal movement, precipitation, and shallow groundwater level changes), the residual gravity changes, which were measured by the relative gravimeter, were subdivided into two types of responses. Gravity changes up to 90 μgal were observed from April 2014 to July 2015. After that, gravity became stable, except for small seasonal changes.

Keywords Gravity change · Hot spring reservoir · Monitoring · Sustainability

J. Nishijima (✉) · K. Naritomi · Y. Sofyan · Y. Fujimitsu
Faculty of Engineering, Kyushu University, Fukuoka, Japan
e-mail: nishijima@mine.kyushu-u.ac.jp

S. Ohsawa
Beppu Geothermal Research Laboratory, Institute for Geothermal Sciences, Graduate School of Science, Kyoto University, Beppu, Oita, Japan

© Springer Nature Singapore Pte Ltd. 2018
A. Endo, T. Oh (eds.), *The Water-Energy-Food Nexus*, Global Environmental Studies, https://doi.org/10.1007/978-981-10-7383-0_5

5.1 Introduction

Beppu city, which is one of the largest and most famous hot spring resort areas, is located in the eastern part of Kyushu island, Japan (Fig. 5.1). The Hohi volcanic zone, which is located in central Kyushu, was dominated by volcanic activity during the Neogene and Quaternary periods (Kamata 1989). The Beppu geothermal field is located in the eastern flank of Mt. Tsurumidake and Mt. Garandake, which are the late Quaternary active volcanic centers. Figure 5.2 depicts the geological map. The geological setting in Beppu comprises Pliocene andesite (Kankaiji andesite), Pleistocene volcanic rock, and fan sedimentary deposits (Hoshizumi et al. 1988). Cretaceous granite is exposed in the north and south areas just outside of the study area (Sasada 1987). Cretaceous granite also crops out locally in this area. Sasada (1984) pointed out that Cretaceous granite is widely spread out throughout northern central Kyushu as well.

Exploitation of Beppu city began as early as the 1880s, and the amount of production from the hot springs is increasing (Yusa et al. 2000). Recently, 50,000 tons of hot spring water was consumed for bathing and sightseeing. Moreover, small-scale geothermal power plants have been introduced in some places since 2014. Hot spring water production has affected the hot spring aquifer, and the number of

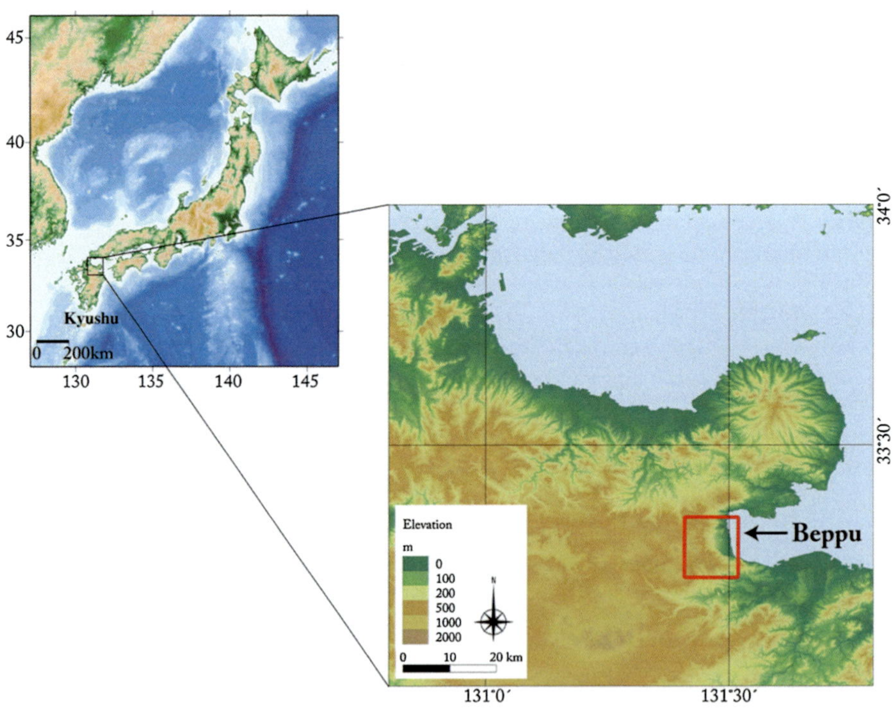

Fig. 5.1 Location of the Beppu geothermal field, central Kyushu, Japan

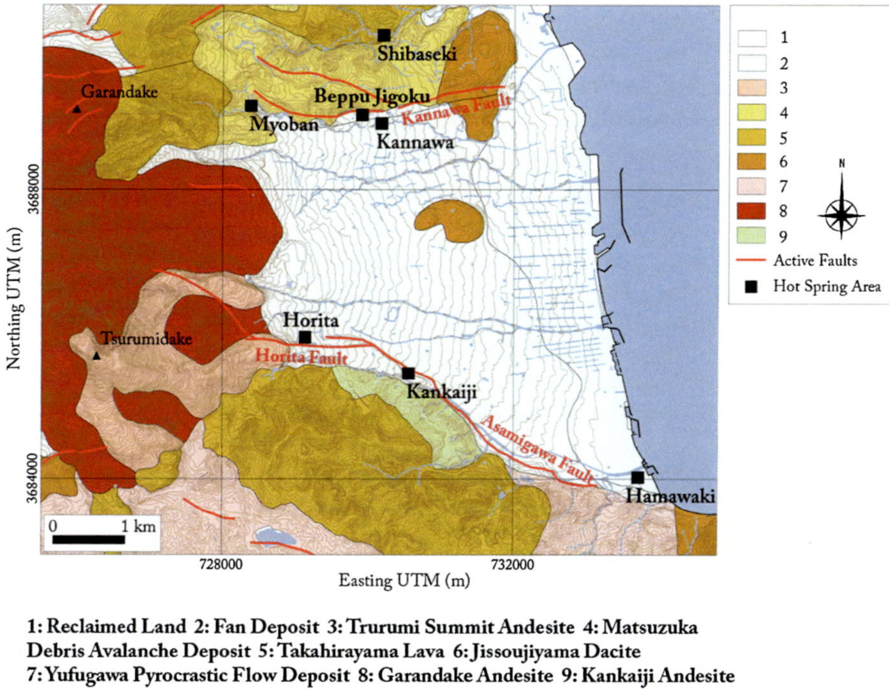

Fig. 5.2 Geological map of the study area and its surroundings. Modified from the Geological Survey of Japan (2014). The red lines and the black squares show the locations of the active faults (Chida et al. 2000) and the hot springs areas, respectively

artesian hot spring wells has decreased since 1960s. Therefore, it is important to establish a hot spring monitoring method for sustainable utilization.

Some previous research has been conducted to monitor hot springs resources using hot springs well properties (temperature and pressure) and chemical component changes. Repeat micro-gravity hybrid measurements is one effective method to detect the water-level change in a hot spring aquifer. This method is applied to reservoir monitoring in some geothermal power stations (Allis and Hunt 1986). The production of hot spring water causes mass changes with a measurable gravity change on the ground surface. In a geothermal power station, gravity monitoring is conducted for geothermal resources monitoring. Previous study results have shown that gravity monitoring can detect the mass balance of a geothermal reservoir and could be an index of sustainability of a geothermal power plant (Nishijima et al. 2010). Yet, little study has been done to apply this method to hot spring aquifer changes. This study was undertaken in order to detect hot spring water level changes using gravity changes.

5.2 Gravity Measurement

5.2.1 *Absolute Gravity Measurement*

The A10 absolute gravimeter manufactured by Micro-g LaCoste Inc. can operate on a 12 V DC power supply, such as a vehicle battery. This gravimeter mainly consists of three parts: a dropper, an interferometer unit, and a controller. The interferometer unit includes a laser, an interferometer, and a long-period inertial isolation device. In order to obtain high accuracy, an atomic clock and laser interferometer were introduced to accurately measure the time and position of the test mass. A test mass was dropped vertically in a vacuum chamber to an average distance of 7 cm.

The raw gravity data were processed using the "g" version 9 software (Micro-g LaCoste Inc.) specifically designed to acquire and process gravity data with the Micro-g LaCoste absolute gravimeter. The software needs the input of some parameters, including the location of the site (latitude, longitude, and altitude) and geophysical corrections. We can correct the effect of the tidal movement, ocean load, barometric pressure, and polar motion when acquiring the gravity data. There were two gravity stations for the absolute gravity measurement. Table 5.1 shows the setting of the measurement. It took about 30 min for the measurements, and the error of measurements is about 10 μgal.

5.2.2 *Relative Gravity Measurement*

The measurement of relative gravity began in 2014 with intervals of three to four months, using a Scintrex CG-5 #549 relative gravimeter. There were eight stations for relative gravity measurements (Fig. 5.3). Because an instrumental drift is included in the measured gravity, it is necessary to remove the instrumental drift with high accuracy. A two-way measurement method was applied to evaluate the instrumental drift precisely. A reference station was set up at the Beppu Geothermal Research Laboratory (BGRL), Kyoto University. Gravity changes at the reference station were monitored using an A10 absolute gravimeter.

The measured gravity data, such as the earth tide, the height of the instrument, and the instrumental drift, were reduced to the necessary corrections. The earth tidal corrections, including the ocean tidal effect, were calculated using GOTIC2 (Matsumoto et al. 2001) instead of the instrument's internal program. The errors of each observation were estimated within 10 μgal.

Table 5.1 A10 absolute gravimeter settings

Drop interval	1 sec
Number of drop/set	100 drops
Set interval	6 min
Number of sets	10

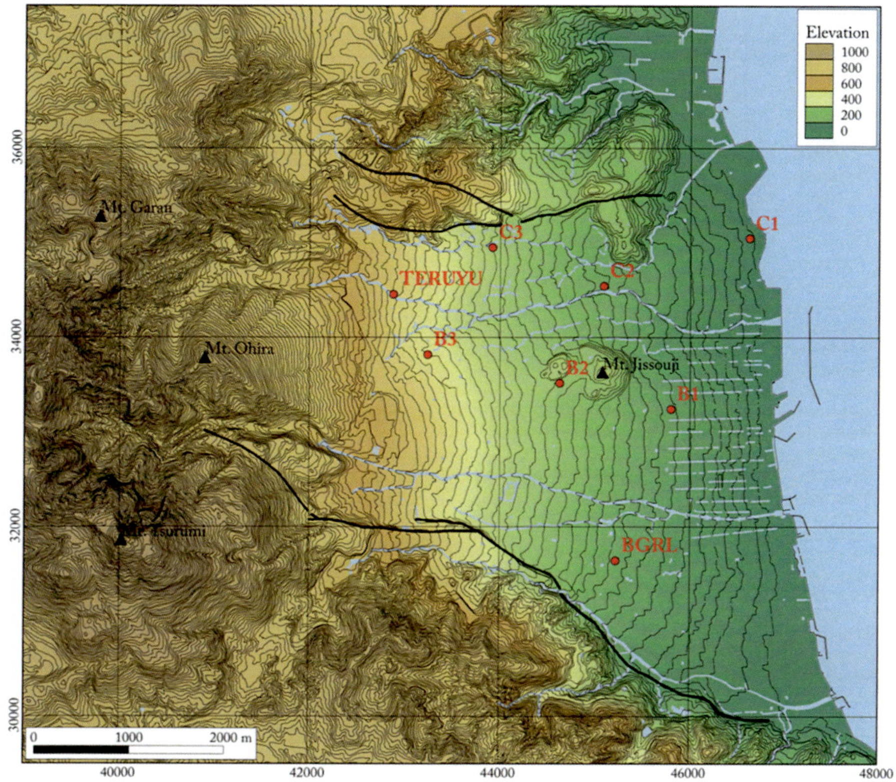

Fig. 5.3 Distribution of the gravity stations

5.3 Result and Discussion

5.3.1 Absolute Gravity Changes

Figure 5.4 shows the results of the absolute gravity measurements at the BGRL station. Because the instrument had trouble with the laser system, the observation period was shorter than the relative gravity measurement (from April 2014 to March 2015). A gravity increase up to 33 μgal was observed in November 2014 at this station. After that, the gravity changes were less than 5 μgal. A comparison between the gravity changes and groundwater level changes shows that there is a good correlation between those data. When the groundwater aquifer assumes the semi-infinite slab, the relationship between the gravity and the groundwater level changes are expressed by

$$\begin{aligned}\Delta g &= 2\pi G\phi\Delta h \\ &= 0.419\phi\Delta h\end{aligned}$$

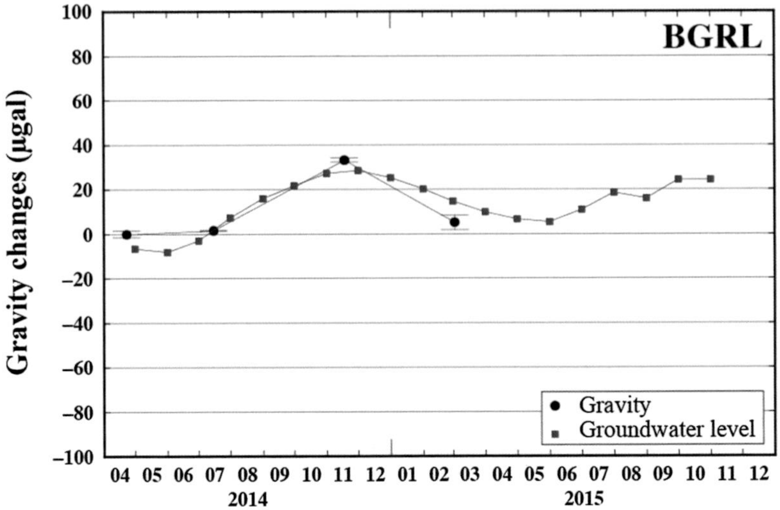

Fig. 5.4 Absolute gravity data at the reference station (BGRL)

Δg: Gravity changes (μgal)
G: Gravity constant (6.673×10^{-11} m^3 kg^{-1} s^{-1})
φ: Porosity
Δh: Groundwater level change (m)

When fitting an approximation line, the calculated gravity fluctuation ($\Delta g/\Delta h_w$) was 9.88 μgal/m and the porosity (%) was 23.6% (Fig. 5.5). Since this value is close to the measured porosity of alluvial fan sediments (20.0%) (Japanese Association of Groundwater Hydrology 1979), it is concluded that the observed gravity changes at the BGRL are caused by the seasonal groundwater level changes.

In order to predict the gravity changes precisely from groundwater level and precipitation, Gwater-1D (Kazama and Okubo 2009) was applied. This method consists of two parts. One simulates the water content changes caused by the soil water infiltration in the unsaturated zone. The other predicts the gravity changes caused by the volume water content change. This method needs some input data, such as precipitation, evapotranspiration, soil parameters, and groundwater level (Fig. 5.6). Evapotranspiration is estimated using the Thornthwaite method (Thornthwaite 1948). Thornthwaite method can estimate an evapotranspiration easily using annual precipitation and monthly average temperature. The soil characteristics, such as saturated permeability and effective porosity, were measured using soil samples (Table 5.2).

Figure 5.7 shows the comparison between the observed and predicted gravity changes. The predicted gravity changes were almost fitted to the observed gravity changes. This means that the observed gravity changes can be explained by the

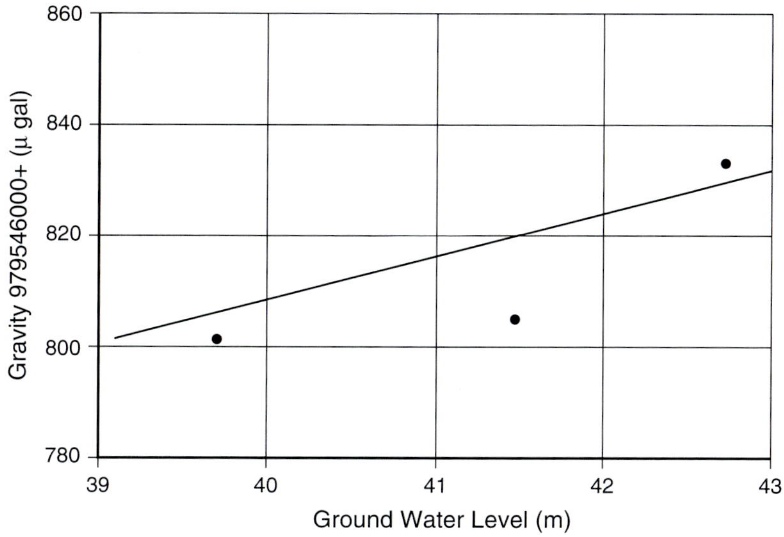

Fig. 5.5 Quantitative comparison between the gravity and water level at the BGRL

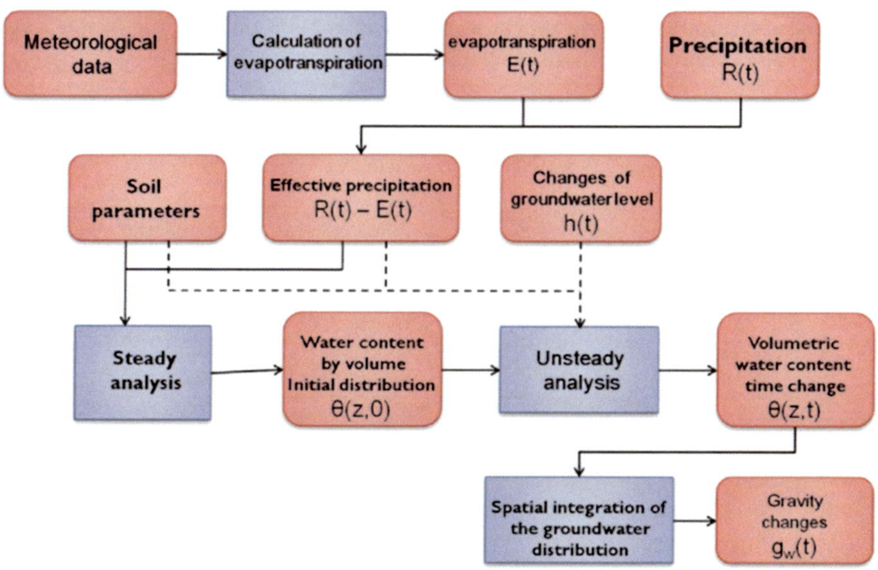

Fig. 5.6 Flowsheet of Gwater-1D (Kazama and Okubo 2009)

Table 5.2 Parameters of soil characteristics

Saturated permeability (Ks)	2.91×10^{-5} (m/s)
Effective porosity (n)	0.53

water content changes in the unsaturated zone. The effect of seasonal gravity changes at the reference station was corrected to the gravity changes, which were measured by relative gravimeter. These seasonal gravity changes were corrected using observed (from April 2014 to March 2015) and predicted (from July 2015 to August 2016) gravity changes.

5.3.2 Relative Gravity Changes

The observed changes at eight stations can be classified into two categories according to the characteristics of the trend (Fig. 5.8). These categories are distributed throughout the northern and central parts of Beppu city as described below.

Northern area (C1, C2, C3, TERUYU) These stations are located in the northern part of Beppu city. Many hot springs and fumaroles are distributed throughout this area. A new small-scale binary geothermal power plant commenced on December 2014 near the TERUYU gravity station. Gravity increased until July 2015 and then became stable, except at the C3 station. The average gravity difference between 2014 and after mid-2015 was 40 μgal. The C3 gravity station is located near the Kannawa hot spring and Beppu-Jigoku, which includes fumaroles and a hot spring area. There are some factors that cause gravity change, such as groundwater level change, mass change in the hot spring aquifer, and more.

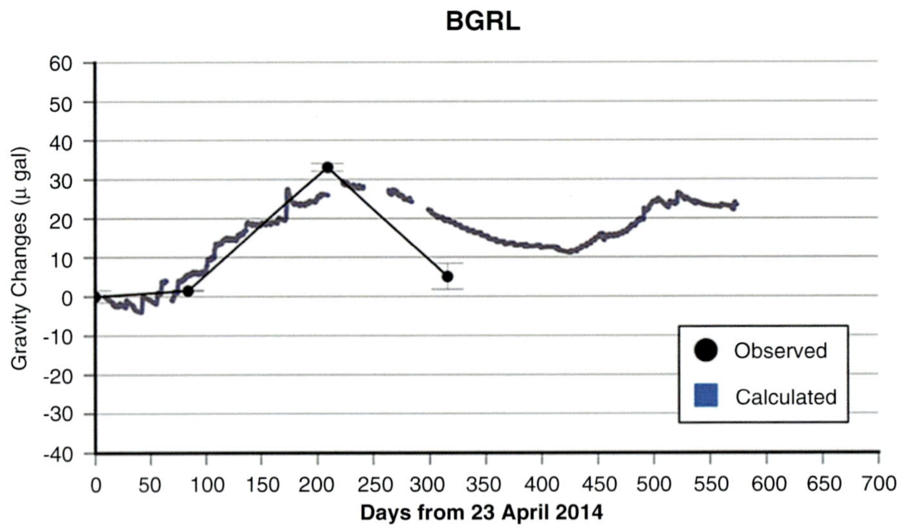

Fig. 5.7 Comparison between observed and predicted gravity changes

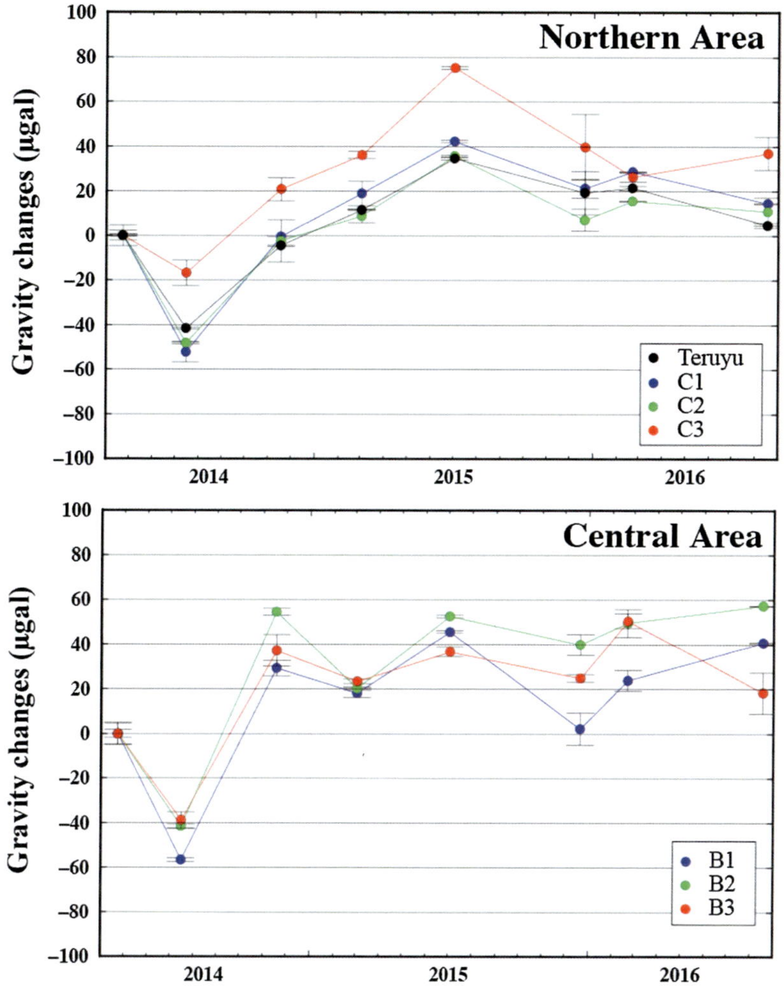

Fig. 5.8 Gravity changes measured using the relative gravimeter

Central area (B1, B2, B3) These stations are located in the central part of Beppu city. A fan deposit is accumulated thickly in this area, and there is no hot spring in the western and southern flanks of Mt. Jissoji-yama. Though a stepped gravity change was observed from July 2014 to March 2015, gravity was basically stable, except for small seasonal changes. The stepped difference of the average gravity change was 50 μgal. According to the geological map, the fan deposit was much thicker than that of the reference station. Though data on the effect of seasonal changes at the reference station were collected, it appears to be different in the groundwater level's seasonal changes.

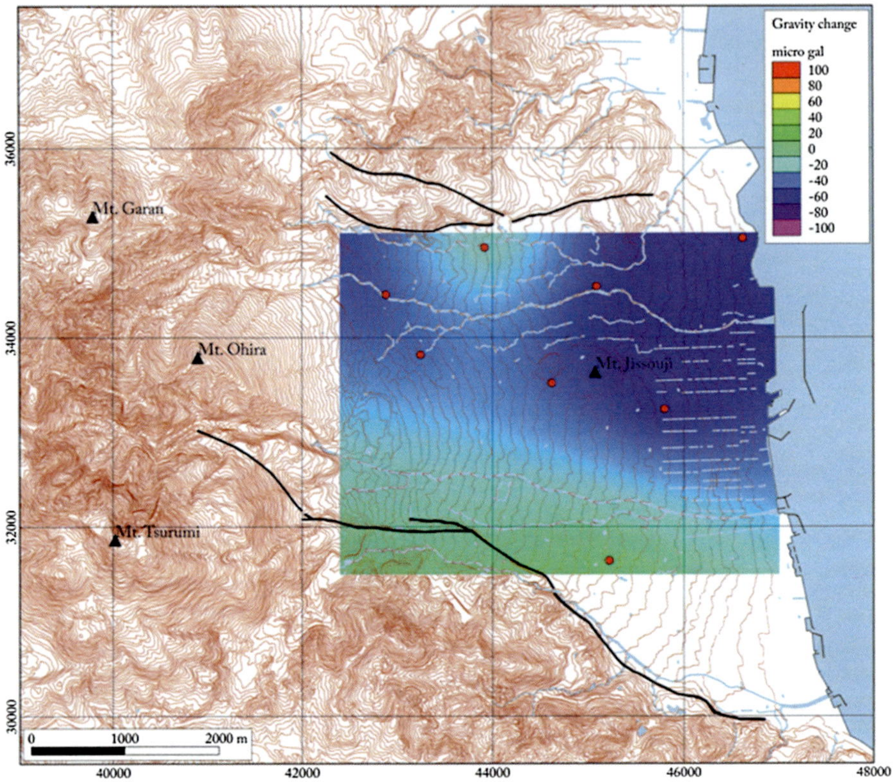

Fig. 5.9 Distribution of gravity changes from April 2014 to July 2014

Figures 5.9 and 5.10 show the distribution of gravity changes before the commencement of the binary geothermal power plant. From July 2014 to November 2014, a large gravity increase was detected in the entire region, while from April 2014 to July 2014, a gravity decrease was seen everywhere except the BGRL. Moreover, there was a huge difference in the amplitude of gravity changes between the central area (B1, B2, B3) and the northern area (C1, C2, C3, TERYU), as seen in Fig. 5.10. Based on a three-dimensional gravity model, it is assumed that these differences are caused by the thickness of alluvial fan deposits, which are widely distributed in Beppu city (Nishijima and Naritomi 2017). Since the porosity in fan deposits is relatively high, the sediments in the central area can store larger amounts of groundwater. Therefore, the local difference of gravity changes occurred in the second period.

Figure 5.11 shows the distribution of gravity changes from July 2014 to July 2015. In this period, the gravity increased on a large scale at all benchmarks, and

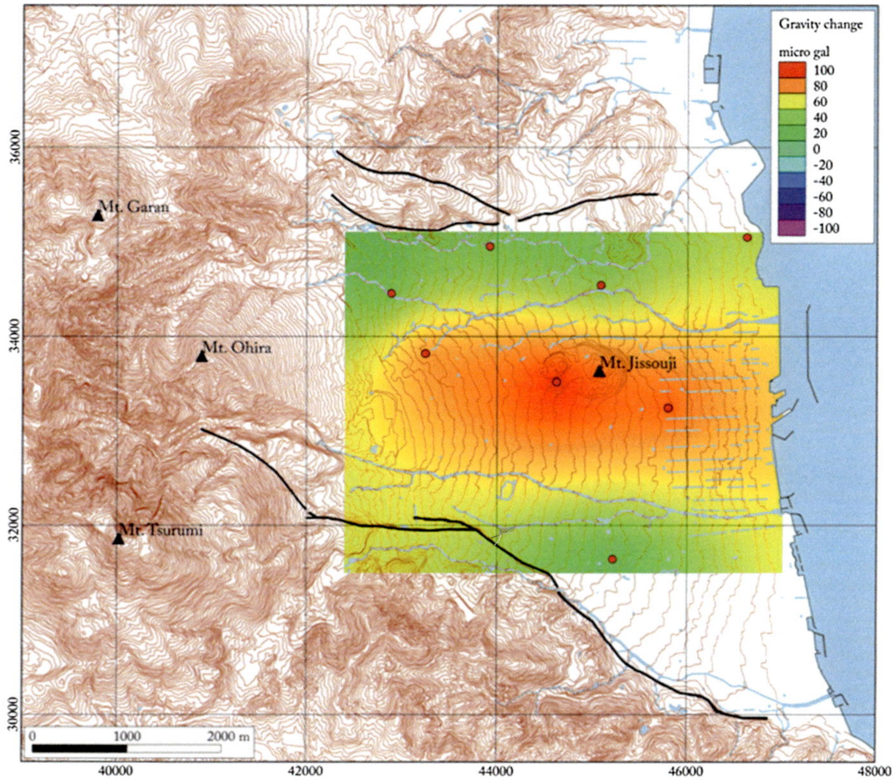

Fig. 5.10 Distribution of gravity changes from July 2014 to November 2014

there were no particular gravity changes near the northwestern area, in which the new binary power plant is located. Figure 5.12 shows the distribution of gravity changes from April 2014 to July 2015. A gravity increase was observed in almost all gravity stations, but the degree of the gravity changes was decreasing. The maximum gravity increase was at the C3 station, which is close to the Kannawa hot spring and the Beppu-jigoku area.

In order to evaluate the sustainability of hot spring use, more gravity and groundwater level data need to be accumulated. Also, modeling of hydrological groundwater disturbances (Kazama and Okubo 2009) is also necessary to detect the gravity changes caused by hot spring water use and the binary geothermal power plant operation.

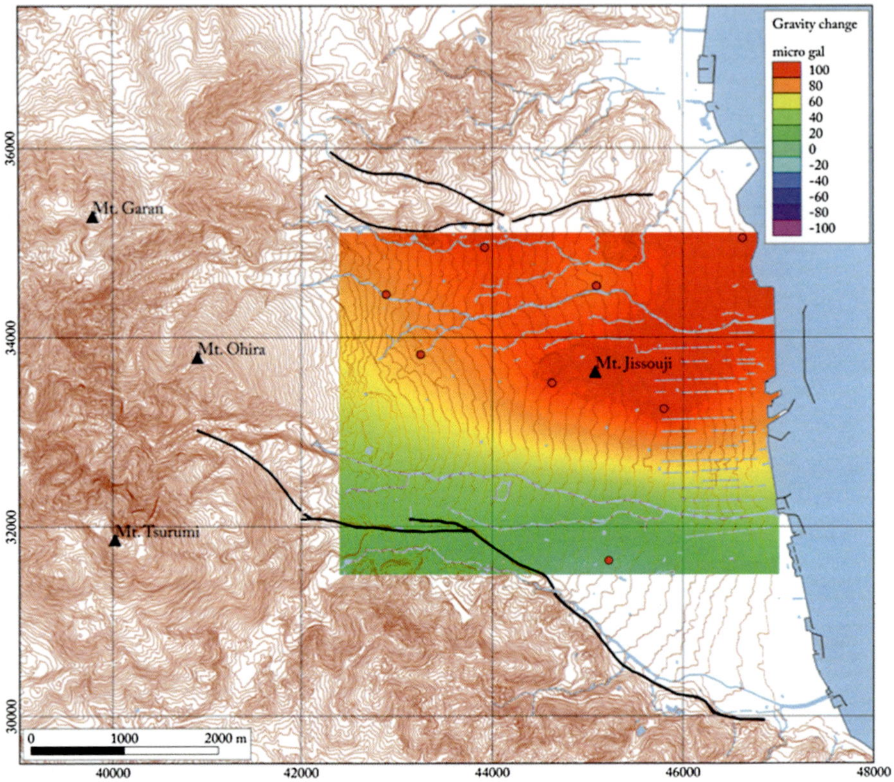

Fig. 5.11 Distribution of gravity changes from July 2014 to July 2015

5.4 Conclusion

Repeat micro-gravity hybrid measurements were conducted in order to evaluate the sustainability of hot spring water use. According to the result of the absolute gravity measurements, the causes of the gravity changes were considered by referring to the groundwater level data as well as the three-dimensional gravity model. Based on qualitative and quantitative explanations, the observed gravity changes at the reference station (BGRL) have a good correlation with the seasonal groundwater level changes. The distribution maps of the gravity changes from July 2014 to November 2014 show a huge difference in the gravity changes between the central area (B1, B2, B3) and the northern area (C1, C2, C3, TERYU) due to the thickness of alluvial fan deposits based on the distribution map of the gravity basement. However, more gravity measurements should be accumulated to detect the gravity changes caused by hot spring water use and the geothermal power station operation in the northwestern part of the study area.

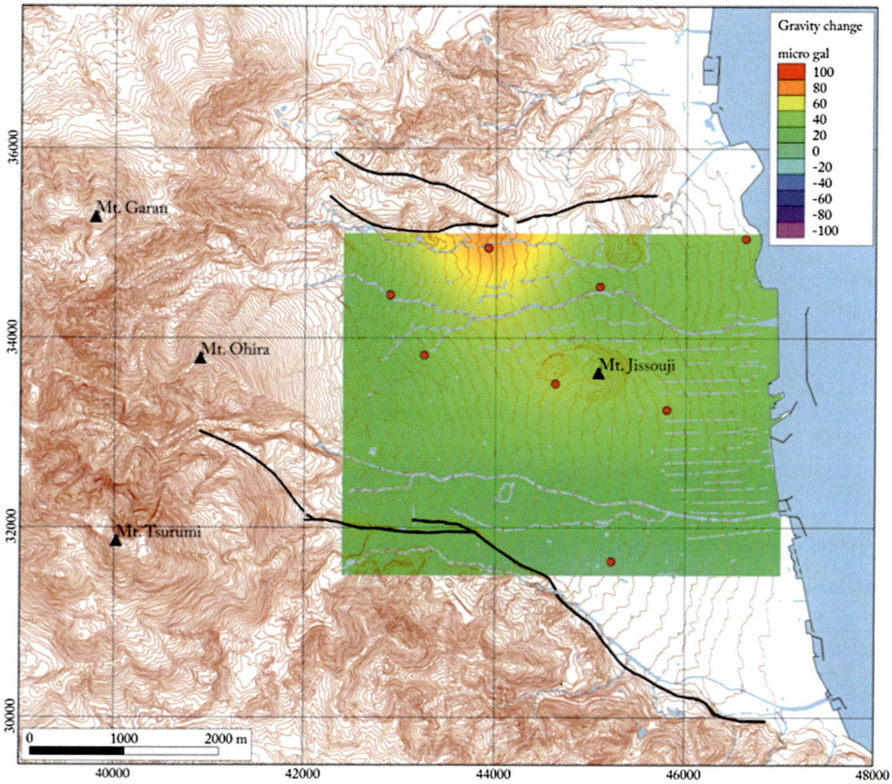

Fig. 5.12 Distribution of gravity changes from April 2014 to July 2015

Acknowledgments This research was financially supported by the project (No. 14200097), entitled "Human-Environment Security in Asia-Pacific Ring of Fire: Water-Energy-Food Nexus," by the Research Institute for Humanity and Nature (RIHN), Kyoto, Japan.

References

Allis RG, Hunt TM (1986) Analysis of exploitation induced gravity changes at Wairakei geothermal field. Geophysics 51:1647–1660

Chida N, Ikeda Y, Nakada T, Okada A, Une H (2000) 1:25000 Active fault map in urban area, Beppu, Geographical Survey Institute Japan. (in Japanese)

Geological Survey of Japan, AIST (ed) (2014) Seamless digital geological map of Japan 1: 200,000. Jan 14, 2014 version. Geol Surv Japan, National Institute of Advanced Industrial Science and Technology

Hoshizumi H, Ono K, Mimura K, Noda T (1988) Geology of the Beppu district with geological sheet map of 1:50,000. Geol Surv Japan (in Japanese with English abstract)

Japanese Association of Groundwater Hydrology (1979) Handbook of groundwater.. (in Japanese)

Kamata H (1989) Volcanic and structural history of the Hohi volcanic zone, central Kyushu, Japan. Bull 51:315–332. https://doi.org/10.1007/BF01056894

Kazama T, Okubo S (2009) Hydrological modeling of groundwater disturbances to observed gravity: theory and application to Asama volcano, Central Japan. J Geophysical Research 114:B08402

Matsumoto K, Sato T, Takanezawa T, Ooe M (2001) GOTIC2: a program for computation of oceanic tidal loading effect. Journal of the Geodetic Society of Japan 47:243–248

Nishijima J, Naritomi K (2017) Interpretation of gravity data to delineate underground structure in the Beppu geothermal field, central Kyushu, Japan. Journal of Hydrology: Regional Studies 11:84–95. https://doi.org/10.1007/BF01056894

Nishijima J, Saibi H, Sofyan Y, Shimose S, Fujimitsu Y, Ehara S, Fukuda Y, Hasegawa T, Taniguchi M (2010) Reservoir monitoring using hybrid micro-gravity measurements in the Takigami geothermal field, central Kyushu, Japan. Proceedings of the World Geothermal Congress 2010

Sasada M (1984) Basement structure of the Hohi geothermal area, central Kyushu, Japan. J Japan Geothermal Energy Association 21:1–11. (in Japanese with English abstract)

Sasada M (1987) Pre-tertiary basement rocks of Hohi area, central Kyushu, Japan. Bull Geol Surv Japan 38:385–422. (in Japanese with English abstract)

Thornthwaite CW (1948) An approach toward a rational classification of climate. Geogr Rev 38:55–94

Yusa Y, Ohsawa S, Kitaoka K (2000) Long-term changes associated with exploitation of the Beppu hydrothermal system, Japan. Geothermics 29:609–625. https://doi.org/10.1016/s0375-6505(00)00025-0

Chapter 6
Evaluation of the Shallow Geothermal Potential for a Ground-Source Heat Exchanger: A Case Study in Obama Plain, Fukui Prefecture, Japan

Hideki Hamamoto, Yuji Miyashita, and Daisuke Tahara

Abstract A ground-source heat exchanger (GHE) is an energy system exploiting shallow geothermal energy that is economical, environmentally friendly, and is rapidly increasing in popularity worldwide. Evaluating the available subsurface heat energy through thermal response tests and/or numerical simulations to design appropriate GHE systems (e.g. deciding the depth and number of boreholes for heat exchange) is important. Geological structures, groundwater properties, and subsurface temperatures are essential input data for such numerical simulations.

In the present study, we demonstrate the application of the GHE potential map, a new method based on the regional geological structure, subsurface temperature, and groundwater flow. Our target area is the Obama Plain in the central part of Japan, which faces the Sea of Japan. Subsurface temperature measurements at four stations in the Obama Plain were used for the present evaluation.The results of the GHE numerical simulations show a linear increase in GHE efficiency of about 35% with a subsurface temperature increase from 15 °C to 20 °C. In addition, GHE efficiency approximately triples when the groundwater flow ranges from 0 to 10 m/year. We estimated the specific heat-extraction for each 100 m × 100 m grid cell. The mapping results indicate a high GHE potential for the central part of the Obama Plain, with a value of more than 100 W/m. The specific heat-extraction ranges from 50 to 110 W/m in most areas of the Obama Plain. This value is sufficient for GHE to cover the potential heat-extraction demand for the Obama Plain. Our evaluation method can be applied to other plains in Japan and around the world.

H. Hamamoto (✉)
Center for Environmental Science in Saitama, Senior Researcher, Saitama, Japan

Y. Miyashita
Hot Springs Research Institute of Kanagawa Prefecture, Senior Researcher, Kanagawa, Japan
e-mail: miyasita@onken.odawara.kanagawa.jp

D. Tahara
Faculty of Marine Bioscience, Fukui Prefectural University, Associate Professor, Fukui, Japan
e-mail: tahara@fpu.ac.jp

© Springer Nature Singapore Pte Ltd. 2018

A. Endo, T. Oh (eds.), *The Water-Energy-Food Nexus*, Global Environmental Studies, https://doi.org/10.1007/978-981-10-7383-0_6

Keywords Shallow geothermal energy · Renewable energy · Ground-source heat exchanger · Heat pump · Obama plain · Groundwater flow · Subsurface temperature

6.1 Introduction

Shallow geothermal energy is a form of renewable energy that originates mainly from solar energy and terrestrial heat flow. To exploit shallow geothermal energy, a ground-source heat exchanger (GHE) system has been developed (also called a "borehole heat exchanger (BHE)" or a "geothermal heat exchanger (GHE)" system). GHE systems have many applications, such as for air conditioning, floor heating, and water heating.

GHE systems are divided into two types, which are commonly used for a heat pump unit: the closed-loop type and the open type (e.g. Sarbu and Sebarchievici 2015). Figure 6.1 shows a schematic of a closed-loop GHE system, where the indirect exchange of geothermal energy is achieved through a fluid flowing in a U-tube,

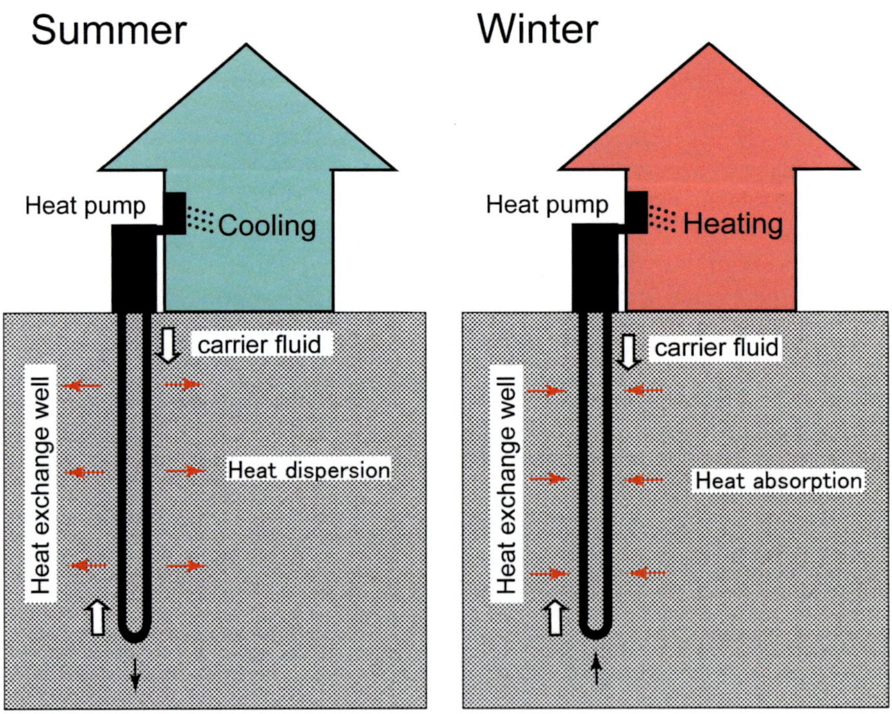

Fig. 6.1 Illustration of the closed-loop type ground-source heat exchanger (GHE) for an air conditioner. The left side is the room cooling mode in summer and the right side is the room heating mode in winter

which is made of a high-density polyethylene pipe. The carrier fluid is water or a mixture of water and anti-freeze fluid. In the open GHE system, the groundwater is pumped directly from the subsurface, where the groundwater is in thermal equilibrium with the surrounding strata. Herein, we investigate the closed-loop GHE system because it is the most commonly used system worldwide.

Lund and Boyd (2015) investigated geothermal energy diffusion in each country. Figure 6.2 shows the capacity of the top nine countries in installed GHE capacity and Japan. The top five countries are the United States, China, Sweden, Germany, and France. Switzerland is the leader in installed capacity per capita. Currently, the total installed capacity is 49,898 MWt, and the annual energy use is 325,028 TJ/year, with a capacity factor of 0.21 (in the heating mode). If we assume that each GHE system contributes 12 kW on average (typical for homes in the United States and Western Europe), then the above total amount corresponds to approximately 4.16 million systems worldwide reported in 2015. There has been 51% and 300% increase in the number of installed units since 2010 and 2005, respectively.

The total installed GHE capacity in Japan is only approximately 100 MWt; 84% are closed-loop GHE, 15% are open GHE, and 1% are both system (used closed-loop GHE and open GHE in one building). GHE usage in northern Japan (Hokkaido and Tohoku) is greater than that in the southern areas of the country (Japan Ministry of the Environment 2012 and 2016). The reason for this is that the cold in the northern areas results in greater heat demand and consequently, higher energy costs (Japan Sustainable Building Consortium 2016). The number of installed GHE

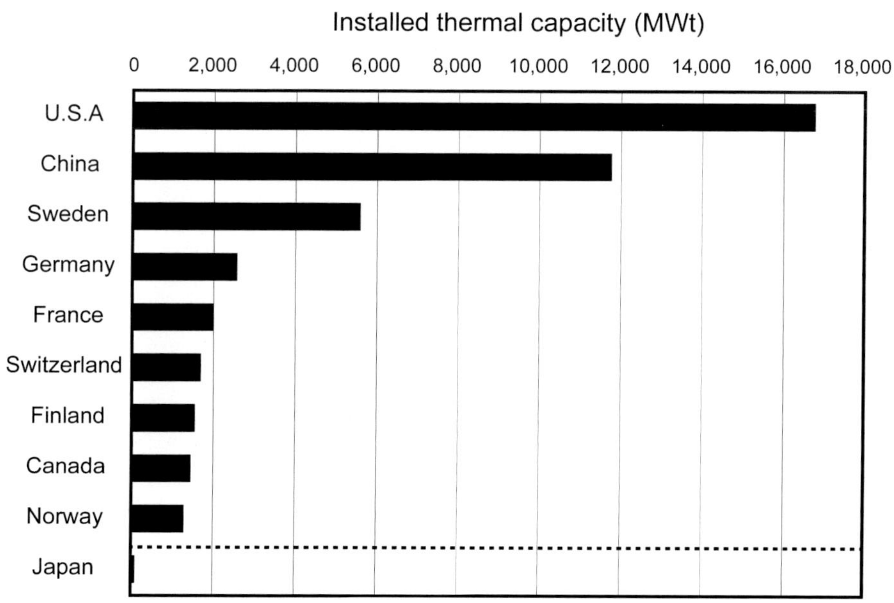

Fig. 6.2 The amount of installed thermal capacity of the top 9 countries and Japan (drawn from Lund and Boyd 2015)

systems is low (Fig. 6.2) because GHE is not well recognized, the systems have a high installation cost, and the local geological conditions are complex, resulting in variable GHE efficiencies depending on the installation site. Therefore, it is important to estimate the specific heat-extraction based on geological information, subsurface temperature, and groundwater flow before installing a GHE system.

The thermal response test (TRT) is a method for obtaining direct measurements of the GHE effective thermal conductivity, which is related to the specific heat-extraction (e.g. Stauffer et al. 2014). In this test, the temperature of the water that flows into and out of the U-tube is monitored for 24 h while the fluid is heated and for another 48 h after the heating is stopped. In a newer method, an electric heating cable and temperature sensors are attached to facilitate measurement. However, these methods incur high costs. Creating a potential map would greatly simplify the process and result in lower costs. Potential maps have been constructed for several areas in the world (Blum et al. 2011, Senate Department for Urban Development and Housing 2015, Shrestha 2015).

Here we propose a new method for constructing GHE potential maps. As an example, we focus on the Obama Plain, which is located in the western part of the Japanese Main Island (Honshu) and faces toward Obama Bay and the Sea of Japan (Fig. 6.3). The Obama Plain is mostly occupied by the city of Obama. The population of Obama is 29,670, and its population density is 127 people/km^2. The total area is 233.09 km^2 (Obama City 2016). The Obama Plain is surrounded by mountains with an elevation of less than about 800 m and has two main rivers: Kitagawa (the north

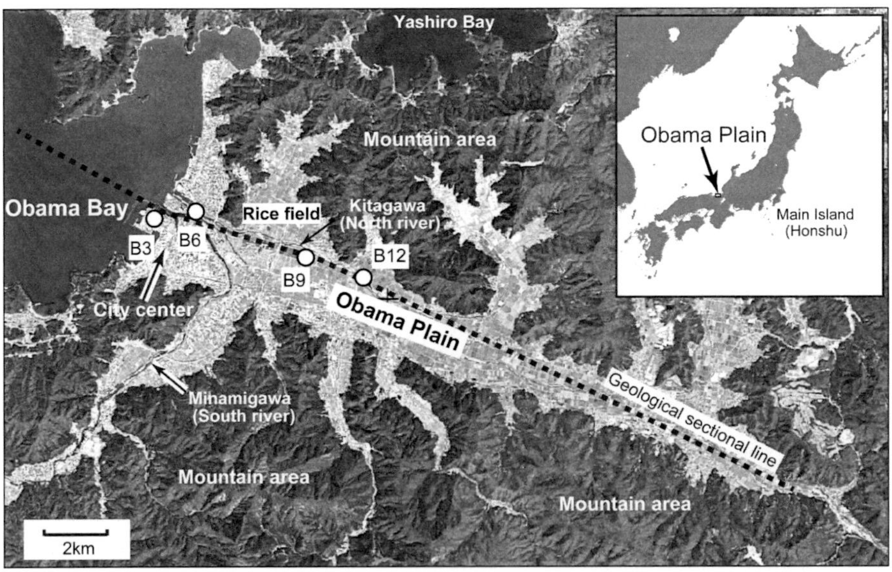

Fig. 6.3 Map of the Obama Plain. Open circles are sites of subsurface temperature measurements. The broken line represents the geological cross section in Fig. 6.4. (Base map is from GSI Maps of Japan)

river) and Minamigawa (the south river). The Obama Plain is about 15 km long and 1.5 km wide along Kitagawa, and about 10 km long and 1 km wide along Minamigawa (Fig. 6.3). It is one of the primary sites of the integrated research project of Human-Environmental Security in Asia-Pacific Ring of Fire: Water-Energy-Food Nexus undertaken by the Research Institute for Humanity and Nature (RIHN) of Japan (Taniguchi et al. 2015). The project aims to maximize human-environmental security and minimize vulnerability by implementing management structures and policies that optimize both the water-food and water-energy connections in the Asian Pacific coastal regions. The target areas in Japan are the Obama, Beppu, and Otsuchi areas. Researchers from disciplines such as Earth Sciences, Agriculture, Social Sciences, and Economics are participating in the project. The Obama plain is just right size for a typical model case of potential evaluation for ground-source heat exchanger. And detailed hydrological model have been constructed by Obama City with Nippon Koei Corporation (Obama City 2017). So, we could conduct new integrated evaluation considered geological information, subsurface temperatures measured in this research, and groundwater properties (groundwater flow and saturated condition) from the hydrological model.

6.2 Subsurface Heat Transfer Theoretical Framework

In the GHE system (closed-loop type), heat is exchanged between the carrier fluid inside the U-tube and the surrounding strata. The relation between the subsurface parameters and the transferred heat is generally described by the following fundamental 3D equation (e.g., Carslaw and Jaeger 1986, Domenico and Schwartz 1997, Turcotte and Schubert 2014).

$$\left(\rho C_p\right)_e \frac{\partial T}{\partial t} = \lambda_e \left(\frac{\partial^2 T}{\partial x^2} + \frac{\partial^2 T}{\partial y^2} + \frac{\partial^2 T}{\partial z^2} \right) - \left(\rho C_p\right)_f \left(u_x \frac{\partial T}{\partial x} + u_y \frac{\partial T}{\partial y} + u_z \frac{\partial T}{\partial z} \right) \tag{6.1}$$

where $(\rho C_p)_e$ is the effective heat capacity of the stratum (J/(m^3·K)); $(\rho C_p)_f$ is the groundwater heat capacity (J/(m^3·K)); T is the subsurface temperature (K); λ_e is the effective thermal conductivity of the stratum (W/m·K); and u_x, u_y, and u_z are the velocities of the groundwater flow (x-, y-, and z-direction, respectively). The relation between $(\rho C_p)_e$, $(\rho C_p)_f$, and the heat capacity of solid soil or rock $(\rho C_p)_s$ is expressed by the following equation (2) using porosity ε;

$$(1-\varepsilon)\left(\rho C_p\right)_s + \varepsilon\left(\rho C_p\right)_f = \left(\rho C_p\right)_e \tag{6.2}$$

When heat is generated, the equation is extended to the following eq. (3) using the heat generation S (W/m^3);

$$\left(\rho C_p\right)_e \frac{\partial T}{\partial t} = \lambda_e \left(\frac{\partial^2 T}{\partial x^2} + \frac{\partial^2 T}{\partial y^2} + \frac{\partial^2 T}{\partial z^2} \right) - \left(\rho C_p\right)_f \left(u_x \frac{\partial T}{\partial x} + u_y \frac{\partial T}{\partial y} + u_z \frac{\partial T}{\partial z} \right) + \mathrm{S} \quad (6.3)$$

Examples of heat generation include radioactive heat generation from subsurface rocks or heat wasting from a heat exchanger well. Because radioactive heat generation is generally low (only a few μW/m^3), it is considered negligible and disregarded when constructing the GHE model. In the above equations, $(\rho C_p)_e$, λ_e, and ε are related to the subsurface soil or rock, and $(\rho C_p)_f$, u_x, u_y, and u_z are used to express the groundwater properties. Thermal conductivity, heat capacity, and porosity measurements are generally difficult to obtain directly without using extremely specialized and expensive instruments. Therefore, we inferred the rock properties from lithology.

6.3 Investigation of the Obama Plain

6.3.1 Geological Setting

The basement of the Obama Plain is composed of sedimentary rocks and an allochthonous rock complex (Public Corporation of Fukui Construction Technology 2010). The sedimentary rocks are Permian–Jurassic sandstones, conglomerates, and shales in alternation. The allochthonous complex comprises greenstone, limestone, and chert. These basement formations are overlain by unconsolidated gravel, sand, and clay. This alluvial cover is especially thick (about 100 m) in the area along Kitagawa.

6.3.2 Groundwater Flow

The Obama City local government conducted an investigation of the quantity of groundwater resources in the Obama Plain using field surveys and groundwater simulations with the Nippon Koei Corporation (Obama City 2017). They performed long-term monitoring of 12 groundwater observation wells, which were built for this investigation. Results revealed that groundwater level decreased by pumping groundwater for snow melting equipment during winter.

In addition, groundwater flow was modeled by combining observations and existing information using the Modflow simulator, which was developed by the United States Geological Survey (USGS) (Harbaugh 2005). The city of Obama and the Nippon Koei Corporation produced a geological model of the Obama Plain using existing lithological information and the geological map of the area (Fig. 6.4). Next, they constructed a 3D hydrogeological model (20 km × 13 km, 13 layers, 100 m × 100 m mesh size) for the simulation. They considered the basement rock as

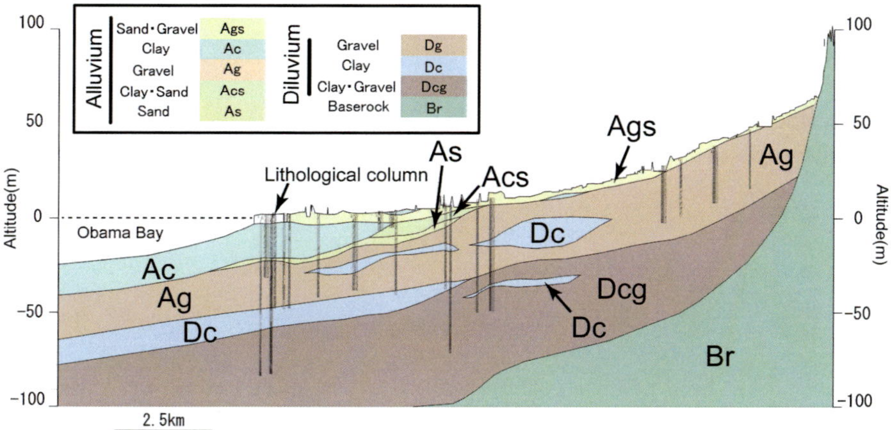

Fig. 6.4 Geological cross section in the Obama Plain. The section is indicated by the broken line in Fig. 6.3. (Modified from Obama City 2017)

an impermeable layer. Hydraulic conductivity ranges were found to be from 1×10^{-8} to 5×10^{-5} m/s for clay and from 1×10^{-4} to 1×10^{-3} m/s for sand.

Then, they performed groundwater simulation of the unsteady state considering the quantity of the pumped water and hydraulic conductivity estimated from geological information. The results were consistent with the observations of the groundwater level. A comparison was based on observational and estimated data obtained in two periods (3 July 2015 in the wet season and 19 March 2016 in the dry season). The relation has a strong correlation of 0.96 of R^2 (coefficient of determination) (Obama City 2017). In GHE analysis, groundwater flow is an important parameter. Based on the average horizontal velocity in each grid (Fig. 6.5), a high-velocity (>10 m/day) zone is located at the center of the plain near B12. The area is located in a groundwater recharge area: it is covered with thick gravel or sand at the surface and a large amount of river water flows into the subsurface here. In the Y area, there is another high velocity zone, which includes a large pumping well named "Yunooka well" for the public water service. The high velocity grids were due to the effect of pumping groundwater. But we must notice that there were some artificial well settings along a boundary between plain and mountain for adjusting the simulation.

6.3.3 *Subsurface Temperature Measurements*

To estimate the regional GHE potential, measuring the spatially variable subsurface temperature is essential because it depends on surface temperature, terrestrial heat flow, and thermal properties. Because subsurface temperature data from our target area was not available, we conducted subsurface measurements at four sites in the Obama Plain (Fig. 6.1). The boreholes had been drilled for groundwater monitoring

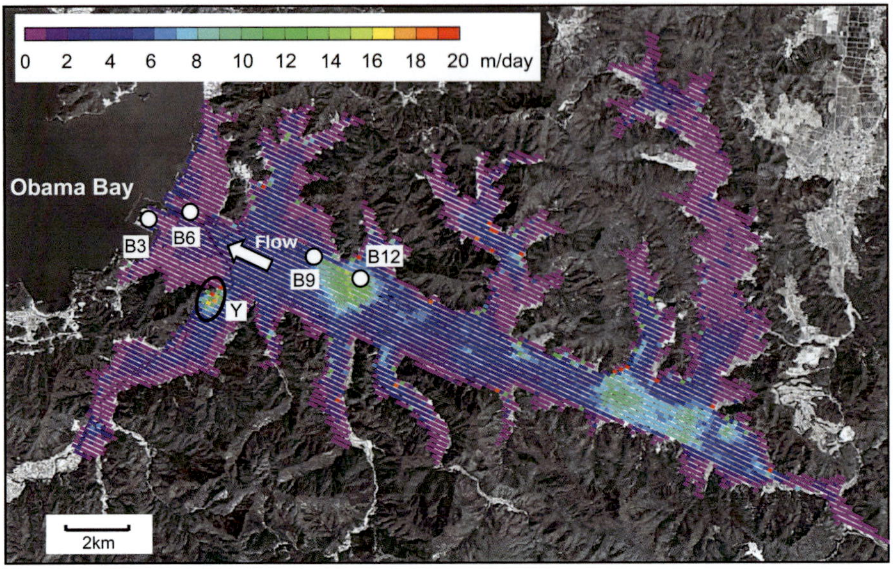

Fig. 6.5 Average horizontal velocity of groundwater flow in the Obama Plain. The grid size is 100 m × 100 m. The arrow indicates the approximate direction of groundwater flow. The open circles are sites of observation wells. (Base map is from GSI Maps of Japan)

by the Obama City local government. Twelve boreholes existed, which were located at four sites (three boreholes at each site) and to different depths to monitor the different aquifers. The maximum depth of all boreholes was about 50 m.

We conducted subsurface temperature measurements from the groundwater level to the bottom of the borehole at the deepest borehole at each site. The temperature sensor is a thermistor with high resolution (0.001 K) (Hamamoto et al. 2014). We monitored real-time subsurface temperature using a digital multimeter connected with a PC. Our measurements had a higher resolution than general subsurface temperature measurements (0.01 K). These high-resolution measurements were expected to detect the effect of groundwater flow and slight differences in thermal properties. The measurements were taken at intervals of 0.5 or 1.0 m. The sensor measured the groundwater temperature directly in the borehole. We assumed that groundwater was in thermal equilibrium with the surrounding strata when there was no pumping of groundwater at the borehole. In general, temperatures at shallow depths vary with seasonal surface temperature variations. Therefore, we measured the subsurface temperature in three or four seasons.

We show the results of the subsurface temperature measurements in Fig. 6.6. Above ~10 m, the temperature is strongly affected by the seasonal changes at all sites in the Obama Plain; the lowest temperature was measured in winter and the highest in summer. Surface temperature is mainly propagated into the subsurface by thermal diffusion. Below 10 m, the temperature remains relatively stable between 14 °C and 16 °C. The temperature at boreholes B9 and B12 may be disturbed by the

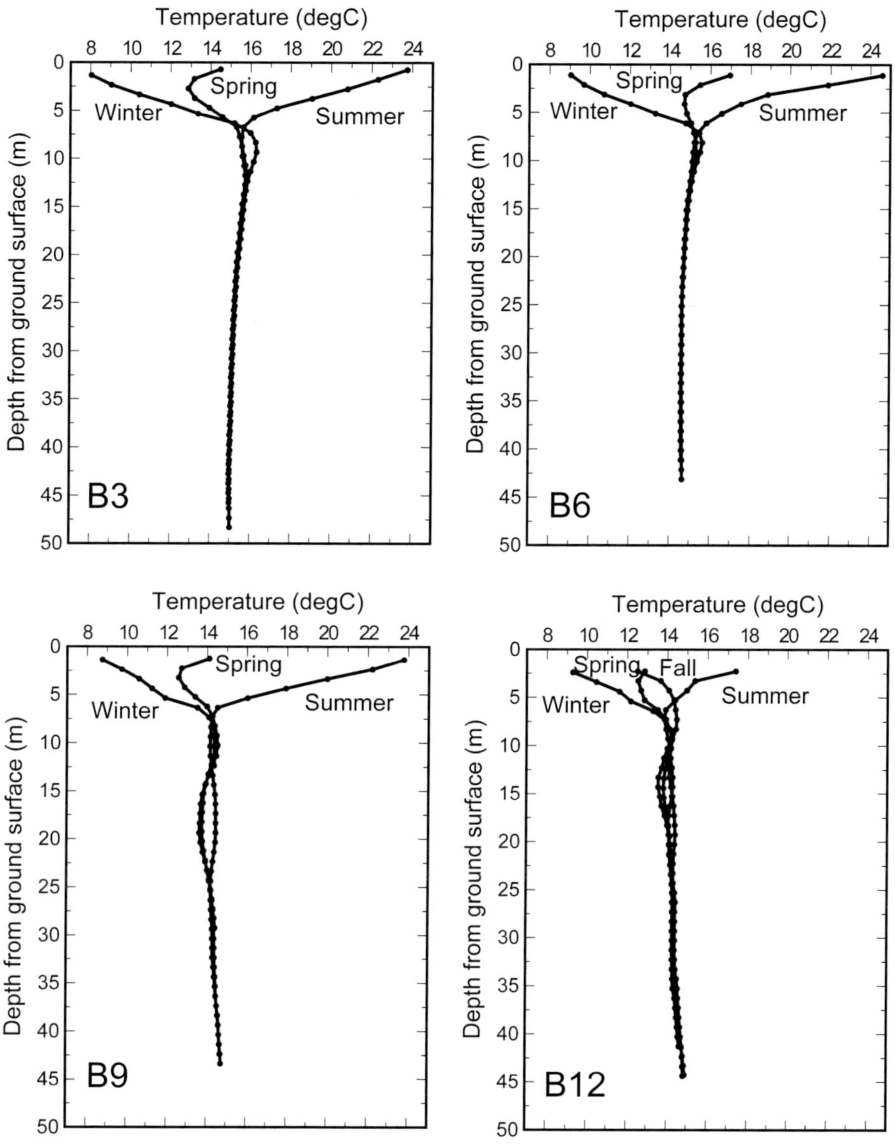

Fig. 6.6 Results of subsurface temperature measurements at four sites in the Obama Plain. The measurements were conducted in three seasons at three sites and four seasons at one site. The locations are described in Fig. 6.3

effect of groundwater flow from 15 to 25 m depth. These sites are affected by high-velocity groundwater flow, as described in Section 6.3.2.

6.4 Potential of GHE System (Closed Type)

The GHE potential map is a useful tool for the initial stage of planning and designing the system, before the detailed investigation, and it includes the thermal response test (TRT). This map is also useful for developing an environmental city design. There are various types of potential maps using indicators such as effective thermal conductivity (W/m·K) and specific heat-extraction (W/m). The effective thermal conductivity is particularly useful for a more detailed design because it depends on the actual thermal properties and can be applied in various scales of a house or a building. However, the specific heat-extraction is useful for determining easily the optimum installation depth for the exchanger, but it limits the model case to a fixed system (The length of heat exchanger ranges from 40 m to 100 m and the installed pipe type is double U-pipes with DN 20, DN 25 or DN 32). We adopt the specific heat-extraction map type for our target area of the Obama Plain.

Our estimation considers the effects of the following:

1. Geology
2. Subsurface temperature
3. Saturated or unsaturated soil or rock
4. Groundwater flow velocity

The easiest way to roughly determine the heat-extraction is using the manual of Verein Deutscher Ingenieure (VDI) (Table 6.1) (Verein Deutscher Ingenieure 2001). This manual provides the specific heat exchange for various geological settings and conditions. The value of the specific heat-extraction can be used for small groundsource heat pumps (<30 kW) with borehole heat exchangers for room heating mode.

To estimate the GHE potential (closed-loop type), we analyzed the effect of subsurface temperature and groundwater flow on specific heat-extraction by the numerical simulator of FEFLOW (DHI-WASY 2010). The simulator has a three-dimensional finite element code for the simulation of heat transport, included with the GHE simulation.

We made a box model for our estimation (200 m × 200 m × 200 m; Fig. 6.7). The GHE well is located at center of the model and considered to be a typical exchange well. The circuit-carrier fluid flows at 43.2 m^3/day with the double U-tube method, installed at a depth of 100 m from the surface. The temperature of the circuit-carrier fluid of the inlet tube going down to the subsurface is 1 °C for room heating mode and 50 °C for room cooling mode. We adopt the average of the specific heat-extraction for the room heating mode for 2400 h. In most Japanese regions including our target area, the heat demand of the room heating mode is higher than that of the room cooling mode.

Table 6.1 Specific heat-extraction of indicated as Verein Deutscher Ingenieure (VDI). Averages of specific heat-extraction for 2400 h by geology or rock type (Modified from Verein Deutscher Ingenieure 2001)

Underground	Specific heat-extraction (W/m) for 2400 h
General guideline values:	
Poor underground (dry sediment) ($\lambda < 1.5$ W/(m·K))	20
Normal rocky underground and water saturated sediment ($\lambda < 1.5$–3.0 W/(m·K))	50
Consolidated rock with high thermal conductivity($\lambda > 3.0$ W/(m·K))	70
Individual rocks:	
Gravel, sand, dry	< 20
Gravel, sand, saturated water	55–65
For strong groundwater flow in gravel and sand, for individual systems	80–100
Clay, loam, damp	30–40
Limestone (massif)	45–60
Sandstone	55–65
Silliceous magmatite (e.g. granite)	55–70
Basic magmatite (e.g. basalt)	35–55
Gneiss	60–70

The values can vary significantly due to rock fabric such as crevices, foliation, weathering, etc

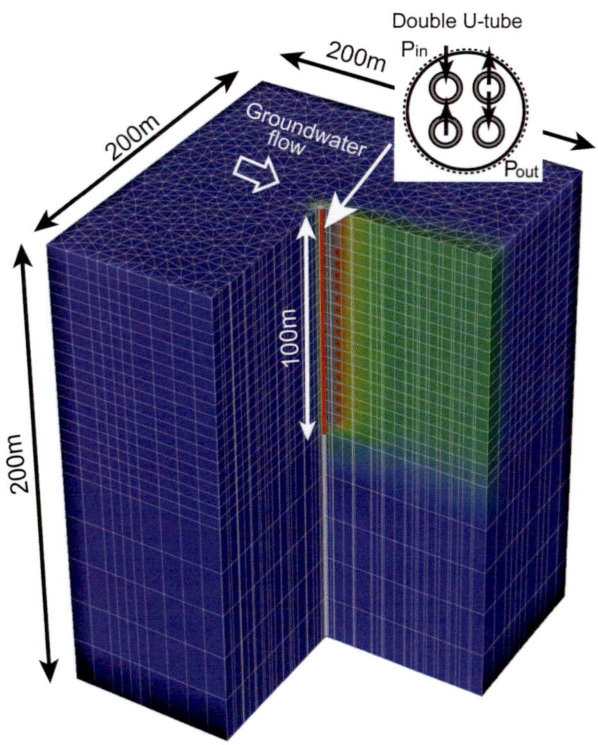

Fig. 6.7 The illustration of the numerical simulation of GHE. The size is 200 m × 200 m × 200 m. The result is an example of the simulation with GHE of the room cooling mode. Carrier fluid circulates at 50 °C from the surface down to the subsurface. The initial subsurface temperature is 15 °C with groundwater flow of 1 m/day

We estimated the effect of subsurface temperature for the specific heat-extraction of GHE. Moreover, we ran various scenarios of initial subsurface temperature values ranging from 10 °C to 30 °C, with an interval of 5 °C. In addition, we used three lithologies for the homogeneous box model: clay, sand, and sandstone. Differences in the thermal properties and the thermal capacity of the three lithologies were examined. The thermal conductivities are 1.8, 2.4, and 2.8 W/m·K, respectively. The thermal capacities are 2.4, 2.5, and 2.2 MJ/m^3·K, respectively. We reveal that the difference in the specific heat-extraction varies with the subsurface conditions (Fig. 6.8). The heat-extraction in the heating mode increases linearly with subsurface temperature. However, the slope of the increase differs between geological settings. For example, the results of the simulations for sand show a linear increase in GHE efficiency by about 35% when subsurface temperatures increase from 15 to 20 °C. The estimated value of sand (32 W/m) is much lower than the VDI value (55–65 W/m). However, we could not obtain more accurate VDI values based on the simple descriptions in the manual. Therefore, we assume that the values are under slight groundwater flow. Groundwater flow is generally possible in the sand layer

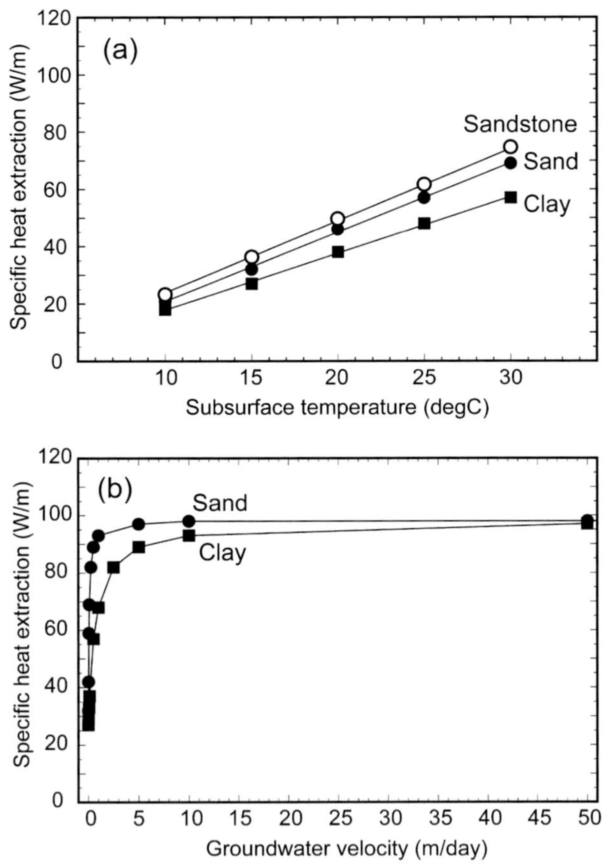

Fig. 6.8 The effect of (**a**) initial subsurface temperature and (**b**) groundwater flow velocity, on the specific heat-extraction

because the sand layer is highly hydroconductive. If there is even slight groundwater flow (about 0.01 m/day), then the two values are consistent with each other.

Furthermore, we estimated the effect of groundwater flow on the specific heat-extraction in the room heating mode. We changed the groundwater flow velocity from 0 to 50 m/day for clay or sand at 15 °C (Fig. 6.8). Independent of lithology, GHE efficiency approximately triples (in sand, an increase from 32 W/m to 98 W/m) when the groundwater flow ranges from 0 to 10 m/day. According to the results of the groundwater simulation, in most clay cells, the horizontal groundwater velocity is lower than 0.01 m/day (Fig. 6.9) in the Obama Plain, which is extremely low. Contrastingly, in most sand cells, the velocity ranges from 1 to 100 m/day, which is much higher than that for the clay cells. Furthermore, we assume that the basement rock is impermeable. If the surface soil cell was unsaturated, we adopted a low value (15 W/m) for the unsaturated layer, in accordance with the VDI recommendations.

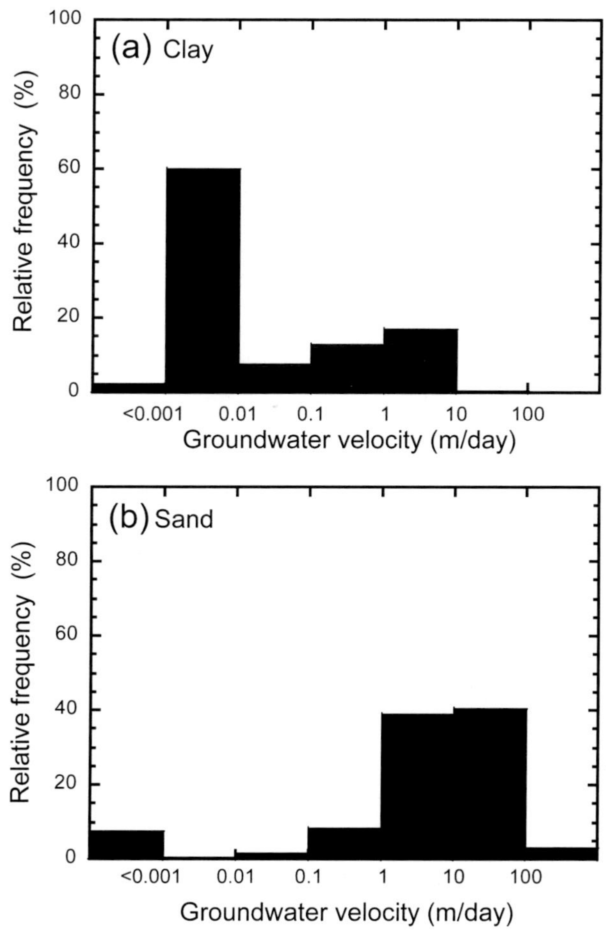

Fig. 6.9 Histogram of the horizontal velocity of groundwater flow in (**a**) the clay layer and (**b**) the sand layer

We calculated the specific heat-extraction for each grid cell (100 m × 100 m) in the Obama Plain. In mountainous areas, basement rocks are exposed to the surface. We did not calculate the specific heat-extraction in the mountainous areas because the population density here is low. We took into account the geology, subsurface temperature, groundwater velocity, and saturation. Geological information was obtained from the 3D geological map of the Obama Plain. We constructed a model of typical subsurface temperature structure based on the results of subsurface temperature measurement at four sites in winter. This was chosen because the room heating mode is used during winter. In the model of typical subsurface temperature, the temperature near the surface is 7 °C and that at a depth of 10 m is 15 °C based on measurement results. The thermal gradient is 0.02 K/m, and the groundwater velocity is adopted only for the horizontal direction (magnitude u_x and u_y); the velocity of the vertical flow is generally one or more orders of magnitude slower than the horizontal flow and it is negligible. In this study, we account only for horizontal groundwater flow. The velocity is obtained from groundwater simulations conducted by the Obama City local government with the Nippon Koei Corporation as described in Section 6.3.2. The condition of soil saturation can be inferred from the level of the groundwater potential. We estimate the specific heat-extraction for each grid using the above data and the subsurface temperature and groundwater flow. Figure 6.10 shows the average heat-extraction map for 2400 h of the room heating mode, which is calculated for each grid. The results show that the central part of the Obama Plain (to the east of observation well B12) is a high heat-extrac-

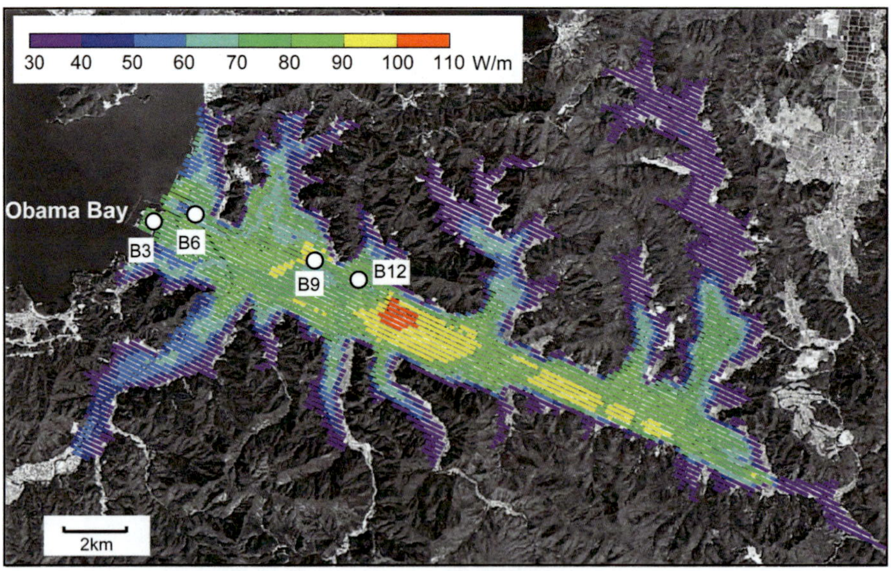

Fig. 6.10 Estimated average heat-extraction for 2400 h of the room heating mode in the Obama Plain. The grid size is 100 m × 100 m. The estimation depth is from the surface to 100 m. The open circles are sites of observation wells. (Base map is from GSI Maps of Japan)

tion area. This area contains a thicker sand layer than the other areas. In the case of groundwater flow in the sand layer, the specific heat-extraction rapidly increases with the flow velocity. There are rice field in the area. The most populated area in the Obama Plain is the coastal area. Though the specific heat-extraction of this area is around 70 W/m, this value is sufficient for usage of a general house as long as a minimum spatial density of 6 m. Therefore, 256 households can use a ground source heat-exchanger per grid cell (100 m × 100 m). The maximum household density of Obama City is about 1200 households /km^2 (approximately 12 households per grid cell). Therefore the ground-source heat exchanger potential is sufficient for the demand in any area of Obama.

6.5 Conclusions

We developed a new method for constructing a potential map of GHE in the Obama Plain. For the potential map, we measured subsurface temperatures at four sites in three or four seasons. The geological model was constructed and the velocity of the groundwater flow was estimated by the Obama City local government with the Nippon Koei Corporation. We calculated the effect of subsurface temperature and groundwater flow on specific heat-extraction. The results showed that the specific heat-extraction increased linearly with increase in subsurface temperature. In contrast, the specific heat-extraction increased rapidly even with a low groundwater flow velocity (lower than about 10 m/day). We constructed the potential map with an indicator of specific heat-extraction. The potential map shows that the central part of the Obama Plain has a high specific heat-extraction. Finally, we revealed that the potential for GHE is sufficient for home heating in the Obama Plain (up to 256 households per grid cell); although the specific heat-extraction of the coastal area of the Obama Plain, which is densely populated, is around 70 W/m.

Most existing methods have used only geological information. Our evaluation considers geology, subsurface temperature, and groundwater flow. Therefore our method is relatively precise with simple data requirements. This method may be expected to find applications in other plains in Japan as well as in other countries.

Acknowledgments This research was mainly conducted as a part of the integrated research project of "Human-Environmental Security in Asia-Pacific Ring of Fire: Water-Energy-Food Nexus by the Research Institute for Humanity and Nature (RIHN)". A portion of the analyses in this work was supported by Japan Society for the Promotion of Science (JSPS) KAKENHI Grant Number JP16K00595 and it was conducted as the research project of the Center for Environmental Science in Saitama. We thank Dr. M. Taniguchi, Dr. A. Endo, and Dr. M. Fujii, who lead the NEXUS project of RIHN, for their invaluable advices for our investigation and further analysis. Dr. S. Hachinohe helped in the investigation of subsurface temperature in the Obama Plain. We thank Ms. T. Okamoto, who provided accounting support. The groundwater simulation information was provided by the Obama City local government and the Nippon Koei Corporation.

References

Blum P, Campillo G, Kölbel T (2011) Techno-economic and spatial analysis of vertical ground source heat pump systems in Germany. Energy 36(5):3002–3011

Carslaw HS, Jaeger JC (1986) Conduction of heat in solids second edition. Oxford University Press

City O (2016) Report of Obama city statistics. Obama City, Fukui

Obama City (2017) The Obama Plain Groundwater Survey Summary Report,. http://www1.city.obama.fukui.jp/category/page.asp?Page=2574. Accessed 26 June 2017

DHI-WASY (2010) User Manual. DHI-WASY GmbH, Germany

Domenico PA, Schwartz FW (1997) Physical and chemical hydrogeology, 2nd edn. Wiley, New York

Hamamoto H, Shiraishi H, Hachinohe S, Ishiyama T, Satake K, Miyakoshi A (2014) Synthesis of subsrface temperature information and evaluation of potential for setting up ground heat exchangers in Saitama prefecture. BUTSURI-TANSA 67(2):107–119. (in Japanese)

Harbaugh AW (2005) MODFLOW-2005, The U.S. Geological survey modular ground-water model—the ground-water flow process. Techniques and methods, US geological survey 6–A16

Japan Ministry of the Environment (2012) Report of investigation of ground source heat energy use.. http://www.env.go.jp/press/15945.html. Accessed 20 Sept 2017. (in Japanese)

Japan Ministry of the Environment (2016) Report of investigation of ground source heat energy use.. http://www.env.go.jp/press/103827.html. Accessed 20 Sept 2017. (in Japanese)

Japan Sustainable Building Consortium (2016) Data-base for Energy Consumption of Commercial buildings.. http://www.jsbc.or.jp/decc_english/index.html. Accessed 26 June 2017

Lund JW, Boyd TL (2015) Direct utilization of geothermal energy 2015 worldwide review. Paper presented at the world geothermal congress 2015, Melbourne, Australia, 19–25 Apr 2015

Public Corporation of Fukui Construction Technology (2010) Geological map of Fukui prefecture. Public Corporation of Fukui Construction Technology, Fukui

Sarbu I, Sebarchievici C (2015) Ground-source heat pumps; fundamentals, experiments and applications. Elsevier, Amsterdam

Senate Department for Urban Development and Housing (2015) Geothermal potential: Specific conductivity and specific extraction capacity.. http://www.stadtentwicklung.berlin.de/umwelt/umweltatlas/ei218.htm. Accessed 26 June 2017

Shrestha G, Uchida Y, Yoshioka M, Fujii H, Ioka S (2015) Development of potential maps for ground source heat pump systems with a new approach. J Geotherm Res Soc Japan 37(4):133–141

Stauffer F, Bayer P, Blum P, Giraldo NM, Kinzelbach W (2014) Thermal use of shallow groundwater. CRC Press, Boca Raton

Taniguchi M, Masuhara N, Burnett K (2015) Water, energy, and food security in the Asia Pacific region. J Hydrol Reg Stud 11:9–19. https://doi.org/10.1016/j.ejrh.2015.11.005

Turcotte D, Schubert G (2014) Geodynamics, 3rd edn. Cambridge University Press

Verein Deutscher Ingenieure (2001) VDI 4640 – blatt 2. Thermal use of the underground ground source heat pump systems. Verein Deutscher Ingegneure, Düsseldorf

Part III
Water-Food Nexus (Water for Coastal Ecosystem Conservation)

Chapter 7
Lacustrine groundwater discharge in southern Laguna de Bay, Philippines

Erwin Racasa, Ronald Lloren, Michelle Manglicmot, Karen Ann B. Jago-On, Maria Ines Rosana D. Balangue, Makoto Taniguchi, and Fernando P. Siringan

Abstract Although a major potential contributor to water and nutrient budgets, lacustrine groundwater discharge (LGD) is often neglected in most lake studies. Through electrical resistivity profiling surveys, the authors examined the possible occurrence of LGD in southern Laguna de Bay, the largest freshwater lake in the Philippines. Discrete and dispersed LGDs were identified. Discrete LGDs were inferred from narrow highly resistive zones that cut vertically across the lake floor. These discrete LGDs line-up with projections of lineaments on land and are thus deemed to be fault-controlled. Dispersed LGDs, interpreted from wide swaths of resistivity signals cutting across the lake floor, were found to occur more commonly in shallower areas. Findings from radon concentrations, nutrient concentrations, and chlorophyll *a* analyses support the perceived patterns of LGD occurrences. Nutrient input through LGD is probably contributing to the lake's current eutrophic condition however where discrete LGDs occur, the fisheries appear to be enhanced.

Keywords Lacustrine groundwater discharge · Nutrients · Eutrophication

E. Racasa · R. Lloren · M. Manglicmot · F. P. Siringan (✉)
Marine Science Institute, University of the Philippines, Diliman, Quezon City, Philippines
e-mail: fpsiringan@msi.upd.edu.ph

K. A. B. Jago-On
School of Urban and Regional Planning, University of the Philippines, Diliman, Quezon City, Philippines

M. I. R. D. Balangue
National Institute of Geological Sciences, University of the Philippines, Diliman, Quezon City, Philippines

M. Taniguchi
Research Institute for Humanity and Nature (RIHN), Kyoto, Japan

© Springer Nature Singapore Pte Ltd. 2018

A. Endo, T. Oh (eds.), *The Water-Energy-Food Nexus*, Global Environmental Studies, https://doi.org/10.1007/978-981-10-7383-0_7

7.1 Introduction

Unlike submarine groundwater discharge (SGD), lacustrine groundwater discharge (LGD) in lakes appears to have not been given attention in the Philippines. SGD or any flow of water that comes out of the seafloor (Burnett et al., 2006) has received wide recognition in the last decades for its volumetric and chemical importance (Taniguchi et al., 2002; Johannes, 1980) and its impact to ecosystems (Moore, 2010; Mulligan & Charette, 2006). At present, an increasing number of SGD studies have been observed in the Philippines (Taniguchi et al., 2008; Cardenas et al., 2010; Senal et al., 2011) and in Southeast Asia (Moosdorf et al., 2014). On the contrary, local studies on LGD or the freshwater equivalent of SGD (Lewandowski et al. 2013) is still non-existent even though it has been identified as a potentially important nutrient contributor in lakes (Meinikmann et al., 2015; Harvey et al., 2000; Vanek, 1987). Some of the reasons why few delve into LGD studies are (1) difficulty in quantifying volume and nutrient fluxes owing to its nature of spatial and temporal heterogeneities, (2) assumption of its residual or negligible volumetric contribution, and (3) existing paradigm of immobile P in groundwater and little ecological relevance from diffuse transport of P (Meinikmann et al., 2014).

The subject of this LGD study is Laguna de Bay, the largest freshwater lake in the Philippines and the third largest in Southeast Asia (Gomez Jr., 2014). With its sheer size, the lake can support a variety of domestic and socioeconomic activities. Some of its uses include opportunities from fisheries, transportation, power generation, and industrial cooling (Jago-on et al., 2017; Jaraula et al., 2015; LLDA, 2012); agriculture in its lakeshore sites with irrigation coming from lake waters (Gomez, 2014; LLDA, 2012); freshwater resource (Jago-on et al., 2017; Jaraula et al., 2014; Gomez, 2014; Tamayo-Zarafalla et al., 2002); retention basin for flood control and sink for treated and untreated waste (Jaraula et al., 2014). Although the lake has conflicting uses, its importance for the 15 million people (LLDA, 2014) living around the vicinity of the lake could not be discounted.

Despite being the most studied lake in the Philippines (Brillo, 2015), only a few groundwater-related studies have been published (Jago-on et al., 2017). Based on the scanning of readily available published knowledge, this is the first LGD study in Laguna de Bay and in the Philippines with particular focus on the mapping and characterization of the southern region of the bay.

7.2 Description of the Study Area

7.2.1 Physical Description and Geology

Laguna de Bay is located 15 km southeast of Manila and bounded by the provinces of Rizal and Laguna in the northern and southern part; and the cities of Metro Manila in the northeast (Fig. 7.1). Based on bathymetric surveys from 2011 to 2014,

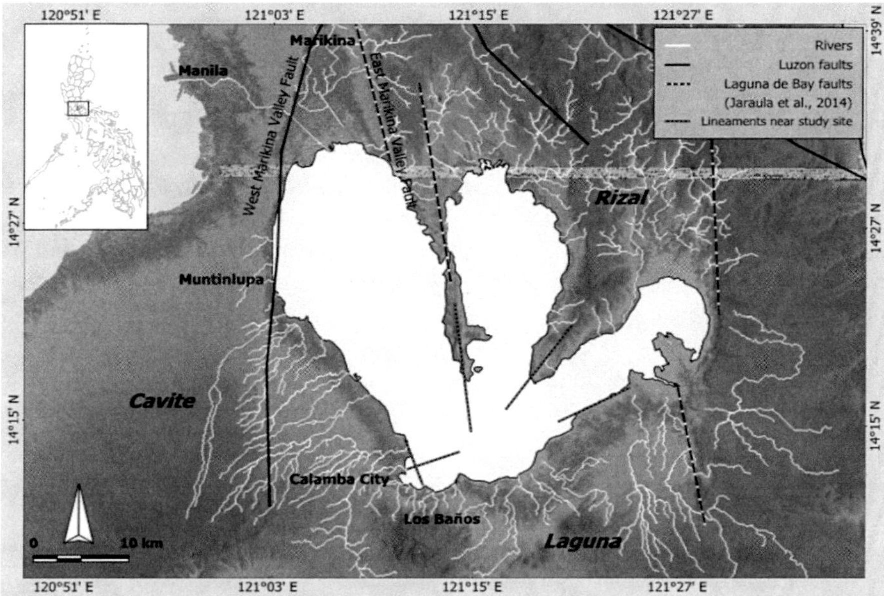

Fig. 7.1 Laguna de Bay in southwestern Luzon island. Inset shows location in the Philippines. Highlighted are the major faults that shaped Laguna de Bay including major Luzon faults and lineaments (Jaraula et al., 2014). The study site is at the southern portion of the bay near Calamba and Los Baños

the average depth of the lake at the mean low lake level is 2.43 m with an equivalent lake volume of 2.03 x 10^9 m^3 (LLDA, 2014). The lake has a surface area of 834.49 km^2 with waters coming from drainage basin of 2980 km^2 spanning the mentioned bordering provinces and cities including towns in Batangas, Cavite, and Quezon (LLDA, 2014). Thirty five tributaries drain the catchment area (Zimmer & Bendoricchio, 2001) with only the 25.5 km Pasig - Napindan Channel as its outlet to Manila Bay.

The tri-lobate, almost heart-shaped lake is visibly divided into three distinct bays – the West Bay, Central Bay, and the East Bay – with the southern portion aptly called the South Bay (LLDA, 2012). The bay is probably a relic of a larger, erupted caldera system (Aurelio & Pena, 2004) and prior to about three thousand years ago, was a marine environment (Jaraula et al., 2014). The Parañaque strip is the resulting landmass of the vertical movement of the West Marikina Valley Fault (Jaraula et al., 2014). Sea level fall and the emergence of the land separating Manila Bay from the lake due to repeated faulting (Fig. 7.1) transformed the marine environment into a fresh water system (Jaraula et al., 2014).

7.2.2 *Study Site*

The study site, located in the southern portion of the lake, straddles Calamba City and the municipality of Los Baños (Fig. 7.1). Pliocene-Pleistocene volcanic centers bound the study area to the south, east and west (Förster et al., 1990). Common "rocks" in the area are scoria cones and tuff cones in addition to monogenetic cones of basaltic composition. Calamba and Los Baños are well-known for their abundant hot springs. Jago-on et al. (2017) reported a rapid increase in the establishment of hotspring resorts in the two adjacent towns in the past decade. A perceived increase in groundwater demand for tourism, population, and continued development is expected to lower the water table and hot spring temperatures (Jago-on et al., 2017) and consequently, the LGD in the study area.

7.3 Methods and Materials

7.3.1 *Mapping LGDs*

Towed electrical resistivity system from the Advanced Geosciences, Inc. was used to map the LGD in September 2014 (Fig. 7.2). The system was equipped with an echo sounder for bathymetry readings. The survey lines were limited by the presence of fish pens in the lake: areas with enough distance between pens and long segments were selected. In March 2015, water samples for radon analysis were acquired in 43 sites - 38 in the lake and 5 on land. Lake water samples were acquired using two, narrow-necked 750 mL glass bottles tied together at 1.3 meters from the end of a ~10 meter pole. The bottles were submerged near bottom with a cap which was released by pulling a string attached to it. A Durridge RAD7 Radon Analyzer was used to determine the radon concentrations in the samples. Only 28 samples were analyzed for radon content (sites 1 to 23 and sites 41 to 45).

SRTM digital elevation models and synthetic aperture radar (SAR) images of Laguna de Bay were utilized to delineate lineaments on land. These lineaments are potential faults and were extrapolated across the lake and correlated to the positions of discrete LGD plumes defined by the resistivity surveys.

7.3.2 *Confirmation of LGD Patterns Through Nutrients and Chlorophyll a*

Water sampling was conducted on September 25, 2014 and October 15, 2015 for nutrient analyses. To determine the nutrient patterns relative to LGD, lake water samples were purposively sampled both near and far from the lineaments that bisect the study site in the South Bay (Fig. 7.2; Stations 5, 7, 9, 13, 15, 16, 18, 19, 34, and

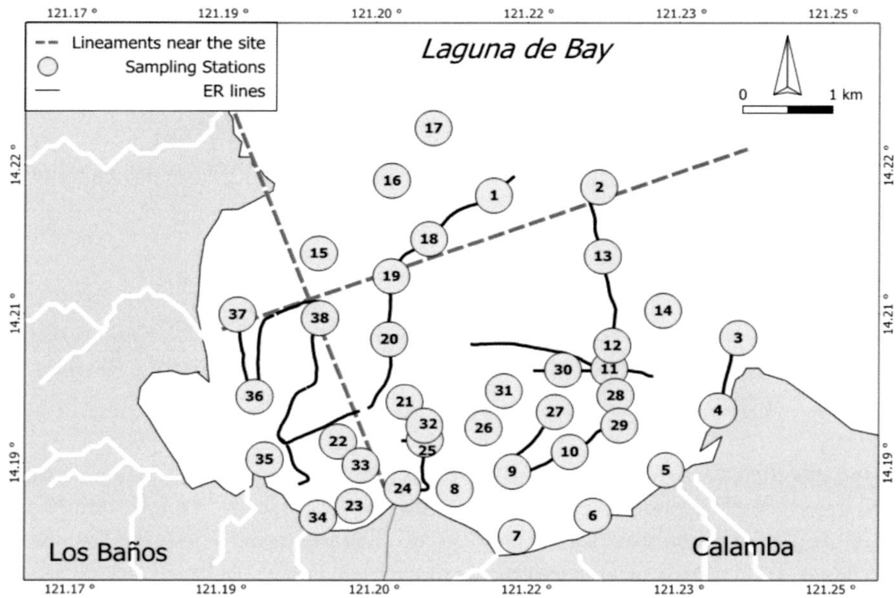

Fig. 7.2 Electrical resistivity tracks (solid lines), lineaments (dashed lines), and radon sampling points (circles with numbers). Lake samples were taken during September 2014 and October 2015. River samples were taken in October 2015 only

37). These numbered sampling sites were the same as with the radon sampling sites. In addition, riverine water samples were also taken in October 2015 for comparison with its LGD counterparts. Physico-chemical parameters such as pH, salinity, conductivity, and total dissolved solids were also measured in situ using YSI field probes from all the water samples.

Lake water samples were collected in 250 mL polypropylene bottles at 0.5 m from the water surface. Filtration using a 0.45 μm cellulose acetate-syringe membrane filter into 15 mL centrifuge tubes was performed on-site or at the laboratory within three days after collection. Duplicate samples were made for each of the following nutrients – dissolved silica ($Si(OH)_4$), phosphate (PO_4^{3-}), nitrate and nitrite ($NO_3^- + NO_2^-$), and ammonia (NH_3). Samples were chilled during transport and were frozen at storage except for dissolved silica samples which were refrigerated at approximately 4 °C. Laboratory methods for the spectrophotometric analysis of the nutrients were performed according to Strickland and Parsons (1972). A Shimadzu Mini UV-Vis 1240 spectrophotometer was used to read the color absorption of the samples and standards. For the analyses of NH_3 in October 2015 and $NO_3^- + NO_2^-$ in both September 2014 and October 2015 surveys, a Skalar San ++ auto-analyzer was used due to issues of detection limits.

LGD observations were also confirmed by measuring levels of chlorophyll *a* (Chl *a*) in lake waters. Nutrient-rich waters are also expected to have high levels of Chl *a*, a proxy for algal biomass and indirectly, primary productivity. For this study, nutrient sampling stations were reoccupied for Chl *a* measurements. The analysis was performed following Parsons and others (1984). 500 mL colored Nalgene bottles were used as sample containers, wherein only 150 mL to 200 mL water samples were filtered due to high suspended solids. Glass fiber filters of 1 μm size were used and were disrupted by sonication before the samples were let to stand in acetone overnight. A Turner fluorometer was used to measure the Chl *a* from the samples.

7.4 Results and Discussion

7.4.1 Lacustrine Groundwater Discharge as Determined by ER Profiles

Using electrical resistivity (ER) profiles, the occurrence of LGD in Laguna de Bay was determined, discrete and dispersed LGDs were observed. Discrete springs of LGD were interpreted from the highly resistive plumes cutting across the lake floor on the right end of panel CC' and left side of panel DD' (Fig. 7.3b). Dispersed LGD was interpreted to occur in panel GG' where zones of various resistivities along a wide swath cut across the lake floor. The size of sediments in the area appears to play a role in the occurrence of the observed LGD types. Shallower areas near the lakeshore tend to have dispersed LGD, probably due to sandier sediments. In the deeper parts of the lake, dispersed LGD is limited by the thicker blanket of muddy sediments as evidenced by the bluish resistivity values as exemplified by the RR' section.

Discrete LGD, springs coming out directly from the lake floor, are likely to be fault-controlled. Projection over the lake of a lineament identified onshore coincide with the discrete LGD in panels DD' and CC' (Fig. 7.3b). Results were according to the observations of Lewandowski et al. (2013) - strongest LGD fluxes occur at the nearshore margins in lakes with uniform geology and highly localized LGD in fractures of fractured-rock settings.

7.4.2 Radon Concentration Confirms Discrete LGD Occurrences

The highest radon concentration among the bottom lake water samples were measured from regions with discrete LGDs (Fig. 7.3a) as indicated by the ER profiles. Radon concentration values amounted to as much as 11.9 dpm L^{-1}. The high radon measurements near lineaments further confirm the occurrence of discrete LGD and its fault-controlled nature. Measured radon concentration from deep wells and springs inland exceeded 236 dpm L^{-1}.

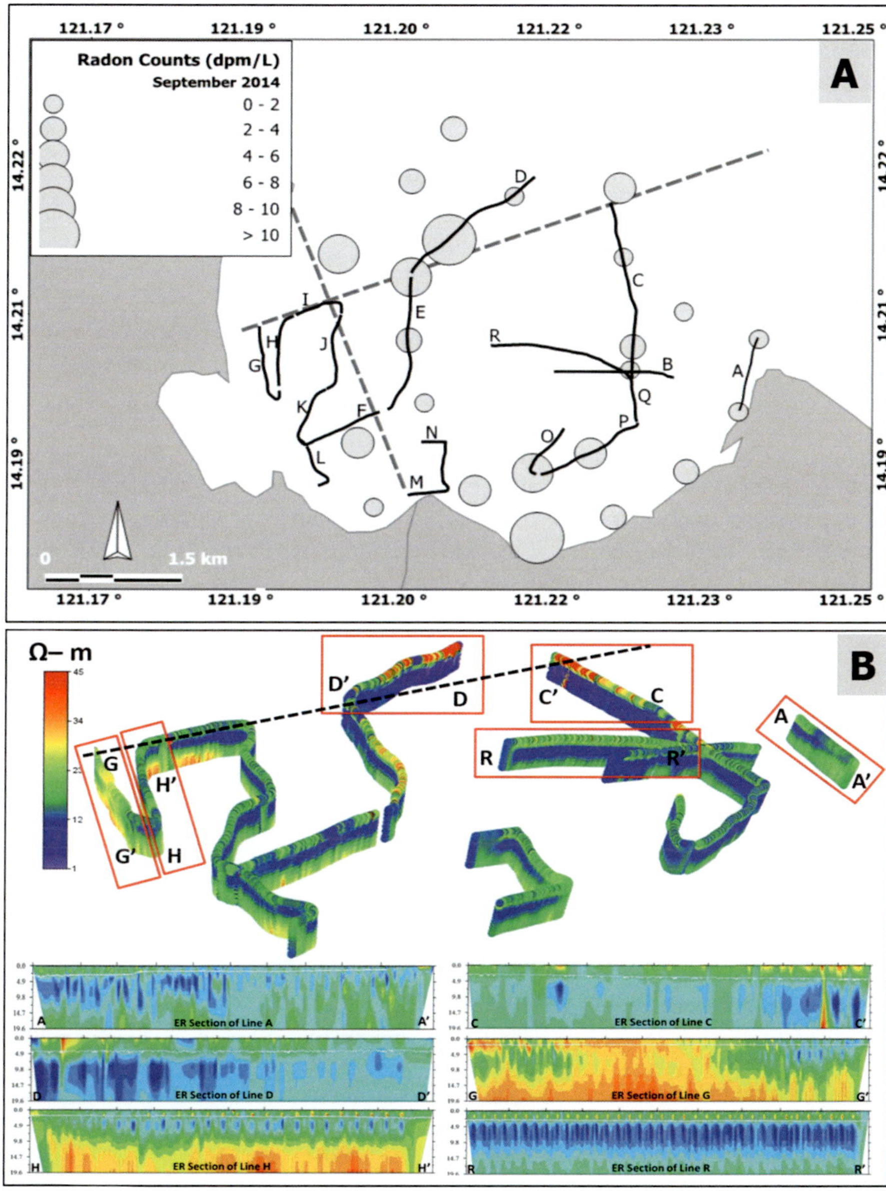

Fig. 7.3 (**a**) Radon concentration in lake water samples. (**b**) Surveyed electrical resistivity lines and their corresponding resistivity profiles in Ω-m. The red boxes illustrate the resistivity profiles of sections tracks A, C, D, F, G, and H.

Table 7.1 Nutrient levels in lake and river waters sampled in September 2014 and October 2015.

Mean	$Si(OH)_4$	PO_4^{3-}	NH_3	$NO_3^- + NO_2^-$	DIN
Range	(μM)	(μM)	(μM)	(μM)	(μM)
Sept 2014	136	2.36	1.02	*0.41	1.43
Lake	71.3–187	1.30–3.67	0.28–2.34	0.25–0.87	0.60–3.21
Oct 2015	191	1.02	*0.59	*0.39	0.98
Lake	135–265	0.40–3.35	0.13–0.88	0.23–1.21	0.38–1.70
Oct 2015	1410	13.0	*6.13	*7.09	13.2
River	1010–1640	3.75–27.1	0.13–29.2	0.27–32.0	0.51–61.1

*Nutrient analyses were performed using Skalar San ++ Auto-analyzer

7.4.3 Spatial Relationship of LGD and Nutrients

Nutrient availability limits primary production of phytoplankton. Diatoms, which are at the base of the fish food chain, need dissolved silica for their frustules. They are important as they are a major food resource for freshwater microorganisms and animal larvae. Higher dissolved silica and other nutrients in the water could boost diatom populations which could enhance local fish populations. Local fisherfolks' anecdotal accounts of bubbling warmer-than-ambient waters were reported in some regions of the lake. These areas where some locals come to fish could be the discrete LGD springs delineated using the ER profiles and radon concentrations. Nutrient-laden discrete LGD could increase the primary productivity and fisheries in these areas and may be the reason why locals fish there.

Nutrient levels of surface lake waters in September 2014 and October 2015, as well as river waters sampled in October 2015, were determined to establish the relationship of LGD and nutrients. Dissolved silica (Si(OH)4), phosphate ($PO4^{3-}$), and inorganic nitrogen (NH_3 + NO_3^- NO_2^-) concentrations were one (1) order of magnitude higher in rivers than in lake waters for both September 2014 and October 2015 sampling (Table 7.1). This is expected since rivers are the main pathways for nutrients to drainage basins such as lakes and coastal areas. A relative comparison of the nutrient concentrations with the available data from the nearest monitoring station (Station VIII) of Laguna Lake Development Authority (LLDA) reveal comparable values as expected (Table 7.2).

Measured nutrients of the sampling stations in September 2014 and October 2015 (Fig. 7.2) were plotted with the lineament projections to determine their spatial relationship with LGD (Fig. 7.4). The following trends were observed for the four tested nutrients for September 2014: 1) high concentrations near the coast, 2) high concentrations for stations 18 and 19 (center of the plots), and 3) relatively lower concentrations in all the other stations. During the October 2015 sampling, the same trend of high nutrient concentrations near the lineament was observed (Fig. 7.4). In addition, station 15 showed consistently high nutrient concentrations. Rivers have grossly higher concentrations as discussed previously. These observed nutrient trends are further evidences of the occurrence of discrete LGD in the southern part of Laguna de Bay.

Table 7.2 Comparison of nutrient concentrations from present study with LLDA monitoring stations

(in µM	Ammonia			Nitrate			Phosphate		
	Present study		LLDA	Present study		LLDA	Present study		LLDA
	Sept 2014	Oct 2015	Q4 2016 Ave (station VIII)	Sept 2014	Oct 2015	Q4 2016 Ave (station VIII)	Sept 2014	Oct 2015	Q4 2016 Ave (station VIII)
Lake samples	1.02	0.59	1.03	0.41	0.39	3.12	2.36	1.02	1.54
San Cristobal River	–	0.24	–	–	0.27	0.55	–	16.4	5.62
San Juan River	–	29.2	--	--	32.0	94.3	--	27.1	8.42

7.4.4 LGD and **chlorophyll *a***

As with the nutrients, Chl *a* measurements were also plotted with lineament projections to determine its spatial relationship with LGD (Fig. 7.5). Patterns of Chl *a* measurements were similar with nutrients - higher concentrations near lineaments. Readings were also high near Alligator Lake at the southern part of the study site. When compared with radon concentrations, Chl *a* distribution matches with the former (Fig. 7.3a). This qualitative spatial relationship of radon concentration and Chl *a* could be further strengthened by its Pearson-product moment of correlation value. Initial calculations showed a correlation value of 0.1083, signifying little correlation between the two variables (Fig. 7.6). However, visual analysis revealed an apparent linear trend if zero and high radon values are removed. A high positive correlation value of 0.8732 can thence be calculated. The occurrence of high Chl *a* measurements in regions with high radon concentration and high nutrients is additional proof for discrete LGD in the study site.

7.4.5 Implications of LGD to Laguna de Bay Ecology

SGD, the marine counterpart of LGD, is increasingly being recognized as an important material pathway between the land and sea that could drive geochemical budgets and ecosystem change (Burnett et al., 2006; Swarzenski et al., 2004; Taniguchi et al., 2003). In the Philippines, Taniguchi et al. (2008) found that SGD contributes as much dissolved inorganic nitrogen as the Pasig and Pampanga rivers - the major rivers that drain into Manila Bay. In Bolinao, Pangasinan, eutrophication of the oligotrophic reefs is expected because of the increased nutrient influx from the seemingly fault-controlled SGD in the area (Senal et al., 2011; Cardenas et al., 2010). Nitrogen (N) and phosphorus (P) fluxes were one to two orders of magnitude

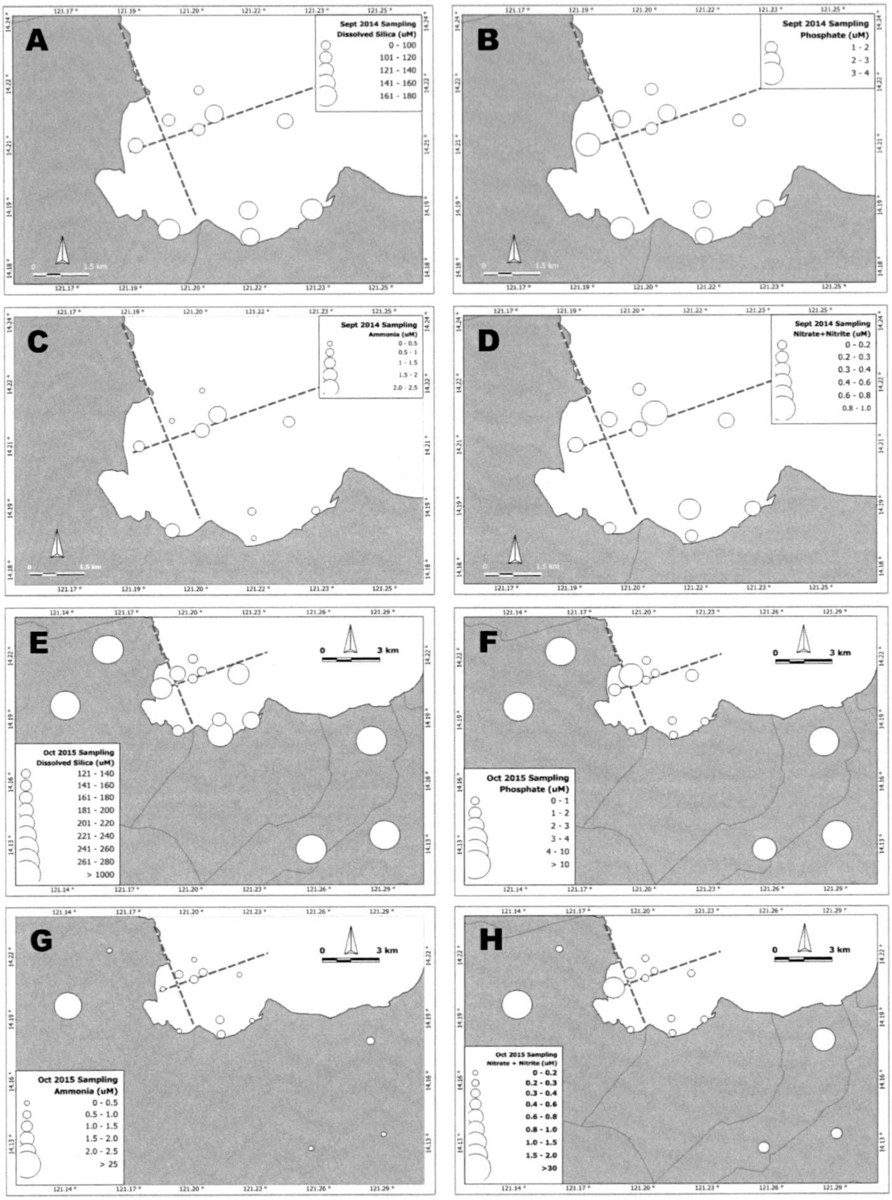

Fig. 7.4 Nutrient concentrations of sampling stations in September 2014 (A-D) and October 2015 (E-H)

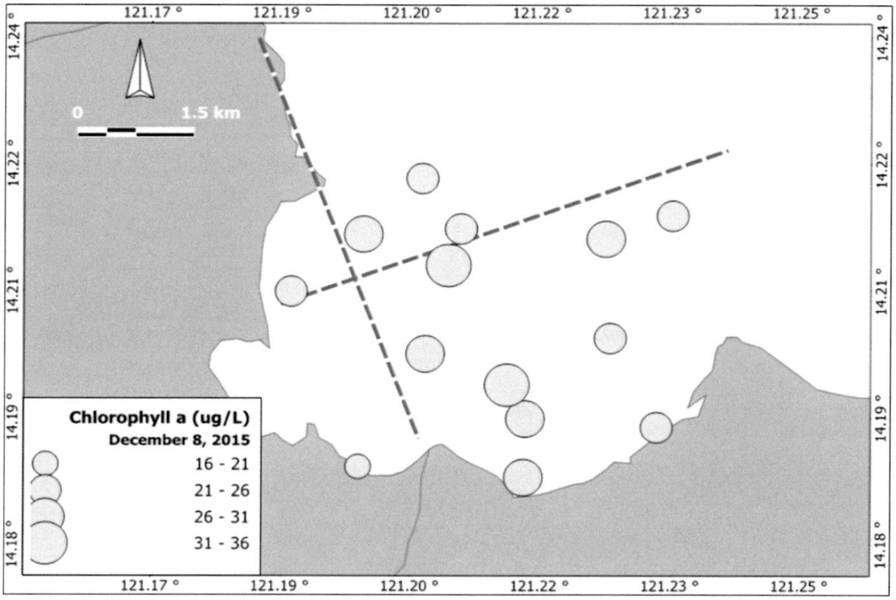

Fig. 7.5 Chlorophyll *a* concentration of surface waters of re-occupied stations

higher than ambient seawater, an apparent result of the increased anthropogenic activities on shore which find its way to the reefs through the fault-controlled SGD (Senal et al., 2011).

In lakes, eutrophication has largely been attributed to the input of phosphorus from surface runoff. P has traditionally been assumed to exist in low concentrations in groundwater since it is adsorbed in the soil and sediment matrix (Meinikman et al., 2015). Evidence however suggests that groundwater can contribute the same or higher amounts of phosphorus as river waters leading to eutrophication. P input of groundwater to Lake Arendsee, Germany was quantified at 53% of all external inputs which makes it the lake's main driver of ongoing eutrophication (Meinikmann et al., 2015). The oligotrophic Lake Hampen in Denmark gets 67% of the total N and 85% of the total P inputs from groundwater. These excess nutrients are expected to shift the lake into a eutrophic state (Ommen et al., 2012). A much earlier study by Loeb & Goldman (1979) identified that groundwater, although just 16% of Ward Creek which flows into Lake Tahoe, contributed 49% of the total nitrate and 44% of the total soluble phosphorus loads in the lake. In the case of Laguna de Bay, previous studies have shown that nitrogen may be the main eutrophication factor of the lake (Santos-Borja and Nepomuceno, 2006).

As such, LGD occurrence in Laguna de Bay has implications to both its water and nutrient budgets and hence its ecology. Faults create preferential pathways for groundwater flow as inferred in this study and previously by Cardenas and others (2010) and Senal and others (2011) in Bolinao, Pangasinan. Since there are numerous faults identified in the watershed of Laguna de Bay, several of which cut across the lakeshores, possible occurrence of discrete LGDs in the larger region of the bay

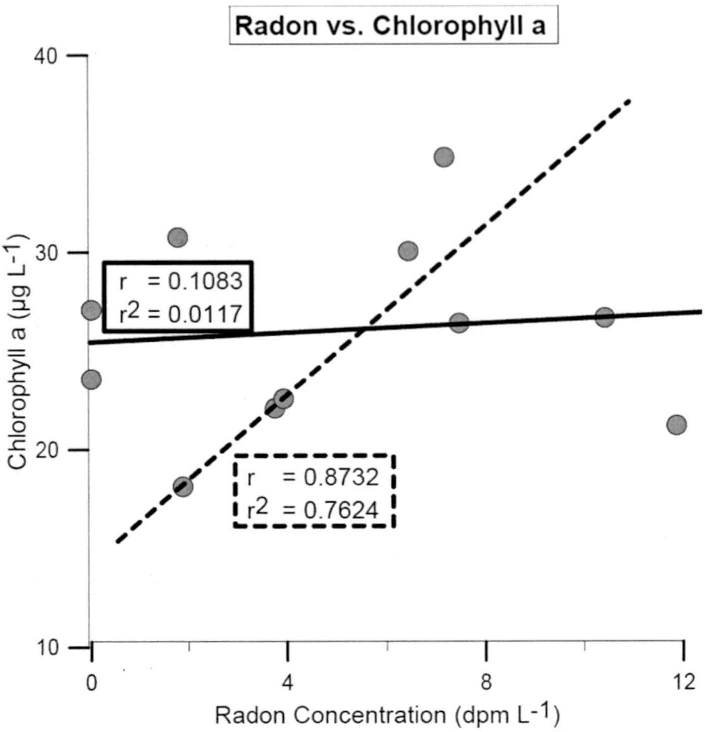

Fig. 7.6 Correlation of radon concentration and chlorophyll *a* measurements. Initial calculations showed a correlation value of 0.1083 (solid line) but application of visual analysis yielded a linear trend with correlation value of 0.8732 (dashed line) once zero and high radon values are removed

is expected. In addition, dispersed LGD near the lakeshore is equally important as discrete springs. Main LGD in Lake Arendsee was observed near urban areas at the lakeshore where most groundwater was heavily contaminated with P. Increasing urbanization and continued development of Laguna de Bay shores, e.g. Calamba, Cabuyao, and San Pedro, and the general lake area will have tremendous impact on the quality of groundwater that eventually goes into the lake through dispersed and discrete LGDs. These previously unknown pathways translate to additional input of water and nutrients into the lake that were never considered in previous studies. This study shows that the overly eutrophic Laguna de Bay source its nutrients not only from surface runoff but also from LGD.

7.5 Summary

Discrete and dispersed LGDs occur in Laguna de Bay. LGD is dispersed in the near shore because of sandier substrate. Further offshore where the bottom sediments are mostly mud, LGD likely seeps out along faults. Where discrete LGD occurs, there is higher algal biomass because of localized additional input of dissolved silica and other nutrients. These sites coincide with areas where some locals come to fish. The contribution of LGDs to the water and nutrient budget of Laguna de Bay can be substantial given the two modes of occurrence: (1) the wide swaths of dispersed LGDs in the lakeshores and, (2) the potentially common occurrence of discrete LGDs due to several faults possibly cutting across the lake.

References

Brillo BB (2015) The status of Philippine lake studies: scholarly deficit in social science and small-lake research. Asia-Pacific Social Science Review 15(1):78–101

Burnett WC, Aggarwal PK, Aureli A, Bokuniewicz H, Cable JE, Charette MA et al (2006) Quantifying submarine groundwater discharge in the coastal zone via multiple methods. Sci Total Environ 367:498–543

Cardenas M, Zamora P, Siringan FP, Lapus MR, Rodolfo RS, Jacinto GS et al (2010) Linking regional sources and pathways for submarine groundwater discharge at a reef by electrical resistivity tomography, 222-Rn, and salinity measurements. Geophys Res Lett 37(16):L16401

Förster H, Oles D, Knittel U, Defant MJ, Torres RC (1990) The Macolod corridor: a rift crossing the Philippine island arc. Tectonophysics 183:265–271

Gomez JEA Jr (2014) Urbanizing at the lake's edge: watershed planning and governance lessons from the Laguna de bay region, Philippines. Journal in urban and regional. Planning 1(1):34–50

Harvey FE, Rudolph DL, Frape SK (2000) Estimating ground water flux into large lakes: application in the Hamilton Harbor, western Lake Ontario. Ground Water 38:550–565

Jago-on KAB, Siringan FP, Balangue-Tarriela R, Taniguchi M, Reyes YK, Lloren R, Peña MA, Bagalihog E (2017) Hot spring resort development in Laguna Province, Philippines: challenges in water use regulation. Journal of Hydrology: Regional Studies 11:96–106

Jaraula CMB, Siringan FP, Klingel R, Sato H, Yokoyama Y (2014) Records and causes of Holocene salinity shifts in Laguna de Bay, Philippines. Quat Int 349:207–220

Johannes RE (1980) The ecological significance of the submarine groundwater discharge of groundwater. Marine Ecology - Progress Series 3:365–373

Lewandowski J, Meinikmann K, Ruhtz T, Poschke F, Kirillin G (2013) Localization of lacustrine groundwater discharge (LGD) by airborne measurement of thermal infrared radiation. Remote Sens Environ 138:119–125

Laguna Lake Development Authority (LLDA) (2012) 2009 to 2012 Annual Water Quality Report on the Laguna de Bay and its Tributary Rivers.. Retrieved from http://www.llda.gov.ph/index.php?option=com_content&view=article&id=218&Itemid=67.. Accessed 7 Mar 2016

Laguna Lake Development Authority (LLDA) (2014) 2013–2014 Laguna de Bay Environment Monitor. Study finds Laguna Lake shallower, Section

Loeb SL, Goldman CR (1979) Water and nutrient transport via groundwater from Ward Valley into Lake Tahoe. Limnol Oceanogr 24(6):1146–1154

Meinikmann K, Hupfer M, Lewandowski J (2015) Phosphorus in groundwater discharge – a potential source for lake eutrophication. J Hydrol 524:214–226

Moore WS (2010) The effect of submarine groundwater discharge on the ocean. Annu Rev Mar Sci 2:59–88

Moosdorf N, Stieglitz T, Waska H, Dürr HH, Hartmann J (2014) Submarine groundwater discharge from tropical islands: a review. Grundwasser – Zeitschrift der Fachsektion Hydrogeologie

Mulligan AE, Charette MA (2006) Intercomparison of submarine groundwater discharge estimates from a sandy unconfined aquifer. J Hydrol 327:411–425

Ommen DAO, Kidmose J, Karan S, Flindt MR, Engesgaard Nilson B, Andersen F (2012) Importance of groundwater and macrophytes for the nutrient balance at oligotrophic Lake Hampen, Denmark. Ecohydrology 5:286–296

Parsons TR, Maita Y, Lalli CM (1984) A manual of chemical and biological methods for seawater analysis. Pergamon Press, Oxford, p 173

Santos-Borja A, Nepomuceno DN (2006) Laguna de Bay: Experience and lessons learned brief.. Retrieved 7 May 2017.. Accessed from http://www.worldlakes.org/uploads/15_Laguna_de_Bay_27February2006.pdf

Senal MS, Jacinto GS, San Diego-McGlone M, Siringan F, Zamora P, Soria L et al (2011) Nutrient inputs from submarine groundwater discharge on the Santiago reef flat, Bolinao, northwestern Philippines. Mar Pollut Bull 63:195–200

Strickland J, Parsons T (1972) A practical handbook of seawater analysis. Fisheries Research Board of Canada, Ottawa

Swarzenski P, Bratton J and Crusius J (2004) Submarine groundwater discharge and its influence on coastal processes and ecosystems. Retrieved 10 Feb 2013, from USGS Science for a Changing World: http://soundwaves.usgs.gov/2004/06/research4.htm

Tamayo-Zarafalla M, Santos RAV, Orozco RP, Elegado GCP (2002) The ecological status of Lake Laguna de bay, Philippines. Aquat Ecosyst Health Manag 5(2):127–138

Taniguchi M, Burnett WC, Cable JE, Turner JV (2002) Investigation of submarine groundwater discharge. Hydrol Process 16:2115–2129

Taniguchi M, Burnett WC, Dulaiova H, Siringan F, Foronda J, Wattayakorn G et al (2008) Groundwater discharge as an important Land-Sea pathway into Manila Bay, Philippines. J Coast Res 24(1A):15–24

Taniguchi M, Burnett WC, Smith CF, Paulsen RJ, O'Rourke D, Krupa SL et al (2003) Spatial and temporal distributions of submarine groundwater discharge rates obtained from various types of seepage meters at a site in the northeastern Gulf of Mexico. Biogeochemistry 66:33–53

Vanek V (1987) The interactions between lake and groundwater and their ecological significance. Stygologia 3:1–23

Zimmer V, Bendoricchio G (2001) Nutrients and suspended solid loads in the Laguna de bay, Philippines. Water Sci Technol 44(7):77–86

Chapter 8
Submarine Groundwater Discharge and its Influence on Primary Production in Japanese Coasts: Case Study in Obama Bay

Hisami Honda, Ryo Sugimoto, and Shiho Kobayashi

Abstract We report the relationship between submarine groundwater discharge (SGD) and primary production in the nearshore coast of Obama Bay, Japan, using three approaches. First, we conducted high-resolution mapping of ^{222}Rn and biogeochemical properties along the coast. The eastern part of the bay was strongly influenced by groundwater through several direct and indirect pathways. Lower $\delta^{15}N$ values in seaweed collected from the eastern area were indicative of larger influences of groundwater. Second, we measured the vertical distributions of ^{222}Rn, salinity, and chlorophyll-a (Chl-a) concentrations along two transects from onshore to offshore at two sites (Tomari and Kogasaki) located on the eastern coast of the bay. In Tomari, Chl-a concentrations were higher in the surface layer in the nearshore coastal area where ^{222}Rn and salinity showed higher and lower values, respectively, due to terrestrial spring water and SGD in the intertidal zone. In contrast, higher ^{222}Rn and Chl-a values were detected in the bottom layer in Kogasaki. This suggested that SGD was composed mainly of recirculated seawater discharge from the seafloor. Finally, temporal variations in multiple parameters related to SGD and phytoplankton production were recorded in Kogasaki in July and November. There was no clear relationship between tide and ^{222}Rn concentrations in either month, but pCO_2 and dissolved O_2 showed clear diurnal variations. The estimated O_2 production rate in July was higher than that in November. This seasonal difference may have been caused by differences in the SGD rate (7.1 cm d^{-1} in July and 3.7 cm d^{-1} in November).

H. Honda
Research Institute for Humanity and Nature, Kyoto, Japan
e-mail: h_honda@chikyu.ac.jp

R. Sugimoto (✉)
Department of Marine Bioscience, Faculty of Marine Bioscience, Fukui Prefectural University, Fukui, Japan
e-mail: sugiryo@fpu.ac.jp

S. Kobayashi
Field Science Education and Research Center, Kyoto University, Kyoto, Japan
e-mail: shihok@kais.kyoto-u.ac.jp

© Springer Nature Singapore Pte Ltd. 2018

A. Endo, T. Oh (eds.), *The Water-Energy-Food Nexus*, Global Environmental Studies, https://doi.org/10.1007/978-981-10-7383-0_8

Keywords Submarine groundwater discharge · ^{222}Rn · Nutrients · Primary production · Coastal seas

8.1 Introduction

8.1.1 Overview

Submarine groundwater discharge (SGD) is defined as terrestrially derived freshwater and recirculated seawater, which includes discharge associated with salt water intrusion along a coastline, tidal pumping, and wave setup, which flow out from the seabed into coastal water through underlying sediments (Destouni and Prieto 2003). Most estimates of terrestrially derived fresh SGD range from 6 to 10% of surface water inputs (Burnett et al. 2003). According to a recent report by Kwon et al. 2014, the total SGD, including fresh terrestrial groundwater and recirculated seawater, is 3 to 4 times greater than river water fluxes into the Atlantic and Indo-Pacific Oceans. Therefore, SGD is non-negligible, even from the perspective of the water cycle.

SGD is an important nutrient source for coastal ecosystems (Johannes 1980; Valiela et al. 1990). Nutrient concentrations in coastal groundwater are relatively higher than those in river water and seawater; therefore, even a small amount of groundwater discharge may contribute to nutrient budgets in coastal ecosystem (Burnett et al. 2003; Moore 2010). For example, nutrients transported through groundwater can support benthic and water column primary production in coastal seas such as sandy beaches and tidal flats (Kamermans et al. 2002; Miller and Ullman 2004; Paytan et al. 2006; Waska and Kim 2010; Waska and Kim 2011). In the coral reef off the west coast of Australia, SGD contributes significantly to reef productivity (Greenwood et al. 2013). In recent years, a direct relationship between SGD and *in situ* phytoplankton primary productivity in nearshore coastal areas in Japan has been clarified (Sugimoto et al. 2017). In addition, nutrient addition bioassay experiments support that SGD acts as a continual nutrient source (Gobbler and Boneillo 2003; Lecher et al. 2015). Conversely, it is well known that excess nutrient loadings via polluted groundwater into coastal seas cause cultural eutrophication and microalgal blooms (Pearl 1997; Gobler and Boneillo 2003).

8.1.2 Study Site

Obama Bay is a semi-enclosed embayment located in the central part of Wakasa Bay facing the Sea of Japan (Fig. 8.1). It has a surface area of 58.7 km^2, and is 17 km wide from east to west, 6 km wide from major river mouth to the bay mouth, and the bay mouth is 2.5 km wide. It has a volume of 0.74 km^3 with a mean depth of 13 m. Shallow areas (<5 m depth) occupy 16% of the surface area (Sugimoto

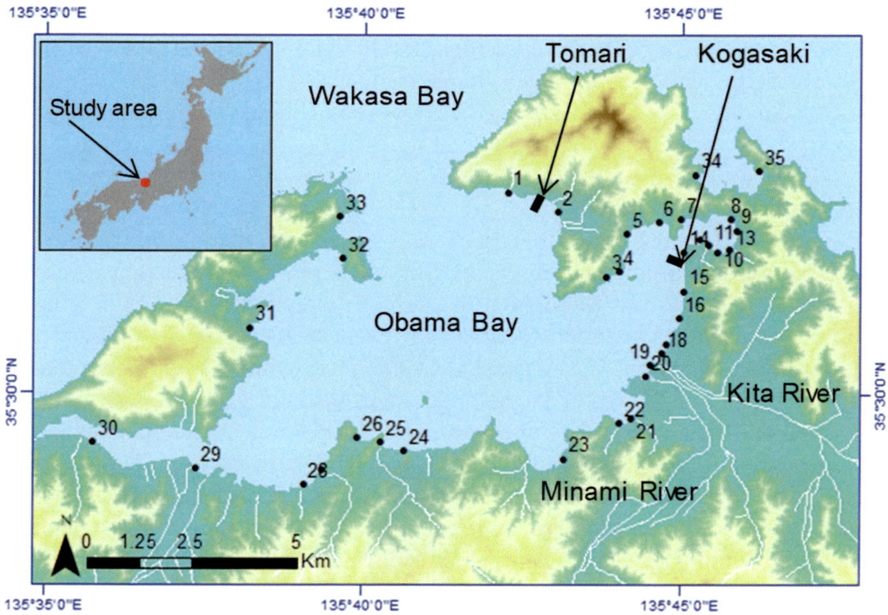

Fig. 8.1 Location of the study site, Obama Bay, Japan
We conducted the towing survey along the coast in March 2013. The dots and numbers indicate the *Ulva pertusa* sampling points in May 2012. We conducted the transect survey at two transects, as indicated bold lines, in Tomari and Kogasaki in July 2012. We conducted the mooring survey in Kogasaki in July and November 2013

et al. 2016). The volume of bay water exchange with the adjacent Wakasa Bay is about 2000 $m^3\ s^{-1}$ in summer, and the mean water residence time in summer is approximately 4 days (Sugimoto et al. 2016). Two major rivers, Kita River and Minami River, flow into the eastern part of the bay with a mean river discharge of approximately 10 $m^3\ s^{-1}$. The annual precipitation of the watershed is >2000 mm $year^{-1}$ (Sugimoto and Tsuboi 2017). Meanwhile, the alluvial fan (i.e., Obama Plain) is formed along the Kita River, and there are substantial groundwater resources on this plain (Sasajima and Sakamoto 1962; Obama City 2017). More than 100 flowing artesian wells are found within the Obama Plain near the coast.

The fresh SGD rates flowing into Obama Bay were estimated based on monthly ^{222}Rn and salinity data using a steady state mass balance model. The annual mean rate was $0.36 \times 10^6\ m^3\ day^{-1}$ (= 0.62 cm d^{-1}) with large intra-annual variability from 0.05×10^6 to $0.77 \times 10^6\ m^3\ day^{-1}$, which was relatively high in spring (March–April), when snowmelt water was predominant, and in the rainy summer season (June–September) (Sugimoto et al. 2016). The mean fresh SGD rate corresponded with the result based on seepage meter measurements in the nearshore coast (0.41 cm d^{-1}) and the total SGD rate was approximately 4.6 cm d^{-1} (Kobayashi et al. 2017). Comparing the nutrient concentrations in terrestrial groundwater with those in river water, dissolved inorganic nitrogen (DIN) and phosphorus (DIP) concentrations in groundwater were about 2 times and 7 times higher than those of river water, respectively. The resulting mean fluxes of DIN, DIP, and dissolved silica (DSi) via SGD were 42.8 kg day^{-1} (range: 64.7–947.1 kg day^{-1}), 38.0 kg day^{-1} (5.6–81.4 kg day^{-1}), and 914.2 kg day^{-1} (133.6–1955.6 kg day^{-1}), respectively. The mean fractions of groundwater in the total terrestrial fluxes of DIN, DIP, and DSi were 42.4% (9.0–72.2%), 65.3% (26.1–87.6%), and 33.4% (6.6–61.6%), respectively (Sugimoto et al. 2016).

DIP-enriched groundwater discharge could represent a non-negligible nutrient source for primary production in Obama Bay, since primary production is mostly restricted by phosphorous throughout the year (Sugimoto et al. 2016). Sugimoto et al. (2017) revealed the direct relationship between SGD and *in situ* primary production of phytoplankton. They conducted *in situ* measurements of primary productivity using a stable ^{13}C tracer method and environmental parameters (e.g., ^{222}Rn concentration, light intensity, temperature, and nutrient concentrations) at six stations in summer. They found a significant relationship between primary productivity and ^{222}Rn concentration. Although light intensity and water temperature differed at each station and in each month, nutrient concentrations limited primary productivity. Honda et al. (2016) also determined the sudden groundwater discharge and associated chlorophyll peak in the bottom layer (depth: 18 m) around 2 km offshore from the river mouth in June 2011, one week after a notable flood (Fig. 8.2). These results indicate that the nutrient supply from SGD has a crucial impact on primary production in Obama Bay.

8.1.3 Objectives

In this chapter, we show the relationship between SGD and primary production in the nearshore coastal area of Obama Bay using three approaches based on the above-mentioned method. First, we conducted a spatial assessment of SGD and biogeochemical parameters in the nearshore coastal area using a towing survey. A method of continuously monitoring active ^{222}Rn concentrations in water has been developed and used to visualize the distribution of groundwater discharge into coastal seas (Burnett et al. 2001; Burnett and Dulaiova 2003). Based on this method, in this study, we conducted high-resolution mapping of ^{222}Rn and biogeochemical properties along the coast of Obama Bay. Moreover, we used the δ^{15}N values of seaweed as an indicator of nutrient inputs through SGD, as they assimilate and accumulate nutrients from the water column, integrating continuous and pulsed nutrient loadings.

Second, transect surveys from nearshore to offshore at two sites (Tomari and Kogasaki in Fig. 8.1) were conducted. The eastern coast of the bay has a few streams, but freshwater springs on land and groundwater discharge at the intertidal zone exist along the coast of these areas.

Finally, we evaluated the diurnal variation in SGD and primary production in summer and autumn at Kogasaki (Fig. 8.1). Recent technological advances (i.e.,

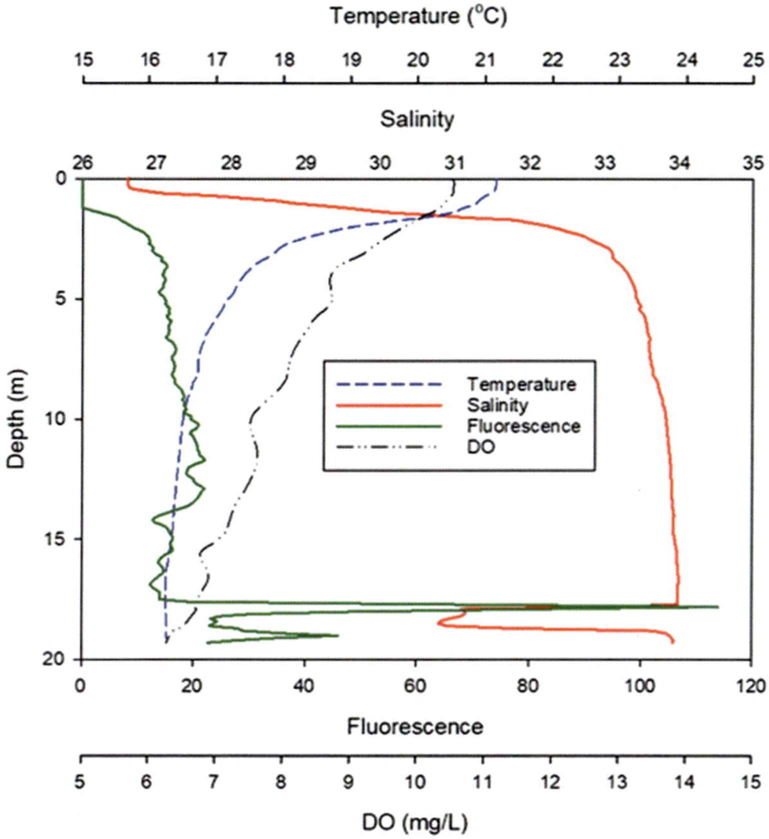

Fig. 8.2 Vertical profiles of temperature, salinity, chlorophyll fluorescence, and dissolved O_2 (DO) in the central part of Obama Bay on 7 June 2011. (Modified from Honda et al. 2016)

automation) have increased the ability to assess SGD in coastal ecosystems using natural tracers such as ^{222}Rn. Simultaneous monitoring of ^{222}Rn with indicators of primary production such as pCO_2, dissolved O_2 and chlorophyll-a (Chl-a) enabled us to determine the relationship between these processes.

8.2 Materials and Methods

8.2.1 Towing Survey

To detect SGD and to estimate the biogeochemical influence of SGD on primary producers in the nearshore coastal area of Obama Bay, the spatial distributions of ^{222}Rn, salinity, and nutrient concentrations were investigated along the coast on 13,

15, and 16 March 2013. We applied the multidetector method (Dulaiova et al. 2005; Stieglitz et al. 2010) for continuous ^{222}Rn measurements using three radon detectors (RAD7, Durridge) at a boat velocity of about 1–2 knots with water depths of around 2 m. We used two sets of three radon detectors, with a measurement interval of 10 min for each system; therefore, the average values and standard deviations of the ^{222}Rn concentrations were obtained every 5 min. Temperature and salinity were simultaneously measured every 1 min using the conductivity and temperature logger (AAQ1183, JFE-Advantech). Seawater samples for nutrient concentrations were collected every 10 min. In the laboratory, concentrations of NO_3^-, NO_2^-, PO_4^{3-}, and DSi were measured using an auto-analyzer (QuAAtro, BL-Tech). The concentration of NH_4^+ was measured using a fluorometer (Trilogy, Turner Design). In this study, we defined DIN as the sum of NO_3^-, NO_2^- and NH_4^+, and DIP as PO_4^{3-}.

Seaweed (*Ulva pertusa* Areschoug) was collected from 33 coastal sites in the bay and two sites outside the bay (numbers 1 to 35 in Fig. 8.1). The $\delta^{15}N$ values of *U. pertusa* were analyzed using a PDZ Europa ANCA-GSL elemental analyzer interfaced to a PDZ Europa 20–20 isotope ratio mass spectrometer (Sercon) at the University of California-Davis Stable Isotope Facility.

8.2.2 *Transect Survey*

The vertical distributions of temperature, salinity, and chlorophyll fluorescence were measured in two transects from onshore to offshore in Tomari on 24 July 2012 and Kogasaki on 29 July 2012 (locations shown in Fig. 8.1), using a CTD instrument (AAQ1183, JFE Advantech). Along each transect, six stations were set up at the points where the water depths were 0, 1, 2, 3, 4, and 5 m. Seawater was sampled at the surface and 0.5 m above the sea bottom of each station using a 6-L Van Dorn water sampler (RIGO Co., Ltd.). Seawater samples were filtered through 0.7-μm glass fiber filter (GF/F, Whatman) and the concentration of Chl-a on the GF/F filters was quantified with a calibrated fluorometer (Trilogy, Turner Design) to calibrate the chlorophyll fluorescence sensor attached to the CTD instrument. The ^{222}Rn concentrations of bottle samples were measured using the RAD7 detector.

8.2.3 *Mooring Survey*

The temporal variations of the parameters related to SGD and phytoplankton production were recorded in Kogasaki from 9 to 11 July 2013 and from 15 to 16 November 2013. To measure ^{222}Rn and pCO_2 simultaneously, seawater was pumped via a submersible pump into a gas equilibration chamber (RAD Aqua, Durridge), where the headspace of the equilibrated gas was measured using mobile ^{222}Rn (RAD7, Durridge) and CO_2 (GMP343, Vaisala) detectors. ^{222}Rn and pCO_2 were measured every 10 min and 5 min, respectively. Temperature, salinity, and dissolved O_2 were measured using a multi-parameter sonde (ProPlus, YSI), and depth and light intensity were measured using data loggers (DEFI-DHG and DEFI-L, JFE Advantech) with a measurement interval of 5 min. The Chl-a concentration was measured every 5 min using a calibrated fluorescence sensor (Cyclops-7, Turner). For the atmospheric conditions, ^{222}Rn and CO_2 concentrations were measured using the aforementioned detectors. In this study, all data were averaged to 1-h data to remove short-term variability.

8.3 Results and Discussion

8.3.1 Towing Survey

The integrated map of the 3-day survey performed in March 2013 showed that the ^{222}Rn ranged from 0 to 347 Bq m^{-3} (Fig. 8.3a). Higher concentrations were measured in the eastern area, relatively high (>150 Bq m^{-3}) concentrations were present in the zones between No.2 to No.14 and No.18 to No.22 (Figs. 8.1, 8.3a). The mean value and standard deviation of ^{222}Rn concentrations in the eastern area of the bay was 162 ± 56 Bq m^{-3}, much higher than that in the western area (203 ± 37 Bq m^{-3}). The spatial variability in salinity ranged from 22.6 to 32.5, and lower salinity was found in the eastern area (Fig. 8.3b). Salinity showed a significant negative correlation with ^{222}Rn ($r = -0.47$, $p < 0.05$). The DIN, DIP, and DSi concentrations were higher in the eastern area, similar to the ^{222}Rn concentrations (Figs. 8.3c, d, e). DIN and DSi concentrations were particularly high in the zone between No.5 and No.10 (Figs. 8.1, 8.3c, e), while DIP concentrations were higher in the zone between No. 15 and No.23 (Figs. 8.1, 8.3c). ^{222}Rn was significantly positively correlated with DIN ($r = 0.47, p < 0.01$) and DSi ($r = 0.34, p < 0.05$), whereas no significant correlation between ^{222}Rn and DIP was observed ($r = 0.08$, $p = 0.57$).

The ^{222}Rn and nutrient concentrations in the eastern area were markedly higher than those in the western area. As noted by Kobayashi et al. (2017), there are two major sources of ^{222}Rn in Obama Bay, river water, which has high ^{222}Rn concentrations, and groundwater discharge offshore areas at the mouths of the Kita and Minami Rivers. The ^{222}Rn concentration in river water was one order of magnitude higher than that in seawater, because groundwater seeped from the riverbed. The ^{222}Rn concentrations along the coast ranged from 0 to 346.7 Bq m^{-3}, while the mean ^{222}Rn concentrations obtained from monthly river sampling were 1800 and 1200 Bq m^{-3}, respectively (Sugimoto et al. 2016). In Obama Bay, river water is dispersed sufficiently into the bay before ^{222}Rn supplied from the rivers has decayed (Kobayashi et al. 2017). In addition, the northward wind enhances the advection of the water high in ^{222}Rn supplied from the river toward the eastern area of the bay (Kobayashi et al. 2017). This effect is significant because March has the highest mean river discharge. Terrestrial groundwater discharge and/or recirculated seawater discharge

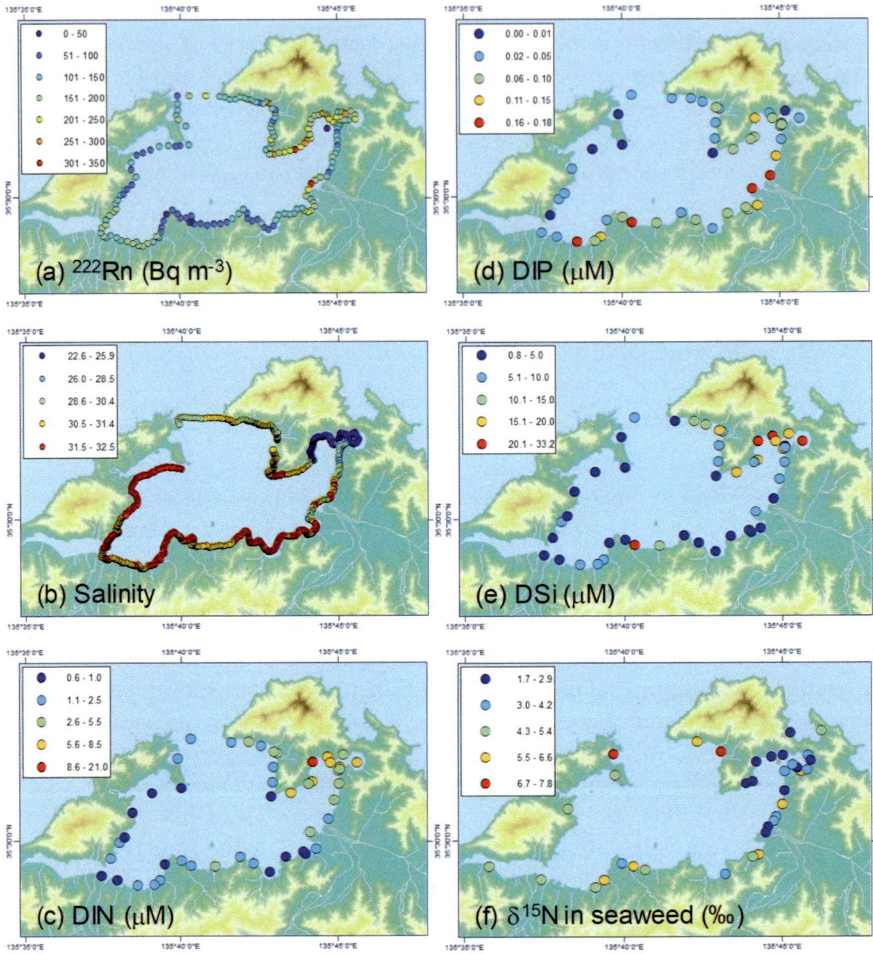

Fig. 8.3 Distributions of (**a**) ^{222}Rn, (**b**) salinity, (**c**) DIN, (**d**) DIP, and (**e**) DSi concentrations in surface water during 13, 15, and 16 March 2013 and (**f**) δ^{15}N in seaweed (*Ulva pertusa*) collected on 7, 8, and 11 May 2012

offshore the estuary represents another presumable source of water high in ^{222}Rn. PO_4^{3-} and ^{222}Rn concentrations were both higher in the zone between No.19 and No.22. This was considered to indicate the presence of significant SGD inputs, because PO_4^{3-} concentrations are higher in terrestrial groundwater than in river water (Sugimoto et al. 2016).

δ^{15}N values of primary producers such as phytoplankton, seaweed, and seagrass reflect the δ^{15}N of substrate nitrogen (Costanzo et al. 2001). The δ^{15}N values of *U. pertusa* ranged from 1.7 to 7.9‰ (Fig. 8.3e). The δ^{15}N values of *U. pertusa* around the eastern part of the bay (2.9 ± 1.1‰) were significantly lower than those in other sites (5.2 ± 1.3‰) ($p < 0.001$, *t*-test). Although we could not differentiate the δ^{15}N

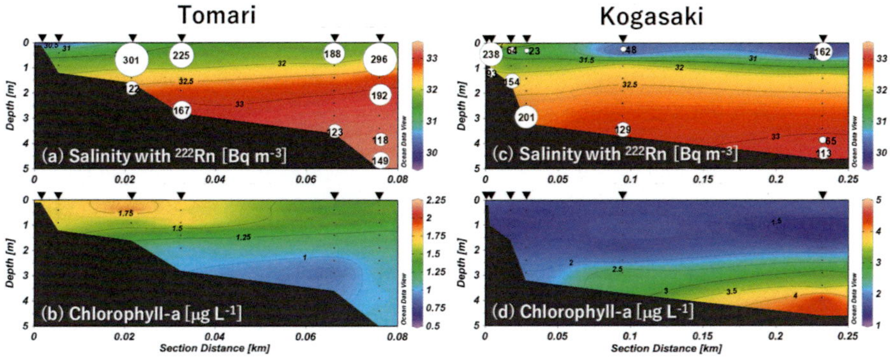

Fig. 8.4 Distributions of salinity (contour) with ^{222}Rn (circle: *top panels*) and Chl-a (*bottom panels*) in the transect of Tomari and Kogasaki on 24 and 29 July 2012. Triangles indicate the locations of the observation points

values of NO_3^- ($\delta^{15}N_{NO3}$) obtained from terrestrial freshwater ($\delta^{15}N_{NO3}$ = 3.7‰ of river, $\delta^{15}N_{NO3}$ = 3.6‰ of spring water; Kobayashi, unpublished), the $\delta^{15}N_{NO3}$ values of terrestrial freshwater were lower than those in oceanic waters of the Sea of Japan (5.4 ± 0.2‰; Nakanishi and Minagawa 2003). These results imply that *U. pertusa* in the eastern part of the bay assimilate nitrogenous nutrients supplied from land, including groundwater.

8.3.2 Transect Survey

The ^{222}Rn concentrations in each zone separated by distance from the land and the vertical distributions of ^{222}Rn, salinity, and Chl-a along the two transects at Tomari and Kogasaki are shown in Fig. 8.1. Along the Tomari transect (Figs. 8.4a, b), the highest ^{222}Rn concentration (301 Bq m^{-3}) was observed around the stations closest to land. ^{222}Rn concentrations were higher at the surface than at the bottom along the transect. The vertical differences in the surface and bottom temperature were around 5 °C, suggesting that the influence of tidal mixing was not significant, although the section was in the shallow zone (>5 m). Salinity was low in the surface layer but high at the bottom, suggestive of river water flowing into the sea from the land, although there was no river or stream near this section. The Chl-a concentrations were relatively high in the surface layer, corresponding to the distribution of low-salinity water and ^{222}Rn.

Along the Kogasaki transect (Figs. 8.4c, d), ^{222}Rn concentrations were higher in the bottom layer than in the surface layer, except for the station closest to land. Salinity was lowest at the station farthest from land in the surface layer, suggestive of the influence of the advection of low-salinity water from the main rivers (Kobayashi et al. 2017). The highest Chl-a concentrations were observed in the bottom layer at the station farthest from land, and Chl-a concentrations were higher in the bottom layer than in the surface layer, corresponding to the distribution of ^{222}Rn.

The relatively high ^{222}Rn concentrations and low salinity at the surface in Tomari indicated the influence of inflow from groundwater springs on land or SGD in the intertidal zone. In contrast, the relatively high concentrations of ^{222}Rn at the bottom unaccompanied by low salinity in Kogasaki suggested the influence of SGD mainly consisting of recirculated SGD, which has been observed in this area using a seepage meter (Kobayashi et al. 2017). In addition to river water, SGD is known to supply nutrients that are needed for phytoplankton growth in Obama Bay (Sugimoto et al. 2016; Honda et al. 2016).

Sugimoto et al. (2017) and Kobayashi et al. (2017) identified significant correlations between ^{222}Rn concentration and the primary production rate and ^{222}Rn and Chl-a, respectively, in the entire Obama Bay. The correlation between ^{222}Rn and Chl-a concentrations along the transects from the land toward offshore obtained in this study also suggest that groundwater influences primary production in the shallow zone along the coast. Although the link between ^{222}Rn and the primary production rate and the influence of the seawater residence time on the distributions of both ^{222}Rn and Chl-a must be clarified in future studies, the results of this survey provide persuasive evidence of the influence of SGD on primary production in the shallow zone of Obama Bay.

8.3.3 Mooring Survey

There were no significant differences in the mean water depth and tidal ranges in July and November, although the water temperature and salinity in July were considerably higher than those in November (Fig. 8.5). The ^{222}Rn concentration of 62 ± 9 Bq m^{-3} (mean ± standard deviation) (range: 42–87 Bq m^{-3}) in July was significantly lower than that in November (71 ± 11 Bq m^{-3}, range: 54–92 Bq m^{-3}) ($p < 0.001$, t-test). There was no clear relationship between tide and ^{222}Rn concentration in either month.

Variations in the time-series ^{222}Rn record from the mooring survey were used to estimate the SGD flux with a non-steady-state mass balance model, as follows (Burnett and Dulaiova 2003; Hosono et al. 2012):

$$F_{\text{benthic}} - \lambda I_{\text{exRn}} - F_{\text{atm}} \pm F_{\text{hor}} = 0 \tag{8.1}$$

where F_{benthic} is the combined advective and diffusive flux of ^{222}Rn to the overlying water column, λ is the decay constant of ^{222}Rn (= 0.181 d^{-1}), I_{exRn} is the inventory of excess ^{222}Rn concentration (= ^{222}Rn – ^{226}Ra), F_{atm} is the flux of ^{222}Rn to the atmosphere, and F_{hor} is the horizontal mixing of ^{222}Rn into or out of the mooring site.

The excess ^{222}Rn activities (unsupported by ^{226}Ra) in the water column were estimated by subtracting the single measured ^{226}Ra concentration of offshore water (7.7 Bq m^{-3}; Sugimoto et al. 2016). Decay was not considered because the fluxes

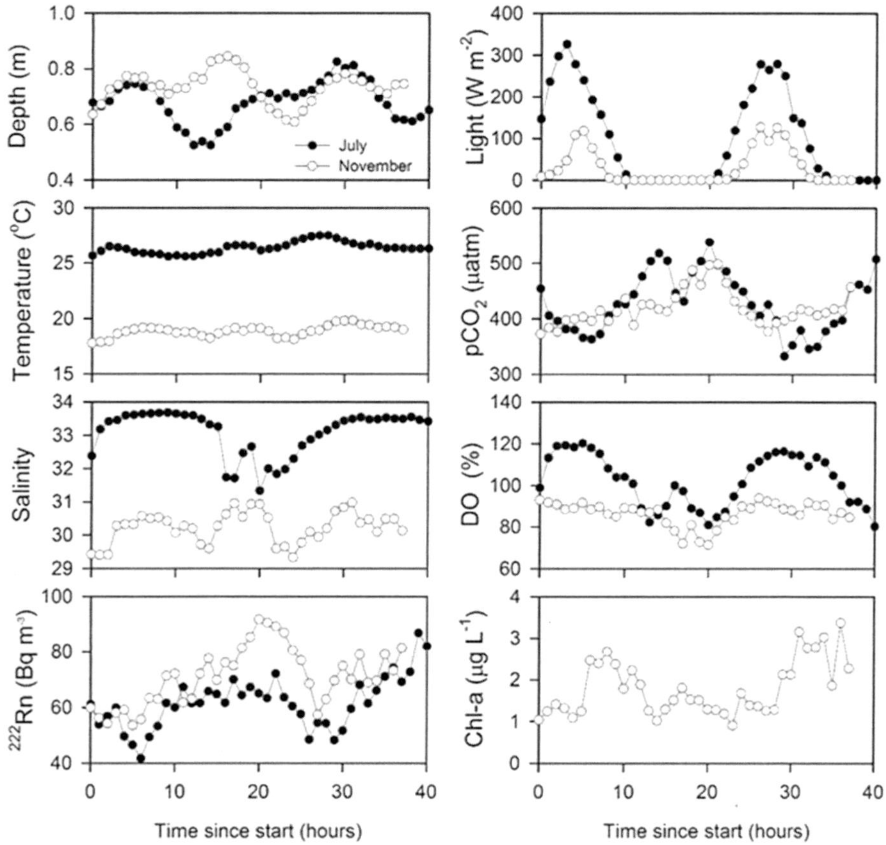

Fig. 8.5 Time series of water depth, temperature, salinity, ^{222}Rn concentration, light intensity, pCO_2, dissolved O_2, and Chl-a concentration in July and November 2013. All data are shown beginning at 09:00 LST

were evaluated on a very short (1-h) time scale relative to the half-life of ^{222}Rn. F_{atm} was determined based on molecular diffusion and turbulent transfer models (MacIntyre et al. 1995; Wanninkhof 1992; Turner et al. 1996), and detailed calculations referred to those in Sugimoto et al. (2016). Atmospheric ^{222}Rn was assumed to be zero, and air temperature and wind speed data were derived from Japan Meteorological Agency reports. During the mooring survey period, these values ranged from 23.0 to 33.9°C and 0.6 to 4.0 m s^{-1} in July and from 8.6 to 16.9°C and 0.1 to 3.3 m s^{-1} in November, respectively. To calculate water flux, we simply divided $F_{benthic}$ by ^{222}Rn activity in shallow groundwater around 10 m inland from the coast (3470 Bq m^{-3}). We assumed that advective flux of radon from the sediment dominated the total flux, and diffusive flux was ignored in our calculations.

The estimated SGD rates were 8.3 ± 3.2 cm d^{-1} in July and 6.3 ± 3.3 cm d^{-1} in November. In contrast to ^{222}Rn concentrations, the SGD rates in July were higher than those in November (p = 0.017, t-test), because apparent ^{222}Rn loss to atmo-

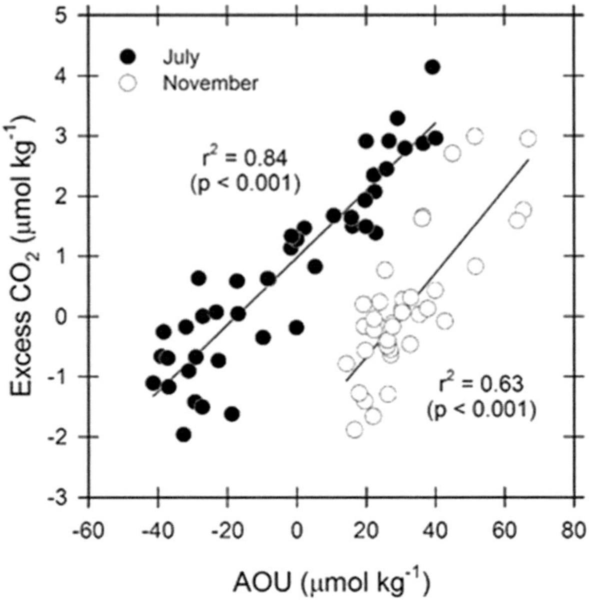

Fig. 8.6 Plots of apparent oxygen utilization (AOU) versus excess CO_2 in July and November 2013

sphere was higher in July. The SGD fluxes estimated from ^{222}Rn were similar to those estimated with a seepage meter at the same site (4.6 cm d^{-1}; Kobayashi et al. 2017). The difference in SGD rates in July and November could affect primary production in the overlying water column.

pCO_2 and dissolved O_2 showed clear diurnal variations, and were lower and higher in daytime than at night, respectively (Fig. 8.5). These inverse trends are suggestive of the influence of phytoplankton photosynthesis and respiration. To facilitate the discussion, we defined the free dissolved CO_2 from atmospheric equilibrium as excess CO_2 (Eq. (8.2) and the oxygen as apparent oxygen utilization (AOU) as Eq. ((8.3), following the method of Zhai et al. (2005).

$$\text{Excess}\,\text{CO}_2 = \left[\text{CO}_2^{\,*}\right] - K_{\text{H}}^{\ \text{CO2}} \times p\text{CO}_2\left(\text{in air}\right) \tag{8.2}$$

where $[CO_2^*]$ is the concentration of total free CO_2 (i.e., $[CO_2^*] = [CO_2] + [H_2CO_3] = K_H^{CO2} \times pCO_2$ in water) and K_H^{CO2} is the solubility coefficient of CO_2.

$$\text{AOU} = \left[\text{O}_2\right]_{\text{eq}} - \left[\text{O}_2\right] \tag{8.3}$$

where $[O_2]_{eq}$ is the DO concentration at equilibrium with the atmosphere and $[O_2]$ is the *in situ* DO concentration.

AOU and excess CO_2 showed positive correlations in both months (Fig. 8.6). These clear trends support that the processes of photosynthesis and respiration control the temporal variations in pCO_2 and dissolved O_2. However, the variability of CO_2 and O_2 due to photosynthesis and respiration in seawater is generally assessed

Table 8.1 Average data of O_2 production rate, SGD, and limitation factors of temperature (F_T) and light (F_I) from 05:00 to 11:00 LST in July and November 2013

	O_2 production (μmol kg^{-1} h^{-1})	SGD (cm d^{-1})	F_T	F_I
July	11.2	7.1 ± 2.9	1.00 ± 0.00	0.61 ± 0.31
November	7.9	3.7 ± 1.4	0.96 ± 0.00	0.44 ± 0.41

Optimum temperature and light are assumed to be 25°C and 104.7 W m^{-2}

by examining the relationship between total CO_2 (sum of CO_2^*, HCO_3^-. and CO_3^{2-}) and O_2 (DeGrandpre et al. 1997). To evaluate the stoichiometric relationship between total CO_2 and O_2, additional direct measurements such as alkalinity are needed for future studies.

In this study, we evaluated photosynthesis as the O_2 production rate using the AOU data. From 05:00 to 11:00 LST (integrated time from 21 to 27 of each month), when light intensity increased from near zero to the maximum, dissolved O_2 showed a linear increasing trend in both months (Fig. 8.5). The slopes in this period were −11.2 μmol AOU kg^{-1} h^{-1} in July ($r^2 = 0.98$) and −7.9 μmol AOU kg^{-1} h^{-1} in November ($r^2 = 0.91$). The SGD rate during the same period in July was 7.1 cm d^{-1}, which was considerably higher than that in November (Table 8.1). As noted by Kobayashi et al. (2017), nutrients are major limitation factor at the mooring site. Therefore, nutrients supplied from SGD could induce the difference in O_2 production rates (photosynthesis) in July and November. Since seasonal differences in light availability and water temperature may affect the difference in the O_2 production rate in July and November, we evaluated the limitation factors of light (F_I) and temperature (F_T) following the method of Sugimoto et al. (2017). As a result, light availability (0.61 in July and 0.44 in November, Table 8.1) may be a non-negligible environmental factor of lower O_2 production rate in November. However, simultaneous measurements of multiple parameters related SGD and primary production in future studies will lead to a better understanding of the roles of SGD on coastal ecosystems.

8.4 Conclusion

In this chapter, we presented the relationship between SGD and primary production in the nearshore coastal area of Obama Bay, Japan, using three approaches: towing, transect, and mooring surveys. Although the results collected in each approach support earlier findings and are thought to be correct, the results should be considered somewhat preliminary due to their collection within a few days and deficiency for individual use. Our recent papers (i.e., Honda et al. 2016; Kobayashi et al. 2017; Sugimoto et al. 2017) provide more comprehensive results. Considering that nearshore coastal areas have high productivity and biodiversity and many environmental issues, such as harmful algal blooms and alteration of the coast, occurs at this active area, the results in this chapter are highly suggestive and show the need for specific surveys in nearshore coastal sites. In the near future, SGD studies will be expected to become considered an essential component of understanding coastal ecosystems.

Acknowledgment The transect survey from onshore to offshore in two transects were made in collaboration with Wakasa High School. We are grateful to the diving club members and their teachers, Dr. Yasuyuki Kosaka and Mr. Hiroaki Hirayama, and the headmaster of Wakasa High School for their assistance. This research was financially supported by the R-08-Init Project, entitled "Human-Environmental Security in Asia-Pacific Ring of Fire: Water-Energy-Food Nexus" the Research Institute for Humanity and Nature (RIHN), Kyoto, Japan.

References

Burnett WC, Dulaiova H (2003) Estimating the dynamics of groundwater input into the coastal zone via continuous radon-222 measurements. J Environ Radioact 69:21–35

Burnett WC, Kim G, Lane-Smith D (2001) A continuous monitor for assessment of 222Rn in the coastal ocean. J Radioanal Nucl Chem 249:167–172

Burnett WC, Bokuniewicz H, Huettel M, Moore WS, Taniguchi M (2003) Groundwater and pore water inputs to the coastal zone. Biogeochemistry 66:3–33

Costanzo SD, O'Donohue MJ, Dennison WC, Loneragan NR, Thomas M (2001) A new approach for detecting and mapping sewage impacts. Mar Pollut Bull 42:149–156

DeGrandpre MD, Hammar TR, Wallace DWR, Wirick CD (1997) Simultaneous mooring-based measurements of seawater CO2 and O2 off Cape Hatteras, North Carolina. Limnol Oceanogr 42:21–28

Destouni G, Prieto C (2003) On the possibility for generic modeling of submarine groundwater discharge. Biogeochemistry 66:171–186

Dulaiova H, Peterson R, Burnett WC, Lane-Smith D (2005) A multi-detector continuous monitor for assessment of 222Rn in the coastal ocean. J Radioanal Nucl Chem 263:361–365

Gobler CJ, Boneillo GE (2003) Impacts of anthropogenically influenced groundwater seepage on water chemistry and phytoplankton dynamics within a coastal marine system. Mar Ecol Prog Ser 255:101–114

Greenwood JE, Symonds G, Zhong L, Lourey M (2013) Evidence of submarine groundwater nutrient supply to an oligotrophic barrier reef. Limnol Oceanogr 58:1834–1842

Honda H, Sugimoto R, Kobayashi S, Tahara D, Tominaga O (2016) Temporal and spatial variation in primary production in Obama Bay. Bull Jpn Soc Fish Oceanogr 80:269–282. (in Japanese with English abstract)

Hosono T, Ono M, Burnett WC, Tokunaga T, Taniguchi M, Akimichi T (2012) Spatial distribution of submarine groundwater discharge and associated nutrients within a local coastal area. Environ Sci Technol 46:5319–5326

Johannes RE (1980) The ecological significance of the submarine discharge of groundwater. Mar Ecol Prog Ser 3:365–373

Kamermans P, Hemminga MA, Tack JF, Mateo MA, Marbà N, Mtolera M, Stapel J, Verheyden A, Van Daele T (2002) Groundwater effects on diversity and abundance of lagoonal seagrasses in Kenya and on Zanzibar Island (East Africa). Mar Ecol Prog Ser 231:75–83

Kobayashi S, Sugimoto R, Honda H, Miyata Y, Tahara D, Tominaga O, Shoji J, Yamada M, Nakada S, Taniguchi M (2017) High-resolution mapping and time-series measurements of 222Rn concentrations and biogeochemical properties related to submarine groundwater discharge along the coast of Obama Bay, a semi-enclosed sea in Japan. Prog Earth Planet Sci. https://doi.org/10.1186/s40645-017-0124-y

Kwon EY, Kim G, Primeau F, Moore WS, Cho HM, DeVries T, Sarmiento JL, Charette MA, Cho YK (2014) Global estimate of submarine groundwater discharge based on an observationally constrained radium isotope model. Geophys Res Lett 41:8438–8444

Lecher AL, Mackey K, Kudela R, Ryan J, Fisher A, Murray J, Paytan A (2015) Nutrient loading through submarine groundwater discharge and phytoplankton growth in Monterey Bay, CA. Environ Sci Technol 49:6665–6673
MacIntyre S, Wanninkhof R, Chanton JP (1995) Trace gas exchange across the air-sea interface in freshwater and coastal marine environments. In: Matson PA, Harris RC (eds) Biogenic trace gases: measuring emissions from soil and water. Blackwell Science, pp 52–97
Miller DC, Ullman WJ (2004) Ecological consequences of ground water discharge to Delaware Bay, United States. Groundwater 42:959–970
Moore WS (2010) The effect of submarine groundwater discharge on the ocean. Annu Rev Mar Sci 2:59–88
Nakanishi T, Minagawa M (2003) Stable carbon and nitrogen isotopic compositions of sinking particles in the northeast Japan Sea. Geochem J 37:261–275
Obama City (2017) http://www1.city.obama.fukui.jp/category/page.asp?Page=3741
Paytan A, Shellenbarger GG, Street JH, Gonneea ME, Davis K, Young MB, Moore WS (2006) Submarine groundwater discharge: an important source of new inorganic nitrogen to coral reef ecosystems. Limnol Oceanogr 51:343–348
Pearl HW (1997) Coastal eutrophication and harmful algal blooms: importance of atmospheric deposition and groundwater as "new" nitrogen and other nutrient sources. Limnol Oceanogr 42:1154–1165
Sasajima S, Sakamoto K (1962) Subsurface geology and groundwater of Obama plain, Fukui pref., central Japan. Part 2: groundwater of Obama plain. Memoirs of the Faculty of Liberal Arts, University of Fukui. Ser. II Natural science (in Japanese with English abstract)
Stieglitz TC, Cook PG, Burnett WC (2010) Inferring coastal processes from regional-scale mapping of 222 Radon and salinity: examples from the great barrier reef, Australia. J Environ Radioact 101:544–552
Sugimoto R, Tsuboi T (2017) Seasonal and annual fluxes of atmospheric nitrogen deposition and riverine nitrogen export in two adjacent contrasting rivers in central Japan facing the sea of Japan. J Hydro Reg Stud 11:117–125
Sugimoto R, Honda H, Kobayashi S, Takao Y, Tahara D, Tominaga O, Taniguchi M (2016) Seasonal changes in submarine groundwater discharge and associated nutrient transport into a tideless semi-enclosed embayment (Obama Bay, Japan). Estuar Coasts 39:13–26
Sugimoto R, Kitagawa K, Nishi S, Honda H, Yamada M, Kobayashi S, Shoji J, Ohsawa S, Taniguchi M, Tominaga O (2017) Phytoplankton primary productivity around submarine groundwater discharge in nearshore coasts. Mar Ecol Prog Ser 563:25–33
Turner SM, Malin G, Nightingale PD, Liss PS (1996) Seasonal variation of dimethyl sulphide in the North Sea and an assessment of fluxes to the atmosphere. Mar Chem 54:245–262
Valiela I, Costa J, Foreman K, Teal JM, Howes B, Aubrey D (1990) Transport of groundwater-borne nutrients from watersheds and their effects on coastal waters. Biogeochemistry 10:177–197
Wanninkhof R (1992) Relationship between wind speed and gas exchange over the ocean. J Geophys Res Oceans 97:7373–7382
Waska H, Kim G (2010) Differences in microphytobenthos and macrofaunal abundances associated with groundwater discharge in the intertidal zone. Mar Ecol Prog Ser 407:159–172
Waska H, Kim G (2011) Submarine groundwater discharge (SGD) as a main nutrient source for benthic and water-column primary production in a large intertidal environment of the Yellow Sea. J Sea Res 65:103–113
Zhai W, Dai M, Cai WJ, Wang Y, Wang Z (2005) High partial pressure of CO2 and its maintaining mechanism in a subtropical estuary: the Pearl River estuary, China. Mar Chem 93:21–32

Chapter 9
Relationships Between Submarine Groundwater Discharge and Coastal Fisheries as a Water-Food Nexus

Jun Shoji and Osamu Tominaga

Abstract We introduced the results of investigations conducted in Japan (Yamagata, Fukui, Hiroshima and Oita Prefectures) and discussed the contribution of SGD to the production of fishery resources and their prey (meiobenthic communities). Our recent surveys provided evidence of the contribution of SGD to the production of fishes and their prey. These results strongly support the hypothesis that the contribution of nutrients from SGD to the coastal fishery resources production is much higher than previously suggested. The negative ecological impacts of SGD on meiofaunal abundance were confirmed at the high seepage areas of Kamaiso where the small spatial scale presented strong heterogeneity in SGD environmental conditions. However, the meiofaunal abundance was not low at the low seepage area located slightly apart from the high seepage area. Therefore, the effect of SGD on meiofaunal communities could be limited. However, the environmental heterogeneity significantly enhanced the conservation of the meiofaunal diversity. Thus, SGD determined the property of benthic communities.

Keywords Submarine ground water · Fishery resources · Meiofauna · Production

9.1 Introduction

At various locations globally, submarine groundwater discharge (SGD) from the sea floor enhances the production and species diversity of fishes and shellfishes because the SGD is richer in nutrients compared to river water (Chap. 8). However, there has been limited research on the processes in which SGD affects the growth of fishery

J. Shoji
Costal Ecology, Hiroshima University, Higashi-hiroshima 739-8528, Hiroshima, Japan
e-mail: jshoji@hiroshima-u.ac.jp

O. Tominaga (✉)
Marine Science and Technology, Fukui Prefectural University,
Katsumi, Obama 917-0116, Fukui, Japan
e-mail: tominaga@fpu.ac.jp

© Springer Nature Singapore Pte Ltd. 2018

A. Endo, T. Oh (eds.), *The Water-Energy-Food Nexus*, Global Environmental Studies, https://doi.org/10.1007/978-981-10-7383-0_9

Fig. 9.1 Hiji Castle, Ohita Prefecture (**A**), marbled sole (**B**), Castle Flounder Festival (**C**), and fishes near the submarine groundwater discharge area (**D**)

resources. In Hiji, Oita Prefecture, southwestern Japan, SGD seepages occur at the foot of the Yokokujo Castle (Fig. 9.1A). Marbled flounder, *Pseudopreulonectes yokohamae*, a flatfish (Fig. 9.1B), caught in the surroundings of the castle is highly prized and was presented to the general during the Edo period. Currently, the flounder is traded regionally at high prices, and contributes economically to the Castle Flounder Festival (Fig. 9.1C) every May. It is also known that bivalves, such as oysters and abalones, grow to large sizes near the SGD in various places in Japan (Akimichi 2010; Shoji et al. 2017). Research on the relationships between SGDs and biological communities can elucidate chemical and biological processes in coastal ecosystems. In addition, regarding the conservation of coastal ecosystems exposed to environmental and ecological degradation, and sustainable use of resources, this research will provide important information on the creation of regional brands and revitalization of the economy at local scale and development of the fishery food culture at national and global scales.

The effect of SGD on biological production in coastal waters has been studied in food webs, especially low order biological communities (Miller and Ullman 2004; Valiela et al. 1990; Kamermans et al. 2002). The effect of SGD on freshwater biological communities has been investigated in the Lake Huron of the Great Lakes, United States (Sanders Jr et al. 2011). Based on carbon stable isotope ratios of fish living in the surrounding area and their prey organisms (e.g., crustaceans), it was revealed that nutrients originating from SGD are utilized by secondary consumers. This study comprehensively evaluated the influence of terrestrial nutrients from SGD on the surrounding biological communities at higher trophic levels. In Japan, relationships between SGD and primary production has been reported in the coastal area of Yamagata Prefecture (Hosono et al. 2012), Nanao Bay (Ishikawa prefecture: Sugimoto et al. 2014), Obama Bay (Fukui prefecture: Sugimoto et al. 2016), and Ariake Sea (Shiokawa et al. 2013). Based on the elevated concentrations of chlorophyll-a and phytoplankton in the waters where the radon concentration (as an indicator of SGD) is high, it is possible that the nutrient supply through SGD increases the primary production of coastal waters. However, there are limited studies where the effect of SGD on primary consumers and predators at higher trophic levels in coastal waters has been clarified.

In this chapter, we introduced the results of investigations conducted in Japan (Yamagata, Fukui, Hiroshima and Oita Prefectures) and discussed the contribution of SGD to the production of fishery resources (9.2.) and their prey (meiobenthic communities: 9.3.). Through these results and our ongoing research, it was found that the contribution of nutrients from SGD to the coastal fishery resources production is much higher than previously suggested. In addition, magnitude and mechanism of the effects on marine organisms seem to differ among taxonomic groups.

9.2 Effects of SGDs on Fish communities - Case Studies in Japan

9.2.1 Sandy Beaches in Yamagata Prefecture: Inter-Bay Scale

In Yusa, northern Yamagata Prefecture, near the border of Akita Prefecture, there is abundant freshwater SGD originating from Mt. Chokai (Fig. 9.2A). A local specialty, oysters (*Iwagaki* in Japanese), grow in the surrounding area, with nutrients supplied by the SGD. Sailfin sandfish, *Arctoscopus japonicus*, spawn in the shallow sea area in early winter and possibly migrate to the SGD area (Akimichi 2010). In a previous study investigating the radon concentration in the coastal area, a record high concentration was observed in the coastal area of Yusa (Hosono et al. 2012). Here, we introduced recent findings on the effects of SGD on biological communities and food networks; mainly fish located at higher trophic levels and their prey organisms.

Physical and biological surveys were conducted at two sites which were selected based on a preliminary survey of the horizontal radon concentration distribution. Kamaiso has high SGD with strong freshwater influence and radon concentrations, and Nishihama has low freshwater influence and radon concentrations. The horizontal distance between the sites was approximately 1.5 km, and the sea bottom of both sites was sandy. To examine the appearance of fish, multiple underwater cameras were set at each site with shooting intervals of 1 min for approximately 2 h.

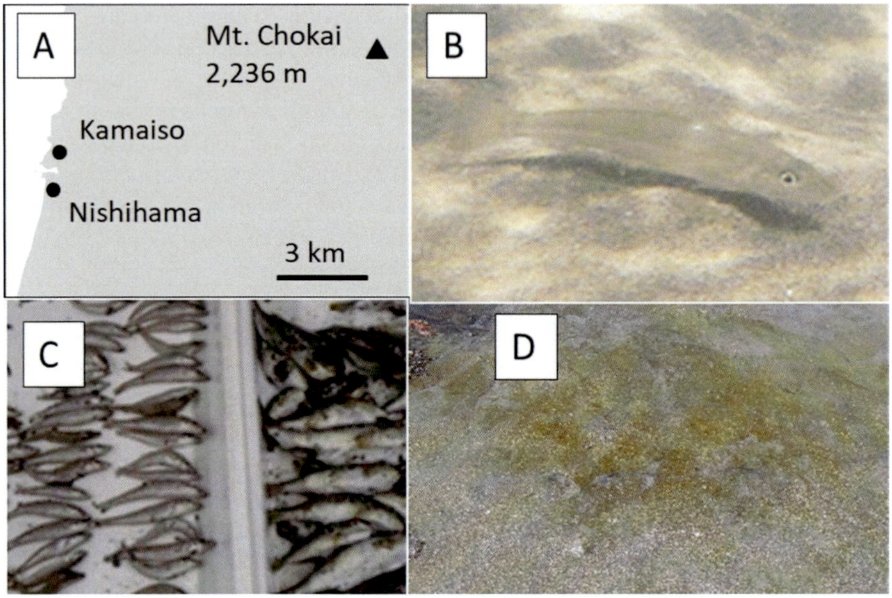

Fig. 9.2 Location of the survey sites in Yamagata Prefecture (**A**), Japanese whiting (**B**), fishes collected at Kamaiso (**C**), and benthic microalgae around submarine groundwater discharge (**D**)

Presence of fish was confirmed in approximately 25% of 600 or more images for analysis. The frequency of appearance of Japanese whiting, *Sillago japonica* (Fig. 9.2B), and mullet (Mugilidae spp.) were high. To reduce errors in identification and overestimation through the image analyses, fish were also collected using nets. Japanese whiting was the most dominant in both number and weight, followed by puffer (*Takifugu porphyreus*), marbled flounder (*Pseudopleronectes yokohamae*) and stone flounder (*Kareius bicoloratus*). The frequency of occurrence of Japanese whiting at Kamaiso was approximately 20 times higher than that at Nishihama.

Stomach contents analyses of the fish collected using nets (Fig. 9.2C) revealed that gammarids (50% by dry weight) were the most dominant prey organisms. Abundance of gammarids sampled by a sledge-net (0.3 x 0.4 m mouth opening, 0.3 mm mesh) at Kamaiso was approximately nine times higher than that at Nishihama. Results of carbon and nitrogen stable isotope ratio analyses supported the results of the stomach content analyses that indicated Japanese whiting fed mainly on gammarids in the surrounding area (Utsunomiya et al. 2017).

Benthic microalgae were abundant around the SGD at Kamaiso (Fig. 9.2D), indicating that nutrients of terrestrial origin supplied through SGD were utilized by gammarids, which feed on benthic microalgae. Gammarids have been suggested as one of the most important components that connect primary producers and fishes (secondary consumers) in the food web in the surrounding area.

9.2.2 Sandy Beaches in Obama Bay, Fukui Prefecture: Within-Bay Scale

Obama City is located in an area where the human population and groundwater are closely related. Previous studies have revealed SGD in Obama Bay resulting in a relatively high total nutrient flux into the bay (> 60% total phosphorus provided through freshwater in river water and SGD: Sugimoto et al. 2016). However, the effect of SGD on biological communities at higher trophic levels in the food web has not been elucidated. In this chapter, we determined the possible effects of SGD on fish and benthic invertebrates at a relatively small spatial scale in Obama Bay (Utsunomiya et al. 2017).

Surveys of the Wakasa district, northeastern Obama Bay, indicated radon concentrations varied between 0.1 and 0.9 dpm/L within a spatial scale of approximately 200 m (Utsunomiya et al. 2017). Regarding the radon concentration gradient in the coastal area, the abundance and biomass of fish and invertebrates were compared between the sites with high (Site 1) and low (Site 2) radon concentrations (Fig. 9.3). The water temperature and salinity were 29.7 °C and 25.4 at Site 1 and 30.5 °C and 28.2 at Site 2, respectively. The sea bottom consisted of sand and small gravel (< 10 cm) in both areas.

Temperate bass (*Lateolabrax japonicus*), black sea bream (*Achanthopagrus schlegelii*), and gobies (Gobiidae) dominated the fish community, and were collected using a small seine (2 x 1 m, 0.3 mm mesh) at the two sites. Fish abundance and biomass at Site 1 were significantly higher than that at Site 2. The abundance of gammarids, a major prey source of the dominant fishes, collected using the sledge net, and those of two gastropods (*Lunella coreensis* and *Pagurus filholi*) surveyed with a quadrat (0.5 x 0.5 m) at Site 1, were also significantly higher than those at Site 2 ($p < 0.05$, Utsunomiya et al. 2017).

9.2.3 Contribution of SGD to the Food Web on a Tidal Flat in Hiroshima Prefecture: Within Bay-Scale

On the tidal flats off the mouth of the Kamo River in Takehara City, Hiroshima Prefecture, marbled flounder larvae settled from February to April, then the juveniles dominated the fish community from April to May. A preliminary survey confirmed a low salinity water mass (< 10) was distributed off the estuary of the Kamo River (Hata et al. 2016), indicating the possibility that fresh water is supplied to the sea via alternative inputs to rivers. We conducted an SGD survey using the sea water radon concentration and found a high radon concentration in the western part of the survey area.

An on-site cultivation experiment of benthic microalgae using plastic plates showed that the production rate was also higher in the western part of the estuary.

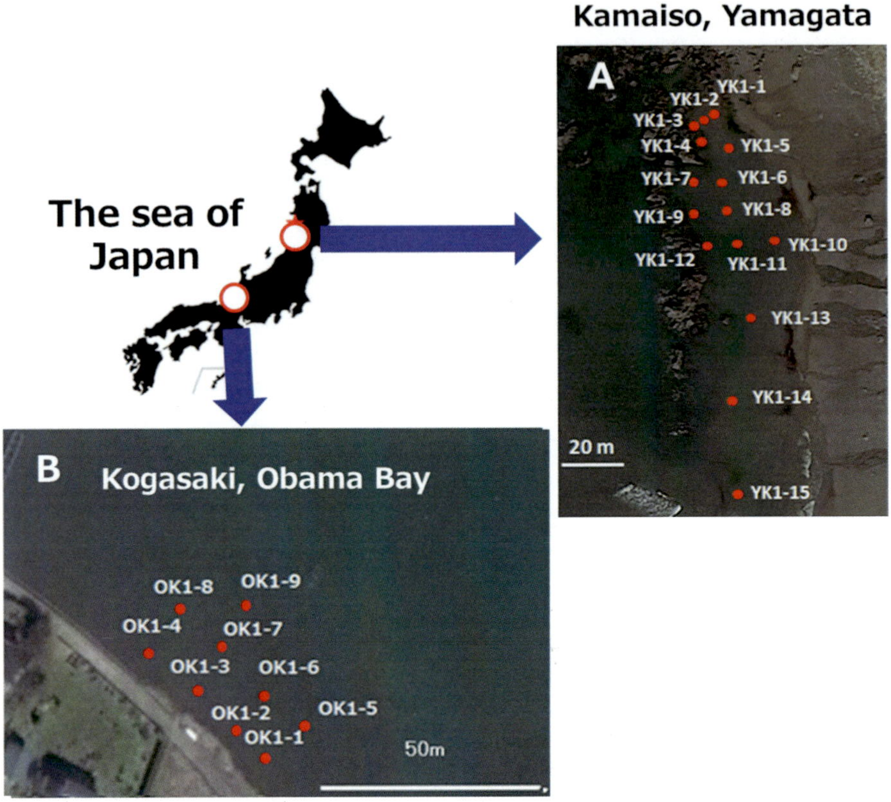

Fig. 9.3 Sampling stations at (**A**) Kamaiso Beach, Yamagata and (**B**) Kogasaki Beach, Obama Bay

The chlorophyll concentration in the water was also high in the same area (Hata et al. 2016). These results suggest that primary production was improved by nutrients of terrestrial origin supplied through the SGD.

An increase in the abundance of juvenile marbled sole and their prey organisms also corresponded with a spatial peak in radon concentrations. Juvenile marbled sole was the most abundant in areas with salinity <10 in the western part of the survey area, and fed mainly on gammarids (Hata et al. 2016). Carbon and nitrogen stable isotope ratio analyses, applied to examine the food web, showed relatively low carbon stable isotope ratios in gammarids. These results were based on laboratory experiments to calculate the half-life of the carbon stable isotope ratio in juvenile marbled sole (Hamaoka et al. 2015), the wild juvenile marbled flounder <40 mm, which fed mainly on gammarids on the tidal flat, depended on nutrients of terrestrial origin provided through the SGD (Hata et al. 2016).

9.2.4 *Influence of SGD on the Reef Fish Community in Beppu Bay, Oita Prefecture: Within Bay-Scale*

In Beppu Bay, Oita Prefecture, the influence of SGD on the reef fish community was investigated. An SGD was found by a preliminary underwater survey using scuba at the bottom of the Hiji Castle. Because the transparency of the sea water in the surrounding area was high, underwater cameras were used to take photographs at 1 min intervals for 2–3 h near the SGD area and in a site where no SGD was observed.

Frequency of occurrence of fish (% photographs with fish) was significantly higher near the discharge area (Fig. 9.1D). The frequency of occurrence of dominant species (damsel fish *Chromis notatus notatus* and Labridae spp.) was also significantly higher at the discharge area (Utsunomiya et al., unpublished). Stomach contents and stable isotope analyses will be conducted to understand if the SGD contributes to the surrounding food web and rock fish community in Beppu Bay.

9.3 Effects of SGDs on Meiobenthic Communities

9.3.1 *Abundance and biodiversity in near-Shore Meiobenthic Community Associated with Submarine Groundwater Discharge*

SGD has been shown to be an important pathway between land and sea for the transport of terrestrial materials (Church 1996; Moore 1996; Moore 2010). SGD results in various ecological changes in coastal environments, such as increased coastal primary production (Sugimoto et al. 2017), eutrophication (Hwang et al. 2005), and ecosystem structure (Waska and Kim 2010). SGD also affects surrounding environments by changing salinity, temperature and sediments characteristics. These changes in environmental condition due to freshwater flow from SGD have both positive and negative effects on coastal biota. In this section, we investigate the effect of SGD on benthic meiofaunal assemblage at two areas which have different environmental conditions of SGD.

9.3.2 *SGD Condition of two Study Sites, Kamaiso and Kogasaki*

The SGD flow rate and the proportion of fresh water to seawater were measured using a seepage meter at Kamaiso Beach along the Mt. Chokai volcanic coast in Yamagata Prefecture Japan in June 2016 (10 sampling stations) and at Kogasaki Beach, Obama Bay, Fukui Prefecture Japan in August 2016 (nine sampling stations)

Fig. 9.4 Photographs of seepage meter used in study areas (**A**) tools to measure the SGD flow rate and freshwater ratio in SGD, (**B**) seepage meter installed in the field, and (**C**) seepage water stored in the plastic bag

(Fig. 9.3). The seepage meter consisted of a 1 L bucket fitted with a sample port and plastic collection bag. The bucket was inserted with the open end down into the sediment (Fig. 9.4). The volume of water in the bag measured over time provided the flux measurements (Q). The salinity of the inflow water to the seepage meters (Ci) was calculated using the following equation:

$$Ci = \frac{Ct \times Tm - \{\frac{-V}{Q} \times Cs \times \left(e^{\frac{-QTm}{V}} - 1\right)\}}{\left(Tm - \frac{-V}{Q} \times e^{\frac{-QTm}{V}} + \frac{-V}{Q}\right)} \tag{9.1}$$

where Ct represented the salinity of water inside the plastic bag, Cs represented the ambient seawater salinity, Tm was the installation time and V was the seepage meter volume (1 L). Therefore freshwater ratio in SGD (FR) was obtained using the following eq. (9.2).

$$FR = \left(1 - \frac{Ci}{Cs}\right) \times 100 \tag{9.2}$$

The area of the study site was approximately 2000 m^2 (20 × 100 m) at Kamiso and approximately 1250 m^2 (25 × 50 m) at Kogasaki. SGD comprised recirculated submarine groundwater discharge (RSGD) and submarine fresh groundwater

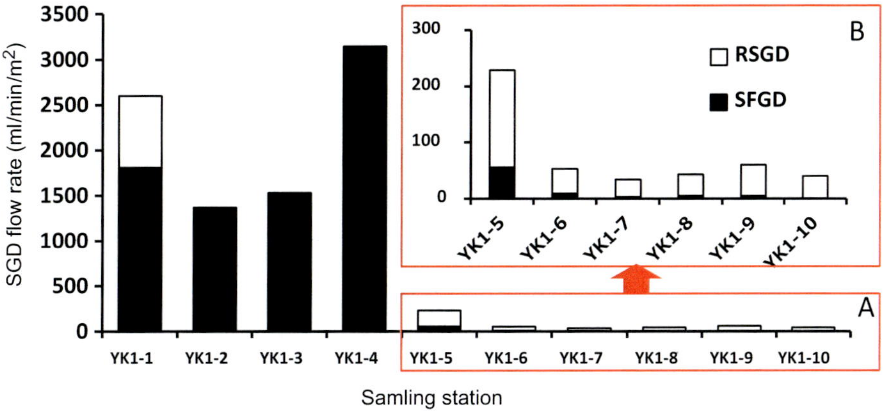

Fig. 9.5 The flow rate of recirculated submarine ground water discharge (RSGD, open bars) and submarine fresh ground water discharge (SFGD, close bars) at 15 sampling stations, Kamaiso Beach, Yamagata Prefecture. The figure **B** is enlarged figure of the part **A**

discharge derived from land (SFGD). Therefore, SGD could be characterized by the SGD flow rate and freshwater ratio (*FR*). In Kamaiso, the survey area was clearly divided into two areas: (1) high SGD flow rate (1369–3144 ml/min/m^2) and high *FR* area (69.5–100%), (2) low SGD flow rate (34–229 ml/min/m^2) and low *FR* area (0–24.2%) (Fig. 9.5). Accordingly, a negative correlation was found between the salinity and SGD flow rate ($R = -0.75$, $p < 0.05$). However, Kogasaki was stable SGD environments and both the SGD flow rate (6–158 ml/min/m^2) and *FR* (0–14.5%) were low overall (Fig. 9.6).The SGD flow rate and *FR* at Kogasaki were similar to those at the low seepage area of Kamaiso.

9.3.3 Effect of SGD on Meiobenthic Abundance

Intensive input of freshwater via SGD could change the abundance and species composition of the benthic assemblage owing to a rapid decrease in salinity. Meiobenthos are useful in assessing environmental disturbances because they are relatively small in size, have a short life span and generally lack a planktonic stage (Austen et al. 1994). Therefore, we sampled meiobenthos to evaluate the ecological influence of SGD on the biota.

Meiobenthos that pass through a 0.5 mm sieve but were retained by a 0.55 mm mesh size, were collected using core samplers (3.5 cm in diameter and 10 cm long) at the same sampling stations of the two study sites where the seepage meters were installed. Three replicate samples were taken and the benthic organisms were identified to the major taxonomic group. The sediment samples were divided into three consecutive layers (0–3, 3–6, 6–9 cm).

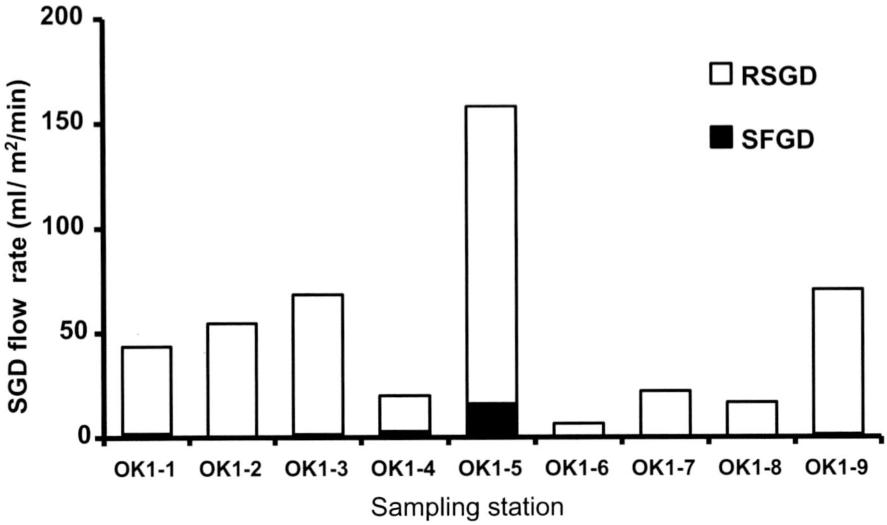

Fig. 9.6 The flow rate of recirculated submarine ground water discharge (RSGD, open bars) and submarine fresh ground water discharge (SFGD, close bars) at 9 sampling stations, Kogasaki Beach, Obama Bay, Fukui Prefecture

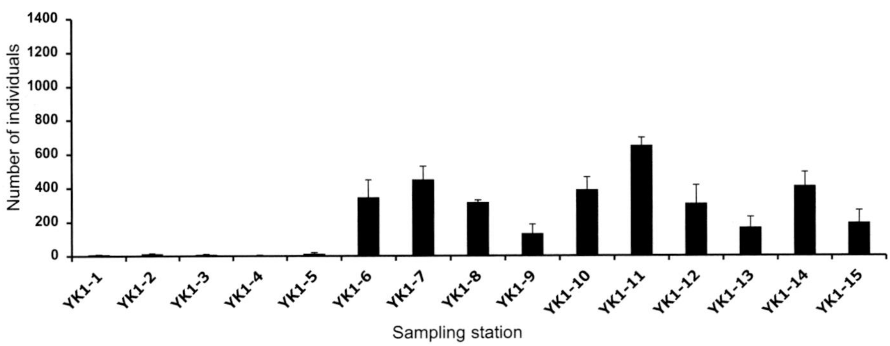

Fig. 9.7 The average meiobenthic numbers of the 0–3 cm layer in the sediment at 15 sampling stations, Kamaiso Beach, Yamagata Prefecture. Error bars indicate SD (n = 3)

The overall average meiobenthic abundance at Kamaiso was 227.4 individuals in the 0–3 cm layer (inds. / layer) and there clear difference between the high seepage (9.6 inds. / layer) areas and low seepage (336.4 inds. / layer) areas (Fig. 9.7, Table 9.1). The average meiobenthic abundance at each station decreased with increased SGD flow rates ($R = -0.70$, n = 10, $p < 0.05$) and there was a significant negative relationship between the average abundance and SGD flow rate within the low seepage area ($R = -0.82$, n = 6, $p < 0.05$) (Fig. 9.8). The meiobenthic abundance

Table 9.1 Average meiobenthic abundance and the coefficient variation of meiobenthic abundance at Kamaiso and Kogasaki

Meiobenthos	Kamaiso			Kogasaki
	High seepage area		Low seepage area	
Abundance (inds./0–3 cm layer)	9.6		336.4	669.2
CV of abundance among stations		0.89		0.39

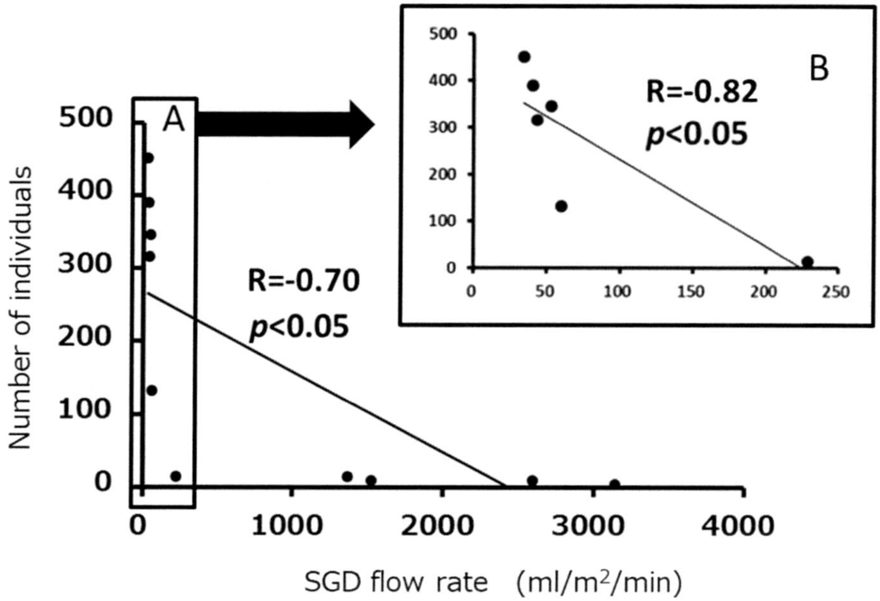

Fig. 9.8 Relationship between the average meiobenthic numbers of the 0–3 cm layer in the sediment and SGD flow rate at Kamaiso Beach. The figure **B** is enlarged figure of the part **A**

of the 3–6 cm layer was similar to that of the 0–3 cm layer. The average meiobenthic abundance at Kogasaki was 406.7–1067.0 inds. / layer (average 669.2) and there was not a significant relationship between the average abundance and SGD flow rate. The coefficient of variation of meiobenthic abundance at Kamaiso and Kogasaki was 0.89 and 0.34, respectively (Table 9.1).

SGD comprised mainly recirculated sea water at Kogasaki. However, FR in SGD at Kamaiso was very different to that within the study area. Particularly, SGD at the high seepage area at Kamaiso comprised mostly freshwater. These results suggest that the environmental conditions at Kamaiso were heterogeneous even at a small spatial scale. Therefore, it could be concluded that intensive freshwater inflow via SGD has a negative ecological effect on the abundance of meiofaunal communities.

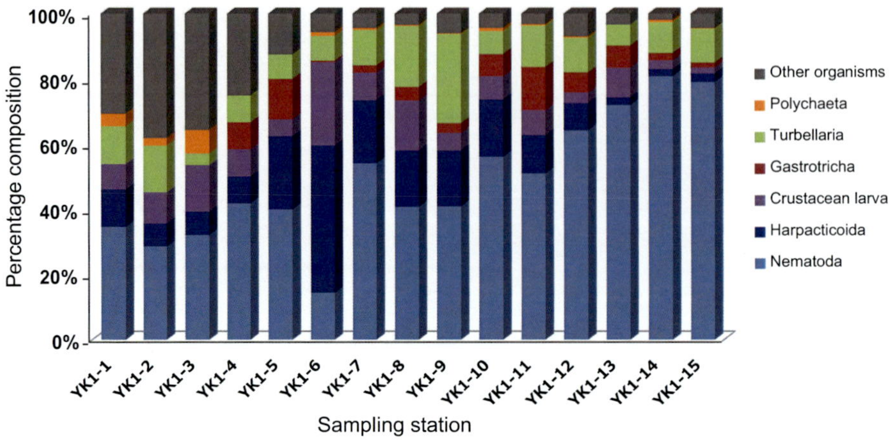

Fig. 9.9 Taxonomic composition of meiobenthic communities at Kamaiso Beach

9.3.4 Effect of SGD on the Taxonomic Composition and Diversity of meiofaunal Communities

A total richness of 14 taxa was sampled at Kamaiso. The most abundant taxa were nematodes, harpacticoids and platyhelminths, which accounted for >80% of the total number of individuals. However, taxonomic compositions were different among sampling sites (Fig. 9.9). The percentage of nematodes of the total number of individuals generally decreased at low seepage areas. Conversely, the percentage of crustaceans including harpacticoids increased at high seepage areas (Fig. 9.9). However, a total of 11 taxa were sampled at Kogasaki and the nematodes were dominant at all sampling stations, followed by the platyhelminths. Very few crustacean species were found (Fig. 9.10).

The average Shannon-Wiener diversity (*H*) and evenness (*E*) indices of meiofaunal communities at 15 sampling stations of Kamaiso were 1.28 and 0.71, respectively and those of the 9 sampling stations of Kogasaki were 0.85 and 0.40, respectively, indicating they were higher at Kamaiso than Kogasaki. Moreover, there was a significant relationship between the *E* of the meiofaunal community and the SGD flow rate at Kamaiso ($R = 0.77$, $n = 9$, $p < 0.01$) (Fig. 9.11). These results suggest that intensive freshwater flow via SGD at high seepage areas suppress the dominance of certain species (e.g., nematodes) and consequently, diversity indices increased.

9.4 Conclusion

Our recent surveys provided evidence of the contribution of SGD to the production of fishes and their prey. These results strongly support the hypothesis that the contribution of nutrients from SGD to the coastal fishery resources production is much

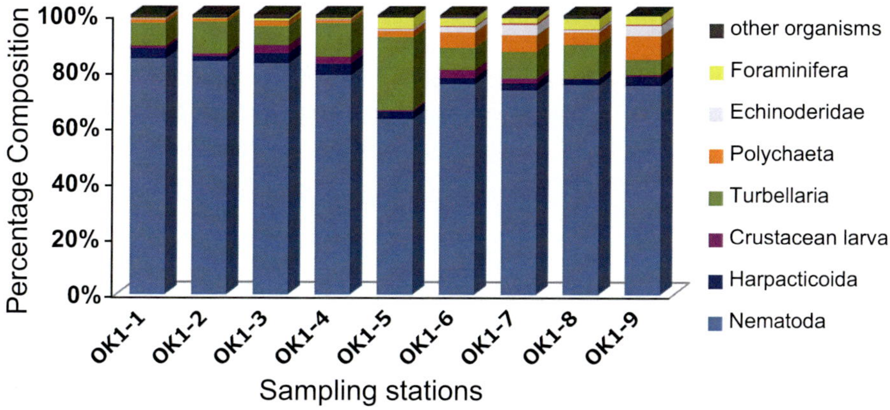

Fig. 9.10 Taxonomic composition of meiobenthic communities at Kogasaki Beach, Obama Bay

Fig. 9.11 Relationship between the evenness index (*E*) and SGD flow rate at Kamaiso Beach

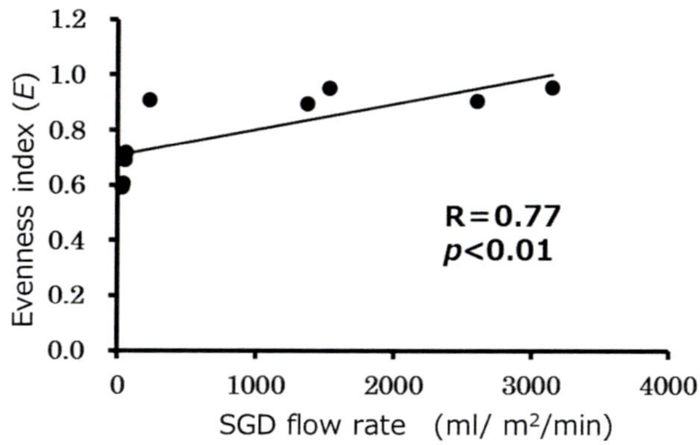

higher than previously suggested. In the future study, it is indispensable to clarify the spatial (i.e. local to global scales) and temporal (diel, seasonal and decadal) variabilities in the processes how SGD affects coastal fishery resources production. In general, fish can move to and feed in multiple habitats within a relatively short temporal scale (e.g., tidal and daily scales). Combination of hydrographic surveys with bio-logging and stable isotope analyses for fish movement and trophic flow in coastal ecosystems would enable accurate estimation of the contribution of SGD to fisheries production. Kotwicki et al. (2014) showed that disturbances caused by SGD affected the number of individuals in meiofaunal communities and decreased meiofaunal densities in seepage areas. They also showed nematodes were present at significantly lower densities at seepage areas. The negative ecological impacts of SGD on meiofaunal abundance were confirmed at the high seepage areas of Kamaiso where the small spatial scale presented strong heterogeneity in SGD environmental conditions. However, the meiofaunal abundance was not low at the low seepage area located slightly apart from the high seepage area. Therefore, the effect of SGD on meiofaunal communities could be limited. However, the environmental heterogeneity significantly enhanced the conservation of the meiofaunal diversity. Thus, SGD determined the property of benthic communities. The effect of groundwater discharge phenomena on benthic communities should be investigated at various coastal areas at different spatial scales.

References

Akimichi T (2010) Water and life around Mt. Chokai – a report from the region. Tohoku Kikaku Publishing, Yamagata

Austen MC, McEvoy AJ, Warwick RM (1994) The specificity of meiobenthic community response to different pollutants: results from microcosm experiments. Mar Pollut Bull 28:557–563

Church TM (1996) An underground route for the water cycle. Nature 380:579–580

Hamaoka H, Shoji J, Hori M (2015) Turnover rates of carbon and nitrogen stable isotopes in juvenile marbled flounder Pleuronectes yokohamae estimated by diet switch. Ichthyol Res 63:201–206

Hata M, Sugimoto R, Hori M, Tomiyama T, Shoji J (2016) Occurrence, distribution and prey items of juvenile marbled sole Pseudopleuronectes yokohamae around a submarine groundwater seepage on a tidal flat in southwestern Japan. J Sea Res 111:47–53

Hosono T, Ono M, Burnett WC, Tokunaga T, Taniguchi M, Akimichi T (2012) Spatial distribution of submarine groundwater discharge and associated nutrients within a local coastal area. Environ Sci Technol 46:5319–5326

Hwang DW, Lee YW, Kim G (2005) Large submarine groundwater discharge and benthic eutrophication in Bangdu Bay on volcanic Jeju Island, Korea. Limnol Oceanogr 50:1393–1403

Kamermans P, Hemminga MA, Tack JF, Mateo MA, Marba N, Mtolera M, Stapel J, Verheyden A, Daele TV (2002) Groundwater effects on diversity and abundance of lagoonal seagrasses in Kenya and on Zanzibar Island (East Africa). Mar Ecol Prog Ser 231:75–83

Kotwicki L, Grzelak K, Czub M, Dellwig O, Gentz T, Szymczycha B, Böttcher ME (2014) Submarine groundwater discharge to the Baltic coastal zone: impacts on the meiofaunal community. J Mar Syst 129:118–126

Miller DC, Ullman WJ (2004) Ecological consequences of ground water discharge to Delaware Bay, United States. Ground Water 42:959–970

Moore WS (1996) Large groundwater inputs to coastal waters revealed by 226Ra enrichments. Nature 380:612–614

Moore WS (2010) The effect of submarine groundwater discharge on the ocean. Annu Rev Mar Sci 2:59–88

Sanders TG Jr, Biddanda BA, Stricker CA, Nold SC (2011) Benthic macroinvertebrate and fish communities in Lake Huron are linked to submerged groundwater vents. Aquat Biol 12:1–11

Shiokawa M, Yamaguchi A, Umezawa Y (2013) Estimation of groundwater-derived nutrient inputs into the west coast of Ariake Bay. Bull Coast Oceanogr 50:157–167

Shoji J, Sugimoto R, Tominaga O (2017) Land-ocean interactions through groundwater/submarine groundwater and human society. Koseisha-koseikaku, Tokyo

Sugimoto R, Honda H, Suzuki T, Ochiai S, Taniguchi M, Nagao S (2014) Effects of submarine groundwater discharge on nutrient concentrations in the western part of Nanao Bay in summer. Bull Jpn Soc Fish Oceanogr 78:114–119

Sugimoto R, Honda H, Kobayashi S, Takao Y, Tahara D, Tominaga O, Taniguchi M (2016) Seasonal changes in submarine groundwater discharge and associated nutrient transport into a tideless semi-enclosed embayment (Obama Bay, Japan). Estuar Coasts 39:13–26

Sugimoto R, Kitagawa K, Nishi S, Honda H, Yamada M, Kobayashi S, Shoji J, Ohsawa S, Taniguchi M, Tominaga O (2017) Phytoplankton primary productivity around submarine groundwater discharge in nearshore coasts. Mar Ecol Prog Ser 563:25–33

Utsunomiya T, Hata M, Sugimoto R, Honda H, Kobayashi S, Miyata Y, Yamada M, Tominaga O, Shoji J, Taniguchi M (2017) Higher species richness and abundance of fish and benthic invertebrates around submarine groundwater discharge in Obama Bay, Japan. J Hydrol Reg Stud 11:139–146. doi.org/10.1016/j.ejrh.2015.11.012

Valiela I, Costa J, Foreman K, Teal JM, Howes B, Aubrey D (1990) Transport of groundwater-borne nutrients from watersheds and their effects on coastal waters. Biogeochemistry 10:177–197

Waska H, Kim G (2010) Differences in microphytobenthos and macrofaunal abundances associated with groundwater discharge in the intertidal zone. Mar Ecol Prog Ser 407:159–172

Chapter 10
Tradeoff Between Hot spring Use and River Ecosystem: The Case of Beppu City, Oita Prefecture, Japan

Makoto Yamada, Hisami Honda, Taketoshi Mishima, Shinji Ohsawa, and Jun Shoji

Abstract In Hirata River at Beppu, Oita prefecture, the inflow of hot spring drainage into the rivers increases the Nile tilapia, an alien species. It is a tradeoff between the hot spring use and river ecosystem. In order to understand details of the mechanisms connecting hot spring use and ecosystem, we investigated the river water quality and the water temperature from the upper reaches to the river mouth in the Hirata River. The results show that the hot spring drainage flowing in a specific section affects the estuary area. Additionally, it is revealed that the current average water temperature of hot spring drainage is very good condition for Nile tilapia in the estuary area. It is suggested that the tradeoffs between the hot spring use and the ecosystems can be resolved by suppressing high-temperature unused hot spring drainage, and/or by collecting hot spring drainage and reduce its average temperature further by using its thermal energy more extensively.

Keywords Hot spring drainage · Hot spring use · River ecosystem · Tradeoff

---

M. Yamada (✉)
Faculty of Economics, Ryukoku University, Kyoto, Japan
e-mail: myam@econ.ryukoku.ac.jp

H. Honda
Research Institute for Humanity and Nature, Kyoto, Japan
e-mail: h_honda@chikyu.ac.jp

T. Mishima
Institute for Geothermal Sciences, Graduate School of Science, Kyoto University, Kyoto, Japan
e-mail: mishima@bep.vgs.kyoto-u.ac.jp

S. Ohsawa
Beppu Geothermal Research Laboratory, Institute for Geothermal Sciences, Graduate School of Science, Kyoto University, Beppu, Oita, Japan
e-mail: ohsawa@bep.vgs.kyoto-u.ac.jp

J. Shoji
Costal Ecology, Hiroshima University, Higashi-hiroshima 739-8528, Hiroshima, Japan
e-mail: jshoji@hiroshima-u.ac.jp

© Springer Nature Singapore Pte Ltd. 2018

A. Endo, T. Oh (eds.), *The Water-Energy-Food Nexus*, Global Environmental Studies, https://doi.org/10.1007/978-981-10-7383-0_10

10.1 Introduction

Hot spring waters are used by people for various purposes such as bathing, drinking and heating. Furthermore, hot springs are important tourism resources. Hot springs have become an important resource that cannot be separated from the life of the local residents of the hot spring area. Hot springs have been used in various applications from ancient times in Beppu, Oita Prefecture. Hot spring waters are used in recent years even for hot spring power generation. In a further diversification of hot springs applications, hot spring demand has increased in this area.

On the other hand, hot spring water after use, i.e., hot spring drainage, is discharged into rivers. Yamada et al. (2017) examined the effects of the hot spring drainage on the river ecosystem near a Beppu area estuary. From the research, they revealed that the inflow of hot spring drainage into the rivers increases the amount of phytoplankton and that hot spring drainage creates a better habitat for Nile tilapia, an alien species, by increasing food availability and water temperature. The increase of the Nile tilapia has disturbed the river ecosystem. Since hot spring drainage will certainly occur if hot springs is used, if tradeoff refers to a relationship in which compatibility is impossible, indicating that taking one means losing the other, we can say that a tradeoff exists between the hot spring use and river ecosystem. Because no fishery activities are performed in the Hirata River, the relation of this tradeoff does not cause a conflict of interest between the hot springs and fisheries. However, in a report of the biological survey in the river of Beppu area, Hiramatsu et al. (1994) described, as a proposal for river environment conservation of the Beppu area, that they hope for restoration of the ecosystem because it has been disturbed by the alien species. In fact, Nile tilapia is listed among "alien species that might damage the ecosystem of our country" (Ministry of the Environment 2015).

From the considerations presented above, probably the present conditions of Hirata River, which include invasion by an alien species, are not in a healthy state from the perspective of biodiversity. As described above, the development of hot spring power generation using hot spring water is being conducted in various Beppu area locations. If hot spring drainage occurs anew at various locations, then a tradeoff between hot spring use and the river ecosystem might occur.

To overcome that tradeoff, one must not only know the causal relation, but also be able to understand details of the mechanisms connecting hot spring use and ecosystem. Therefore, for this study, we investigated the river water quality and the water temperature from the upper reaches to the river mouth in the Hirata River where a tradeoff between the hot spring use and the river ecosystem had already occurred. The purposes of this investigation are the following: to ascertain details of the hot spring drainage inflow situation; to ascertain the expanding influence of hot spring drainage; and to clarify the relation between hot spring drainage temperatures and Nile tilapia. Additionally, we considered how to overcome this tradeoff.

10.2 Site Description and Circumstances of Hot Spring Drainage

The Beppu area is near the center of the southwestern Japan volcanic arc in the northeastern part of Kyushu Island. Beppu is known in Japan as a hot spring resort with numerous hot spring well. In Beppu, there are eight hot spring areas, the "Beppu Hattou": Beppu Onsen, Myoban Onsen, Kankaiji Onsen, Kannawa Onsen, Hamawaki Onsen, Kamekawa Onsen, Horita Onsen, and Shibaseki Onsen (Fig. 10.1a).

The Beppu area has six small river basins (Fig. 10.1a), with rivers generally flowing eastward from the west, eventually flowing into Beppu Bay. The lengths of these rivers, including their tributaries, are 3–6 km. The depths of these rivers are less than 0.5 m. Hirata River is one of these rivers. Except for the upper reaches, the Hirata River stream is lined entirely with concrete. Hirata River flows down to penetrate Kannawa Onsen, a representative hot spring area among those described above. Kannawa Onsen has high-temperature hot springs of about 100 °C. In this area, hot spring drainage is well-known to flow into a river. Local people call the Hirata River the "Yu-no-Kawa (Hot water river)" because steam rises from hot spring drainage flowing into the river.

One reason for discharging hot spring drainage to rivers is the inadequacy of sewer systems. Many undeveloped areas of sewerage dot the mountainside in the Beppu area. Kannawa Onsen area is located on a mountain side of Beppu, where

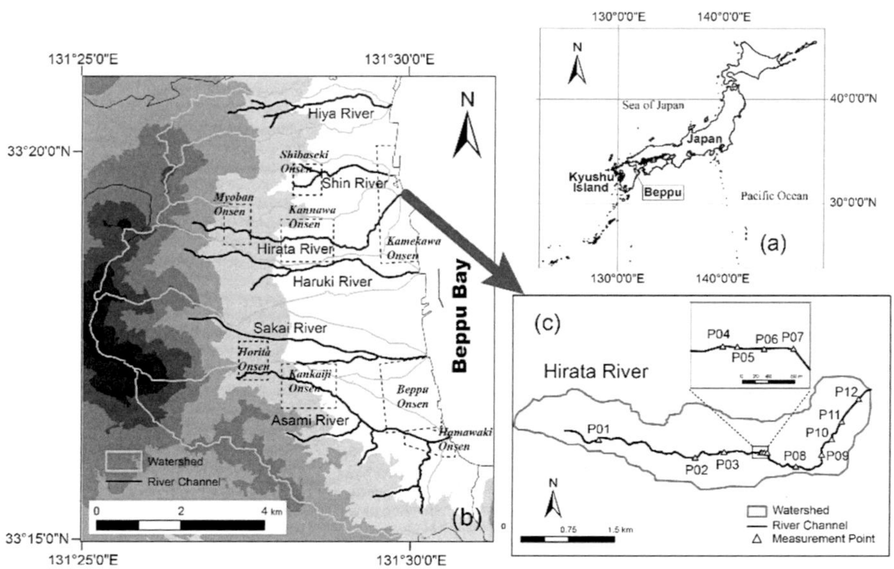

Fig. 10.1 Locations of study areas: (a) location of Beppu area in Japan; (b) watersheds of rivers in the Beppu area and famous Onsen resort area; (c) watershed of Hirata River and observation points

sewers are not fully developed. Another reason is that, by the Beppu City ordinance, hot springs with temperatures higher than 45 °C should not be drained into sewers (Ohsawa et al. 2007). As described above, hot springs in the Kannawa area are extremely hot. For that reason, even if the sewer system is complete, high-temperature hot spring drainage cannot be drained to the sewers. For these reasons, hot spring drainage flows into the Hirata River, irrespective of whether the waters are unused or used.

10.3 Investigation Method and the Results

We conducted a survey in January 2015 at 12 points from the upper reaches to the river mouth in the Hirata River (Fig. 10.1b). Survey items were water temperature, air temperature, electric conductivity, and the flow rate. Because the river is as narrow as 2 m to 5 m at all points, the flow rate was calculated from the flow velocity, water depth, and river width at both the banks and the central part of the river.

Measurement results of water temperature and air temperature demonstrate that a difference between the water temperature and the air temperature has already occurred at P02, the water temperature exceeds 20 °C at P04. Moreover, the influence of the hot spring drain temperature is strongly received from this point (Fig. 10.2). The river water temperature reaches its maximum at P07. Thereafter, although the water temperature is somewhat lower, the water temperature remains

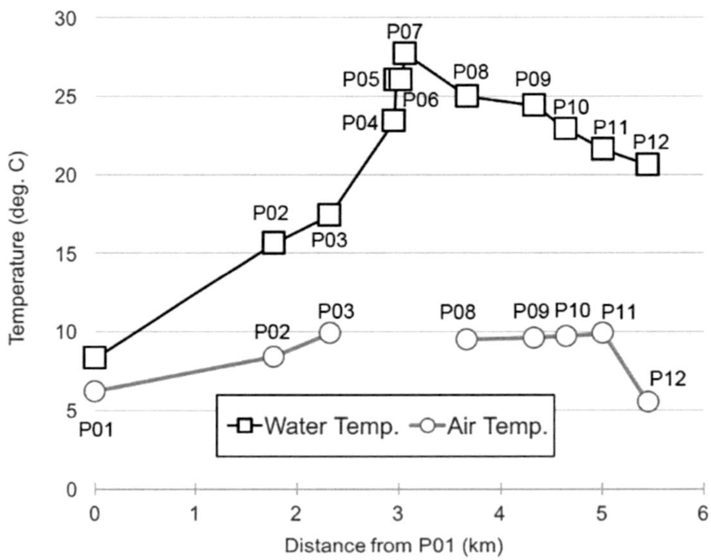

Fig. 10.2 Change of the river water temperature with downward flow and air temperature at each point. Air temperature data from P04 to P07 are not shown

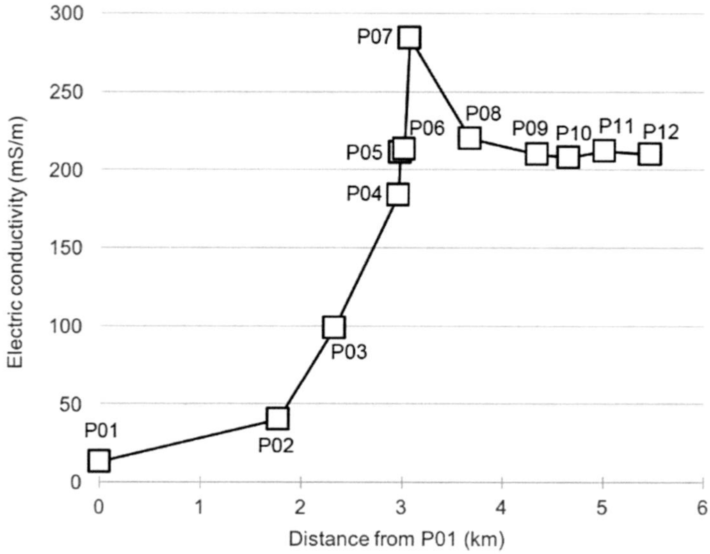

Fig. 10.3 Change of the electric conductivity of river water with downward flow

above 20 °C. River water hotter than 20 °C flows to coastal areas even when air temperatures are lower than 10 °C.

The value of electric conductivity has already risen from P02, but it rises sharply at P04 and becomes maximum at P07 (Fig. 10.3). The value of electric conductivity at all points on the downstream side of P07 was about 210 mS/m, which indicates that almost no influent water changes dissolved chemical components on the downstream side of P07.

The flow rate increased gradually from P01 to P07. Subsequently, it did not change much, but instead decreased slightly near the estuary (Fig. 10.4). Because the flow rate is calculated using three points on the longitudinal section, it is considered that the observation error is not small, but the change of the flow rate trend might resemble that shown in the graph. Therefore, the flow rate measurement result shows that most hot spring drainage flows into rivers between P01 and P07. There is almost no influx of hot spring drainage occurring on the downstream side of P07.

10.4 Inflow Process and Features of Hot Spring Drainage

As described above, the water temperature and the electric conductivity were the maximum at P07. The flow rate also became almost equal to the maximum value. Water temperature and electric conductivity are increased by the influx of hot spring drainage. Therefore, these results indicate that most of the hot spring drainage flows in before reaching P07. Because the same result is obtained from the flow rate

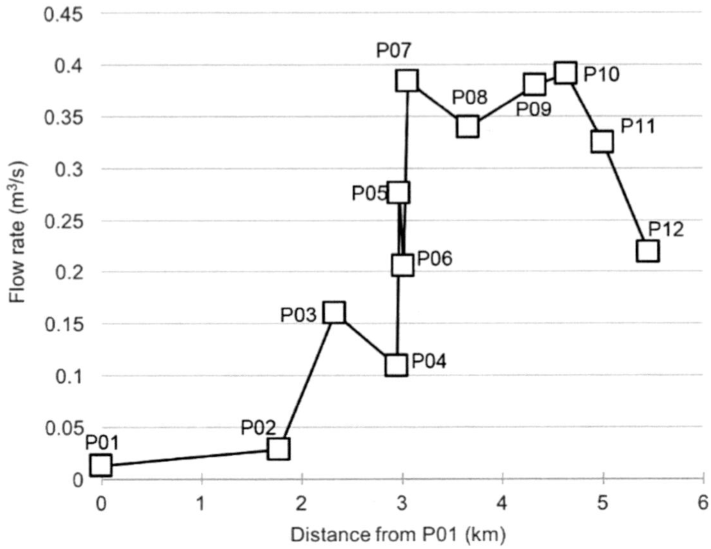

Fig. 10.4 Change of the river water flow rate with downward flow

observations, it is reasonable to infer that the inflow of hot spring drainage is nearly complete before reaching P07. Although water temperatures and electric conductivity increase gradually from P01 to P04, the influx of hot spring drainage starts mainly in the section between P03 and P04 because both water temperatures and electric conductivity are sharply higher at P04. In other words, in the Hirata River, hot spring drainage inflow mainly occurs in the section of P04–P07.

In the Hirata River, hot spring drainage flowing in a specific section affects the estuary area. From data presented before and hereinafter in specific sections, we estimated the average water temperature, mean electrical conductivity and inflow amount of hot spring drainage flowing into the Hirata River. For this estimation, data at P03 (17.4 °C, 98.4 mS/m, 0.16 m^3/s) were used as data of river water flowing into this section. Data at P07 (27.7 °C, 284 mS/m, 0.38 m^3/s) were used as data of river water flowing out of this section. The average of water temperature, electric conductivity and inflow amount of the estimated hot spring drainage were, respectively, 35 °C, 419 mS/m, and 0.22 m^3/s. In this estimate, we do not consider detailed parameters such as heat radiation from the surface of the river at all. However, because these sections are extremely short (about 700 m) and the flow velocity is extremely high (1.3–1.9 m/s), even though detailed parameters are ignored, the estimated value is regarded as roughly correct.

Results of the estimation presented above show that hot spring drainage at temperature of 35 °C is flowing into the river. In reality, however, the drainage of such temperature does not flow in collectively. For this survey, we moved the river in this section in the longitudinal direction and measured several points of hot spring drainage directly. Results show hot spring drainage of 96.2 °C in one place, hot spring drainage of 62.9 °C in another place. Furthermore, the flow rate measurement results indicate that the flow rate increases gradually in this section. In other words, in the Hirata River, not much drainage is flowing in at once, but drainage of various temperatures is flowing in many places: the average temperature of the drainage waters is 35 °C.

10.5 Relation between the Inflow Amount and Water Temperatures of Hot Spring drainage and Nile Tilapia

As described in the previous section, for the Hirata River, it became clear that the hot spring drainage flows in the middle reaches of the river and the river water temperature rises: the river water reaches the estuary without decreasing temperature. For that reason, many Nile tilapia inhabit that estuary. According to FAO data (FAO 2015), Nile tilapia is a tropical species that thrives in shallow water; the lower and upper lethal temperatures for Nile tilapia are 11–12 °C and 42 °C, respectively, although the preferred temperatures are 31–36 °C. Spawning begins when the water temperature reaches 24 °C. Even in the winter, as in this survey, the river water temperature after the hot spring drainage inflow is around 27 °C, so the temperature of the river water supplied to the estuary is likely to be in a favorable condition for Nile Tilapia. The phenomenon by which many Nile tilapia fish inhabit that estuary is based on a balance of the temperature and quantity of hot spring drainage inflowing at the certain section described above (P03–P07).

The balance between the current average temperature and inflow volume create the suitable temperature environment for Nile tilapia. Therefore, changing it will affect Nile Tilapia adversely. To clarify the relation between the average temperature of the hot spring drainage and the inflow amount, using observation data of the river flow rate and water temperature obtained before and after the section (P04–P07) where the hot spring drainage flows in, we derived the relations among the river water temperature, average temperature, and quantity of hot spring drainage. The equation used to derive the relation is the following.

$$V_d = \frac{T_o - T_i}{T_d - T_o} \times V_i \tag{10.1}$$

Therein, V_d stands for the hot spring drainage volume, V_i signifies the river water volume before hot spring drainage flow in (data of P03), T_d denotes the average temperature of hot spring drainage, T_i represents the river water temperature before hot spring drainage flow in (data of P03). In addition, T_o is the river water temperature after hot spring drainage flow (data of P07). Here, the lower limit value of the preferred temperature environment of Nile Tilapia is set as 24 °C. The upper limit value is set as 36 °C. Figure 10.5 portrays the curve calculated using the lower and the upper limit value. It shows the former as a dashed line and the latter as a solid line. The diamond plot in Fig. 10.5 shows the current state. When this plot moves to

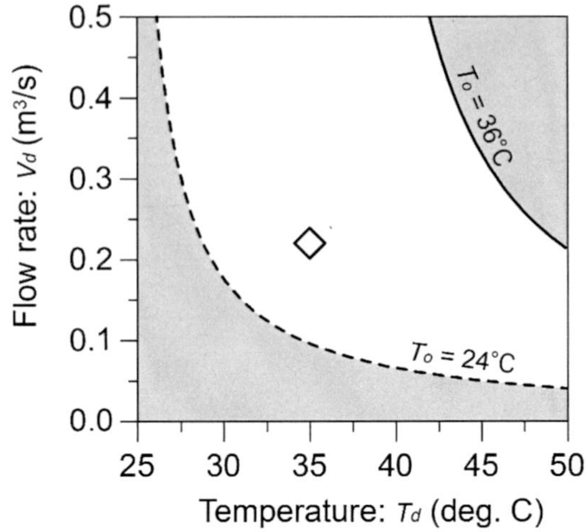

Fig. 10.5 Relations among the river water temperature, average temperature, and quantity of hot spring drainage. Dashed lines and solid lines respectively show the border line calculated using the lower limit value (24 °C) and upper limit value (36 °C) of the preferred temperature environment of Nile Tilapia. The diamond plot in the figure shows the current state

the left side or the lower side of the broken line, the river water temperature becomes an environment unsuitable for Nile Tilapia because of the low temperature. When the plot comes to the right side of the solid line, we find a high temperature for Nile Tilapia. To solve the current relation between river water temperature and Nile tilapia, it seems necessary to change both the amount and average temperature of hot spring drainage in this way.

10.6 Toward Solving Tradeoffs between Hot Spring use and River Ecosystem

Hot spring drainage is not flowing in one place collectively. Rather, it flows in various places and at various temperatures into the Hirata River. The temperature shown in the current state, in Fig. 10.5, is the average value of such hot spring drainage. If the inflow balance of each hot spring drainage were to change, then the average temperature of the hot spring drainage would be expected to change. Accordingly, the position of the plot shown in Fig. 10.5 will also change. As described above, the maximum temperature of the hot spring drainage observed was 96.2 °C. There was also hot spring drainage such as 62.9 °C. Large amounts of hot spring drainage of unused high-temperature waters such as this are known to exist. The average temperature of hot spring drainage is rising by such high temperature and unused hot spring drainage. By suppressing such hot spring drainage, it would be possible to lower the average temperature of hot spring drainage. High-temperature hot spring water is useful for hot spring power generation. It is also useful as a thermal energy source for heating. Therefore, its utility value is very high. Effective use of hot spring water heat is important not only to suppress the hot spring drainage temperature but also to maximize the potential of hot spring resources. Such high temperature hot spring drainage is discharged in a state in which most of it is not used. One might readily imagine that it is extremely difficult to reduce the amount of hot spring water that is normally used because hot spring use is a means of subsistence for local people. However, it would be easy to stop the discharge and waste of unused hot springs because it does not engender restrictions on the activities of local people. By suppressing high-temperature hot spring drainage, it is possible not only to lower the average temperature of the hot spring drainage flowing into the river but also to reduce hot spring drainage overall. Consequently, one would think that the river temperature will become unsuitable for Nile tilapia habitats.

In sections where hot spring drainage gets mixed with river waters, the quantity of thermal energy to raise the river water temperature from its initial (17.4 °C water temperature, 0.16 m^3/s flow rate) to its warmed state (27.7 °C water temperature) is estimated as about 6700 W. Currently, only Nile tilapia in the estuary area benefits from the thermal energy that the hot spring drainage brings to the river. This thermal energy is used only for ecological disturbances. Because the average temperature of the hot spring drainage is about 35 °C, it is difficult to use it because the temperature is slightly low to use the hot spring drainage together. Nevertheless, it is not impossible to use such bulk hot spring drainage: there are examples of using low-temperature hot spring water at around 35 °C for aquaculture of soft-shelled turtle and loach production. One means of resolving the tradeoff between hot spring use and river ecosystems is to collect hot spring drainage and reduce its average temperature further by using its thermal energy more extensively.

10.7 Conclusion

Although hot spring waters benefit residents of hot spring areas, the waters affect ecosystems as described above. This chapter outlined case studies of estuarine areas, but because the rivers are linked directly to coastal areas, the influence is expected to extend to coastal areas. In that sense, hot spring utilization can not only affect river ecosystems, but also present tradeoffs with coastal ecosystems. However, as this research has shown, from the perspective of temperature, one can reduce the load on the surrounding river environments using the hot spring resource without waste, or with much less waste. The same can be said about the effects on coastal areas. One can be fairly certain that tradeoffs between the hot spring use and the ecosystems of the surrounding waters can be resolved by executing quite ordinary measures effectively using resources with much less waste.

References

FAO (2015) FAO fisheries & aquaculture – cultured aquatic species information programme – Oreochromis niloticus (Linnaeus, 1758) http://www.fao.org/fishery/culturedspecies/Oreochromis_niloticus/en

Hiramatsu T et al (1994) Aquatic life in freshwater in the Beppu region. Nat Beppu, pp 323–344. (in Japanese)

Ministry of the Environment (2015) List of alien species that might damage the ecosystem of our country. http://www.env.go.jp/press/files/jp/26594.pdf. (in Japanese)

Ohsawa S et al (2007) The inflow of useful metal elements into rivers: basic research and investigation of untapped hot spring resources. Oita Prefect Hot Spring Res Soc 58:21–30. (in Japanese)

Yamada M et al (2017) Hot spring drainage impact on fish communities around temperate estuaries in southwestern Japan. J Hydrol Reg Stud 11:69–83. https://doi.org/10.1016/j.ejrh.2015.12.060

Part IV
Governance and Management of Resource System

Chapter 11
The Water-Energy-Food Nexus and California's Sustainable Groundwater Management Act

Jason J. Gurdak

Abstract Water, energy, and food are among the most critically important resources for society. The demand for these resources is likely to increase in the coming decades because of global population growth, climate change, and other socioeconomic forces. This chapter provides a brief outline of some potential impacts and feedbacks of how Water-Energy-Food (WEF) Nexus concepts may help local managers and stakeholders in California, USA design optimal groundwater sustainability plans to best meet the diverse interests in groundwater resources while implementing the State's new (2014) Sustainable Groundwater Management Act (SGMA) law. Sustainable groundwater resources will in turn likely increase the resilience of the WEF Nexus across California.

Keywords Water-energy-food Nexus · Sustainable Groundwater Management Act (SGMA) · California

11.1 Introduction

This chapter uses concepts from the Water-Energy-Food (WEF) Nexus to frame sustainable groundwater management, using California's new (2014) Sustainable Groundwater Management Act (SGMA) as a case study. Here, key insight is provided about the following two-way relationship (i.e., impacts and feedbacks) between the WEF Nexus and SGMA; (1) WEF Nexus thinking may have the direct impact of helping local managers and the public design optimal groundwater sustainability plans to better meet the interests of diverse stakeholders in groundwater resources, and in turn, (2) implementation of SGMA and future sustainable

---

J. J. Gurdak (✉)
Department of Earth & Climate Sciences, San Francisco State University, San Francisco, CA, USA
e-mail: jgurdak@sfsu.edu

© Springer Nature Singapore Pte Ltd. 2018

A. Endo, T. Oh (eds.), *The Water-Energy-Food Nexus*, Global Environmental Studies, https://doi.org/10.1007/978-981-10-7383-0_11

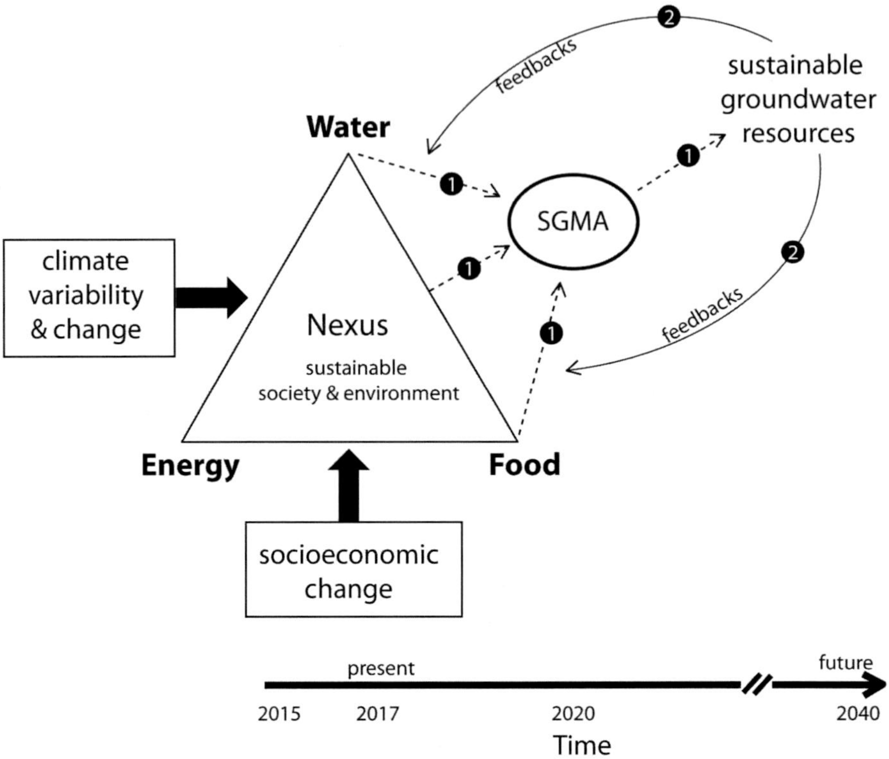

Fig. 11.1 Conceptual diagram of the Water-Energy-Food (WEF) Nexus and relationships and feedbacks with California's Sustainable Groundwater Management Act (SGMA) and sustainable groundwater resources. Climate variability and change and socioeconomic change are the two most significant drivers or perturbations within the WEF Nexus. The dashed arrows (#1) represent the impacts of WEF Nexus concepts on SGMA planning and the solid arrows (#2) represent feedbacks of sustainable groundwater resources on the Nexus. The key dates of SGMA are shown along the x-axis, including 2015 (January 1) when SGMA took effect, 2017 when Groundwater Sustainability Agencies (GSAs) must be formed, 2020 when Groundwater Sustainability Plans (GSPs) must be finalized for high priority basins by the GSAs and begun implementation, and 2040 when the high priority basins must achieve sustainability

groundwater resources will have the positive feedback toward a more resilient WEF Nexus across California (Fig. 11.1). Although the WEF Nexus is not mentioned by name in the regulations that the California Department of Water Resources (DWR) have written to implement the SGMA legislation, the theoretical concepts from the WEF Nexus complement many of SGMA's requirements. Therefore, the purpose of this chapter is to provide some practical considerations for local managers and stakeholders in California to simultaneously meet SGMA regulations and build a more sustainable and resilient WEF Nexus. Many aspects of the case study from California can be applied to other locations within the Asia-Pacific region or the world more broadly.

This chapter is organized as follows. The remaining sections of the Introduction provide brief backgrounds on the WEF Nexus and SGMA. Several key references and resources are also listed for the reader to find additional details about the WEF Nexus and SGMA. Section 11.2 outlines a few key concepts from the WEF Nexus that can be applied to sustainable groundwater management, with specific reference to SGMA regulations. This chapter provides WEF Nexus specific recommendations for local managers and stakeholders that are tasked with implementing SGMA. Section 11.3 briefly envisions how successful implementation of SGMA may transform California's WEF Nexus.

11.1.1 Water-Energy-Food (WEF) Nexus

The nexus between water, food and energy has received considerable attention recently in the international literature (Bazilian et al. 2011; Beisheim 2013; FAO 2014; Lawford et al. 2013; Rahaman and Varis 2005; United Nations ESCAP 2013; Vogt et al. 2010). Such recent articles stress the importance and theoretical framing of the WEF Nexus in sustainability and security research. However, discussions of the practical aspects of policy and management approaches and methods to address the Nexus are relatively new to the literature. Disciplinary and interdisciplinary approaches are required to advance both the theoretical and applied aspects of the Nexus, largely because of the inherent synergies and trade-offs that define the Nexus. For example, evaluating the synergies and trade-offs require analysis of the biophysical and natural systems of the Nexus, but also understanding of the decision-making and human systems.

A fundamental concept of the WEF Nexus is that the use and production, and hence present-day availability and future sustainability of water, energy, and food resources interact in complex ways across multiple spatial and temporal scales (Leck et al. 2015). The Nexus is anchored on the concept that managing water, energy, and food resources collectively is likely to avoid the negative results of more siloed approaches for management and policy of the three resources individually (Rasul and Sharma 2015). Here a common negative result from siloed water management and policy is addressed; overdraft of local groundwater resources and associated undesirable results (i.e., seawater intrusion, declining groundwater quality, subsidence, loss of aquifer storage, etc) in California. For additional details about the WEF Nexus beyond this book, the reader is referred to a recent (2017) special issue in the Journal of Hydrology: Regional Studies titled Water, energy, food nexus in the Asia-Pacific region (volume 11, pages 1–278), and a recent (2015) literature review by Endo et al. (2015) of current methods used in WEF Nexus research.

11.1.2 California's Sustainable Groundwater Management Act (SGMA)

Much has been written about the challenges facing the sustainability of global groundwater resources (e.g., Gleeson et al. 2010; Green et al. 2011; Treidel et al. 2012; Wada et al. 2012). Groundwater is a critically important resource for many local drinking-water supplies, but is also the foundation for irrigated agroecosystems, food production, ecosystem services, energy-extraction (such as shale-gas and hydraulic fracturing) and production, and other manufacturing and industry. As a result of this demand and excessive groundwater extraction, many of the largest and most important aquifer systems, particularly in arid and semi-arid regions, have rapidly declining water levels and substantial loss of storage, including many of California's aquifers such as the Central Valley aquifer (e.g., Konikow 2015). Groundwater overdraft in California is an often cited example of the current 'global groundwater crisis' (Famiglietti 2014).

Groundwater is a valuable resource in California. It provides a substantial portion of California's water supply; about one-third of the water use in average years and more than one-half of the water use in drought years (California Department of Water Resources 2017). The largest volume of extracted groundwater in California is used for irrigation water and food production (Maupin and Barber 2005). Excessive groundwater extraction has caused overdraft in aquifers across California, which has led to wells failing, seawater intrusion and other water quality degradation, land subsidence and related damage to infrastructure, and other negative impacts on baseflow to surface water and other groundwater dependent ecosystems. The importance of groundwater, combined with overdraft conditions and the severe recent drought conditions across California were among the catalysts that enabled SGMA to be signed into law by Governor Brown on September 16, 2014, and became effective on January 1, 2015.

The passage of SGMA, the Governor's Water Action Plan, voter-supported Proposition 1, and the historical drought have prompted some in the popular press to refer to 2014 as *The Year of Water* in California. Quotes from local California newspapers capture this sentiment: "2014 is the most significant year for California in a generation" and "2014 will be viewed as a critical turning point in California water history". However, we will have to wait at least until the year 2040 (the implementation horizon of SGMA) to learn if these statements are true with respect to long-term groundwater sustainability.

SGMA is significant for many reasons, not the least of which is that California is one of the last U.S. States to have comprehensive groundwater regulations. While the State Water Resources Control Board administers the permit system governing appropriative water rights holders, it has not regulated the amount of groundwater used. Because California did not have a statewide groundwater management program, there was only a state constitutional mandate that water not be wasted or put to a non-beneficial use. The new Act is designed to empower local agencies to manage their groundwater resources in a sustainable manner over the long-term, with limited State intervention only when necessary to protect groundwater resources. Under SGMA, the State provides guidance and technical support on how to plan for more sustainable groundwater resources, and only step in when local agencies fail to exercise their responsibilities. Given that groundwater extraction has essentially be un-regulated in many parts of California, there is also consider opposition to SGMA and concerns about socioeconomic damage, largely to businesses and agricultural communities that are heavily reliant on groundwater extraction (Sward 2017). Nevertheless, the passage of SGMA has received widespread support because of the concept that groundwater is best managed locally and optimism that SGMA will help mitigate the challenges associated with sustainable groundwater in California.

SGMA broadly defines groundwater sustainability as avoiding "undesirable results", which are specified in terms of groundwater overdraft, land subsidence, water quality degradation, seawater intrusion, and groundwater-surface water interactions. The undesirable results that occurred before January 1, 2015 are not required to be addressed by the local groundwater sustainability agencies (GSAs), which must be formed by June 30, 2017. GSA are required under SGMA to develop groundwater sustainability plans (GSPs) for basins in critical conditions of overdraft by January 31, 2020 and for all other high- and medium-priority basins not currently in overdraft by January 31, 2022. Twenty years after the adoption of the GSP (2040 and 2042), all high- and medium-priority groundwater basins must achieve sustainability. Additional aspects of SGMA will be discussed in the subsequent sections of this chapter. However, the reader is referred to the SGMA webpages of California Department of Water Resources (DWR) (http://www.water.ca.gov/groundwater/sgm/) for the details about SGMA legislation and regulation.

11.2 Applying WEF Nexus Concepts for Sustainable Groundwater Management

In developing GSPs, the GSAs must work with local stakeholders and will likely have to make controversial decisions (Lund et al. 2015) that may not be supported by all community members or stakeholder groups within a given basin. To minimize potential controversy and increase their effectiveness, GSPs must be based on the best available science regarding the physical and chemical hydrogeologic system, including local climate, surface-water hydrology, land-use activities, and other natural or manmade water inflows and outflows to the basin. Effective GSPs must also include realistic economic, legal, and political conditions of the basins, including information about the neighboring GSPs and their relationships (Lund et al. 2015). Here the argument is made that application of concepts from the WEF Nexus may also help the GSAs and stakeholders develop effective and less controversial GSPs. A few of those key concepts from the WEF Nexus are outline next.

The Nexus approach recognizes the inherent interdisciplinary nature of the water, energy, and food systems and sectors and has a solution oriented perspective on optimizing the trade-offs and synergies (Rasul and Sharma 2015). The most effective GSPs will likely be those that have a scope beyond the immediate physical, socioeconomic, and legal constraints of the aquifer, groundwater resources, and coupled surface-water resources, but extend their scope to explicitly include linkages to the local and regional food systems and energy systems (consumption or production) (Fig. 11.2). Therefore, the data collection and reporting to inform and evaluate the interim objectives and GSPs should include spatial and temporal datasets that extend beyond those used to characterize the basin water balance. Such Nexus datasets might include annual groundwater pumping volumes for irrigated agriculture, groundwater irrigation requirements for each crop type, and other related data characterizing the local and regional food and energy systems. Relevant food system data may focus on production, processing, and possibly distribution, including harvested acres of each irrigated and nonirrigated crop type, crop production per acre, crop prices and total value per crop type, groundwater use for food processing, and energy consumption for distribution. Similar data should be collected that characterizes the water-energy and food-energy nexus linkages.

In developing the GSPs and related management alternatives available to achieve local sustainability objectives and minimize the undesirable results (e.g., Wada et al. 2016), GSAs should consider characterizing and quantifying the linkages, synergies, alterations, and trade-offs related to groundwater sustainability, undesirable results, and the local WEF Nexus (Fig. 11.2). Linkages within the WEF Nexus are the black arrows in Fig. 11.2 that characterize the relationships between the consumption, use, alteration, or production of water, energy, and food. Synergies are those linkages between water, energy, and food where no resource is consumed or degraded in using or producing another resource (e.g., in-stream or run-of-river micro-hydroelectric power), or where there are mutual benefits or co-production of resources (e.g., methane digesters in dairy production).

Alterations are the linkages or interactions between water, energy, and food (Fig. 11.2) where one resource is altered or degraded, but not consumed as a result of producing another resource. An example of alteration is when in-stream water is used for cooling power generation and returned to the river at a higher temperature. Another example of alteration is the irrigation return flow that infiltrated below the root zone and will eventually become recharge, but that often has elevated concentrations of nitrate (NO_3^-) (Gurdak et al. 2016) or other agrichemicals. Groundwater with elevated NO_3^- can still be used to irrigate crops and the NO_3^- may be used as a nitrogen credit when considering fertilization crop requirements. However, irrigation return flow with elevated NO_3^- in groundwater above the drinking water standard (10 mg/L as N) may create situations of trade-offs within the WEF Nexus where the groundwater may not be usable for human (domestic or municipal) consumption.

Trade-offs are linkages that represent consumption of one or more resources during the production of another resource, or where one resource is produced at the expense of another. Trade-offs often represent the primary source of conflict and

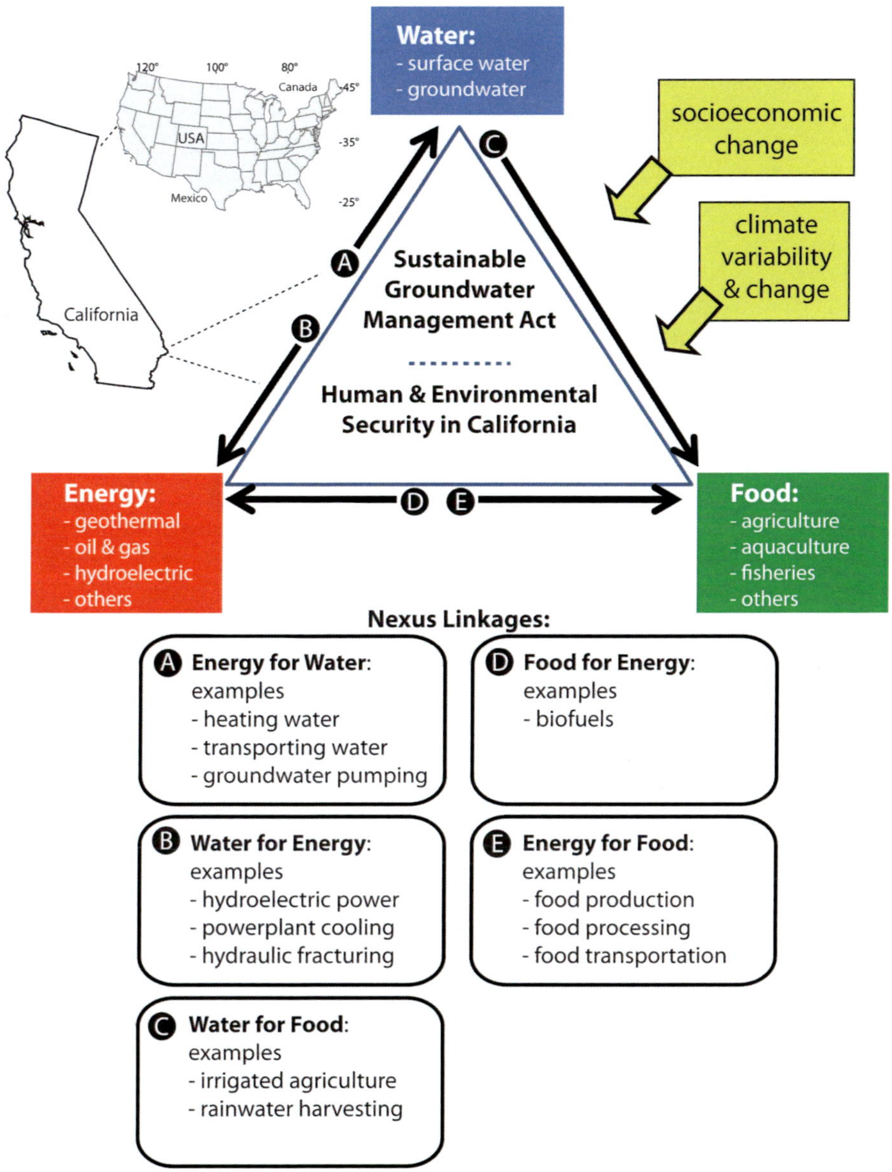

Fig. 11.2 Generalized Water-Energy-Food (WEF) Nexus with the dual goal of implementing the Sustainable Groundwater Management Act (SGMA) and achieving human and environmental security across California. Climate variability and change and socioeconomic change are the two most significant drivers within the WEF Nexus. Examples are listed for the Nexus linkages, including (A) Energy for Water, (B) Water for Energy, (C) Water for Food, (D) Food for Energy, and (E) Energy for Food. Modified from Endo et al. (2015)

disagreement among stakeholders within a system. Examples of trade-offs include the water used in hydraulic fracturing production that is too contaminated to use for irrigated agriculture or other purposes. The use of groundwater for biofuel production as a direct competition for water for agriculture may represent another example of a trade-off. Groundwater overdraft for agriculture and food production is another example of a trade-off because that loss of groundwater storage is not available for energy production or other uses. Careful accounting and public discussions of the linkages, synergies, alterations, and trade-offs during GSP development will create transparency, and will likely lead to greater stakeholder support of the various management alternatives to meet the groundwater sustainability objectives.

The SGMA regulations require that the GSAs document in the communication section of the GSP the opportunities for public engagement and active involvement of diverse social, cultural, and economic interests of the population within the basin. GSAs that embrace an interdisciplinary and transdisciplinary Nexus approach for developing their GSP and related management alternatives will address not only their statutory requirements for public engagement, but will also likely have considerable success in reaching their ultimate groundwater sustainability goals. The literature demonstrates that the complexity of the Nexus and limitations of disciplinary solutions necessitates that solutions to Nexus issues must be approached from an interdisciplinary perspective that includes the natural, social, and human sciences (Taniguchi et al. 2017; 2013). However, the benefits of transdisciplinary approaches often come with additional transaction costs, in the form of additional project costs and longer time to achieve consensus among diverse stakeholders. While implementing a Nexus approach within a GSP, the GSAs will have to evaluate and integrated countermeasures against such possible negative transdisciplinary impacts and costs.

Equally important as the interdisciplinary approach to developing effective GSPs, is the need for a transdisciplinary approach that incorporates the idea of co-construction and integration of different types of knowledge, not just from government officials or so-called experts, but also from the general public and practitioners in groundwater use, agriculture, and energy production (Endo et al. 2015b; Taniguchi et al. 2017). Transdisciplinarity in the WEF Nexus recognizes that in order to identify and implement solutions, the full spectrum of stakeholders and knowledge types must be engaged. For example, identifying stakeholders and agencies that directly manage energy resources and food (agriculture) systems must be engaged within the Nexus approach and outlined in the GSP. The need for transdisciplinarity is motivated by a range of community members that have traditionally been left out of the policy debates. Transdisciplinarity during SGMA activities is needed to identify sustainability goals and management alternatives, but to also build trust and facilitates communication between stakeholders that will help ensure the long-term success of reaching the sustainability goals by the target years of 2040 and 2042.

11.3 Sustainable Groundwater will Transform California's WEF Nexus

The implementation of SGMA will create sustainable groundwater resources at the local scale and across California. Sustainable groundwater resources will likely transform and enhance the resilience of the WEF Nexus in California. A primary reason for SGMA's positive feedback on California's WEF Nexus is that water is the limiting factor in the Nexus. There is no substitution for water during the production of food (agriculture) and energy. Groundwater has a greater spatial distribution than surface-water resources in California, and can often be accessed and used with lower infrastructure and investment costs.

Sustainable groundwater resources will also transform the WEF Nexus in California because of the implications of interannual to multidecadal climate variability and future climate change. California's water supply is highly dependent on climate variability from the El Niño/Southern Oscillation (2–7 year quasiperiodic cycle) (Velasco et al. 2015), atmospheric rivers (ARs), and recent multiyear droughts. Groundwater provides a critical buffer in the State's water supply during drought and dry phases of ENSO and other climate variability. Future climate change will also likely bring more variable and drier conditions to many parts of the State. Thus, there is a direct relationship between the importance of groundwater and increasing climate variability and changes to historical norms caused by future climate change (Green et al. 2011; Gurdak et al. 2015; Taylor et al. 2012; Treidel et al. 2012). SGMA will create a management framework to enable more sustainable groundwater that in turn will directly make food production (irrigated agriculture) and water-intensive energy generation more sustainable as a result.

The importance of groundwater in California's WEF Nexus is clear. Over 6 million Californians rely solely or primarily on groundwater for their water supply, and much of the state's $45 billion agriculture industry is supported by groundwater (Choy and McGhee 2014). The water-energy nexus for the statewide groundwater demand is enormous. An estimated 6000 gigawatt hours is the total energy required to pump groundwater from wells across California, which is a greater energy use than the annual energy requirements for the entire State Water Project (Moran et al. 2014). The city of Fresno alone in 2011 spent $9 million on electricity to pump groundwater (Moran et al. 2014). These energy and financial costs will only increase as groundwater levels continue to decline. Considerable volumes of the state's groundwater resources are also vulnerable to contamination from oil and gas extraction, particularly in the Central Valley aquifer (Kang and Jackson 2016). These are just a few of the many ways that a sustainable groundwater supply will directly increase human and environmental resilience and security within the WEF Nexus of California.

11.4 Conclusions

This chapter highlights a few specific concepts from the WEF Nexus and recommendations for local managers and stakeholders that are tasked with implementing SGMA and developing GSPs for their local groundwater basins. The WEF Nexus is a rich and growing field of study, and many other Nexus concepts than outlined here are relevant for SGMA. GSAs and stakeholders are encouraged to become familiar with and use WEF Nexus concepts to develop effective GSPs and reach their target sustainability goals.

Acknowledgement This research was financially supported by the R-08 Project, entitled "Human-Environmental Security in Asia-Pacific Ring of Fire: Water-Energy-Food Nexus" the Research Institute for Humanity and Nature, Kyoto, Japan. This chapter benefited from insightful discussions about the WEF Nexus with Ellen Doudna and Dr. Leora Nanus from San Francisco State University and Drs. Makoto Taniguchi and Aiko Endo from the Research Institute for Humanity and Nature.

References

Bazilian M, Rogner H, Howells M, Hermann S, Arent D, Gielen D, Steduto P, Mueller A, Komor P, Tol RSJ, Yumkella KK (2011) Considering the energy, water and food nexus: towards an integrated modelling approach. Energy Policy 39:7896–7906. https://doi.org/10.1016/j.enpol.2011.09.039

Beisheim M (2013) The water, energy & food security nexus: how to govern complex risks to sustainable supply? SWP Comments 32:1–8

California Department of Water Resources (2017) SGM Sustainable Groundwater Management. http://www.water.ca.gov/groundwater/sgm/. Accessed 15 Jan 2018

Choy J, McGhee G (2014) Groundwater: ignore it, and it might go away, water in the west.. Stanford Woods Institute for the Environment. http://waterinthewest.stanford.edu/groundwater/overview/. Accessed 15 Jan 2018

Endo A, Burnett K, Orencio P, Kumazawa T, Wada C, Ishii A, Tsurita I, Taniguchi M (2015) Methods of the water-energy-food nexus. Water 7:5806–5830. https://doi.org/10.3390/w7105806

Famiglietti JS (2014) The global groundwater crisis. Nat Clim Chang 4:945–948. https://doi.org/10.1038/nclimate2425

FAO (2014) The water-energy-food nexus: a new approach in support of food security and sustainable agriculture. Food and Agriculture Organization (FAO) of the United Nations, Rome

Gleeson T, VanderSteen J, Sophocleous MA, Taniguchi M, Alley WM, Allen DM, Zhou Y (2010) Groundwater sustainability strategies. Nat Geosci 3:378–379

Green TR, Taniguchi M, Kooi H, Gurdak JJ, Allen DM, Hiscock KM, Treidel H, Aureli A (2011) Beneath the surface of global change: impacts of climate change on groundwater. J Hydrol 405:532–560. https://doi.org/10.1016/j.jhydrol.2011.05.002

Gurdak JJ, Leblanc M, Aureli A, Carvalho Resende T, Faedo G, Green TR, Tweed S, Longuevergne L, Allen DM, Elliot J, Taylor RG (2015) GRAPHIC groundwater and climate change: mitigating the global groundwater crisis and adapting to climate change (position paper and call to action). UNESCO International Hydrological Programme (IHP), Paris

Gurdak JJ, Geyer GE, Nanus L, Taniguchi M, Corona CR (2016) Scale dependence of controls on groundwater vulnerability in the water–energy–food nexus California Coastal Basin aquifer system. J Hydrol Reg Stud 11:126. https://doi.org/10.1016/j.ejrh.2016.01.002

Kang M, Jackson RB (2016) Salinity of deep groundwater in California: water quantity, quality, and protection. Proc Natl Acad Sci 201600400

Konikow LF (2015) Long-term groundwater depletion in the United States. Groundwater 53:2–9. https://doi.org/10.1111/gwat.12306

Lawford R, Bogardi J, Marx S, Jain S, Wostl CP, Knüppe K, Ringler C, Lansigan F, Meza F (2013) Basin perspectives on the water–energy–food security nexus. Curr Opin Environ Sustain 5:607–616. https://doi.org/10.1016/j.cosust.2013.11.005

Leck H, Conway D, Bradshaw M, Rees J (2015) Tracing the water-energy-food nexus: description, theory and practice: tracing the water-energy-food nexus. Geogr Compass 9:445–460. https://doi.org/10.1111/gec3.12222

Lund J, Harter T, Gailey R, Fogg G, Frank R, Dahlke H, Ginn T, Sandoval Solis S, Young T, Fisher AT, Langridge R, Viers J, Harmon T, Holden P, Keller A, Kiparsky M, Greene T, Mehl S, Gurdak JJ, Gorelick S, Knight R (2015) Creating effective groundwater sustainability plans, California water blog. https://californiawaterblog.com/2015/03/04/creating-effective-groundwater-sustainability-plans/. Accessed 15 Jan 2018

Maupin MA, Barber NL (2005) Estimated withdrawals from principal aquifers in the United States, 2000 (Circular No. 1405). U.S. Dept. of the Interior, U.S. Geological Survey ; For sale by U.S. Geological Survey, Information Services, Reston, Va.: Denver, CO

Moran T, Choy J, Sanchez C (2014) The hidden costs of groundwater overdraft, water in the west. Stanford Woods Institute for the Environment. http://waterinthewest.stanford.edu/groundwater/overdraft/. Accessed 15 Jan 2018

Rahaman MM, Varis O (2005) Integrated water resources management: evolution, prospects and future challenges. Sustain Sci Pract Policy 1:1–15

Rasul G, Sharma B (2015) The nexus approach to water–energy–food security: an option for adaptation to climate change. Clim Pol 16:1–21. https://doi.org/10.1080/14693062.2015.1029865

Sward S (2017) Opposing sides in California water wars forced together in groundwater agencies. The Sacramento Bee, Sacramento, California, 24 Mar 2017, http://www.sacbee.com/opinion/california-forum/article140345243.html

Taniguchi M, Allen DM, Gurdak JJ (2013) Optimizing the water-energy-food nexus in the Asia-Pacific ring of fire. EOS Trans Am Geophys Union 94:435. https://doi.org/10.1002/2013EO470005

Taniguchi M, Endo A, Gurdak JJ, Swarzenski P (2017) Water-energy-food nexus in the Asia-Pacific region. J Hydrol Reg Stud 11:1–8. https://doi.org/10.1016/j.ejrh.2017.06.004

Taylor RG, Scanlon B, Döll P, Rodell M, van Beek R, Wada Y, Longuevergne L, Leblanc M, Famiglietti JS, Edmunds M, Konikow L, Green TR, Chen J, Taniguchi M, Bierkens MFP, MacDonald A, Fan Y, Maxwell RM, Yechieli Y, Gurdak JJ, Allen DM, Shamsudduha M, Hiscock K, Yeh PJ-F, Holman I, Treidel H (2012) Ground water and climate change. Nat Clim Chang 3:322–329. https://doi.org/10.1038/nclimate1744

Treidel H, Martin-Bordes JL, Gurdak JJ (eds) (2012) Climate change effects on groundwater resources: a global synthesis of findings and recommendations. CRC Press, London

United Nations ESCAP (2013) The status of the water-energy-food nexus in Asia and the Pacific. United Nations Economic and Social Commission for Asia and the Pacific, Bangkok

Velasco EM, Gurdak JJ, Dickinson JE, Ferré TPA, Corona CR (2015) Interannual to multidecadal climate forcings on groundwater resources of the U.S. west coast. J Hydrol Reg Stud. https://doi.org/10.1016/j.ejrh.2015.11.018

Vogt KA, Patel-Weynand T, Shelton M, Vogt DJ, Gordon JC, Mukumoto CT, Suntana AS, Roads PA (2010) Unpacked: food, energy and water for resilient environments and societies. Earthscan, New York, NY

Wada Y, van Beek LPH, Bierkens MFP (2012) Nonsustainable groundwater sustaining irrigation: a global assessment. Water Resour Res 48(6). https://doi.org/10.1029/2011WR010562

Wada CA, Burnett K, Gurdak JJ (2016) Sustainable agriculture irrigation management: the water-energy-food nexus in Pajaro Valley, California. Sustain Agric Res 5(3):76. https://doi.org/10.5539/sar.v5n3p76

Chapter 12
Pump Tax, Basin Equity Assessment and Sustainability in Groundwater Management: Orange County Water District Experience

Takahiro Endo

Abstract The term "sustainable development" has become very popular among researchers, practitioners and policymakers since the 1980s. The sustainability concept is now applied in the field of groundwater management. Excessive groundwater pumping is taking place in many parts of the world and the means to achieve sustainable groundwater management has become an urgent need. The purpose of this paper is to analyze how to promote sustainable groundwater management, with a focus on Orange County Water District (OCWD), California, USA. OCWD faced excessive groundwater pumping in the 1920s. Since then, the district introduced various policies including a pump tax and basin equity assessment to tackle the problem, and continues to use groundwater as the primary supply. Investigation into the factors that enable OCWD to use groundwater continuously provides useful information to policymakers in other areas who are engaged in groundwater management. In addition, policy implications in terms of sustainability are drawn from OCWD's groundwater management.

Keywords Common-pool resources · Groundwater management · Sustainable development · Sustainability · Pump tax · Basin equity assessment · Orange County Water District · California

---

T. Endo (✉)
College of Sustainable System Sciences, Osaka Prefecture University, Sakai City, Osaka, Japan
e-mail: tte23042@osakafu-u.ac.jp

© Springer Nature Singapore Pte Ltd. 2018

A. Endo, T. Oh (eds.), *The Water-Energy-Food Nexus*, Global Environmental Studies, https://doi.org/10.1007/978-981-10-7383-0_12

12.1 Introduction

The term "sustainable development" has become very popular among researchers, practitioners and policymakers since a report by the World Commission on Environment and Development was published (The World Commission on Environment and Development 1987). The problem the term addresses is how to achieve economic development and environmental conservation at the same time (Pearce and Atkinson 1998). Although there have been controversies regarding the meaning of sustainable development, arguments for sustainability have gradually been directed at social problems in various fields, such as livestock control (Eisler and Lee 2014), fisheries (Heal and Schlenker 2008; Sampson et al. 2015) and transportation policy (Bruun and Givoni 2015).

The sustainability concept is also applied in the field of groundwater management. Excessive groundwater pumping is taking place in many parts of the world and the means to achieve sustainable groundwater management has become an urgent need (Wada et al. 2010). Although a variety of policy options are proposed in groundwater management, their effects remain to be investigated (Green et al. 2011). There are many economic tools to limit water uses such as water rate control, tradable permits and taxation. Taxation of natural resource utilization have been advanced theoretically as early as the 1960s (e.g., Dales 1968; Ruff 1970). However, their application to groundwater management has been limited, with a few exceptional cases such as Bangkok, Thailand (Lorphensri et al. 2011), Indonesia (Braadbaart and Braadbaart 1997), and the Netherlands (Schuerhoff et al. 2013).

The purpose of this paper is to analyze various means of promoting sustainable groundwater management, with a focus on the Orange County Water District, California, USA (hereafter referred to as OCWD). OCWD faced excessive groundwater pumping in the 1920s. Since then, OCWD has introduced various policies including a pump tax to tackle the problem and has continued to use groundwater as the primary supply. Investigation of the factors that enable OCWD to use groundwater continuously will provide useful information to policymakers in other areas who pursue sustainable groundwater management. It has been more than 50 years since OCWD initiated countermeasures. The history of economic tools in groundwater management in OCWD is much longer than the cases mentioned above. Thus, the OCWD case facilitates analysis of long- and short-term policy effects.

A brief history of the groundwater problem in Orange County is presented in the next section. OCWD's policy options including economic tools are explained in the third section. Then, policy implications in terms of sustainability drawn from OCWD's experience are described in the fourth section. Last, an overall summary is given in the fifth section. The paper does not deal with groundwater quality problems because of space limitations. Therefore, the term "groundwater management" herein means "quantitative control."

12.2 Orange County Water District

12.2.1 Brief History of Groundwater use in Orange County

Groundwater is a typical example of a common-pool resource, which has characteristics of difficulty of exclusion and rivalry in consumption. The former implies that controlling a range of beneficiaries through physical or institutional means can be prohibitively expensive, and the latter means that an individual's consumption of the resource reduces the potential consumption of others. Natural resources in general tend to have these common-pool characteristics (Ostrom et al. 1999).

Without adequate arrangements for exclusion, groundwater is left with an open-access status. In such a situation, no one has the incentive to conserve or replenish groundwater, simply because groundwater secured through such efforts can be reaped by someone else. Rather, a user currently has an incentive to pump groundwater as long as its private marginal benefit exceeds private marginal cost. Of course, such conduct generates external costs in the forms of cones of depression and groundwater table decline. Where the number of users is large, external costs caused by a single user may be very small. Nevertheless, the accumulation of such small effects cause clear groundwater depletion, land subsidence and seawater intrusion. Such "tragedy of the commons" (Hardin 1968) actually occurred in OCWD.

Orange County is in the most downstream area of the Santa Ana River, which runs southeast of Los Angeles (Fig. 12.1). This river begins in the San Bernardino Mountains and flows through Prado Dam to Orange County. The climate is dry, and average seasonal rainfall in the OCWD service area over a recent 5-year period (July 1, 2011 through June 30, 2016) was just 9.07 inches (230.38 mm) (OCWD 2017).

Artesian wells were put in use as early as 1868. Pumping machines gradually became popular in the 1890s, which resulted in the expansion of agricultural lands (California Department of Water Resources 1959). As demand for groundwater increased, negative effects were discovered. It was clearly observed in the 1920s that the groundwater table had dropped, resulting in the artesian area shrinking (Lippincott 1925).

12.2.2 Formation of OCWD

In 1931, water users in Orange County brought lawsuits against users in the upstream area. The intention was to make the upstream users curtail water diversion from the Santa Ana River to maintain instream flows, which were considered a primary recharge source of local groundwater. Water users in Orange County organized a body by which they collected property tax to raise revenue for the lawsuit in 1933. This was the origin of OCWD (Weschler 1968).

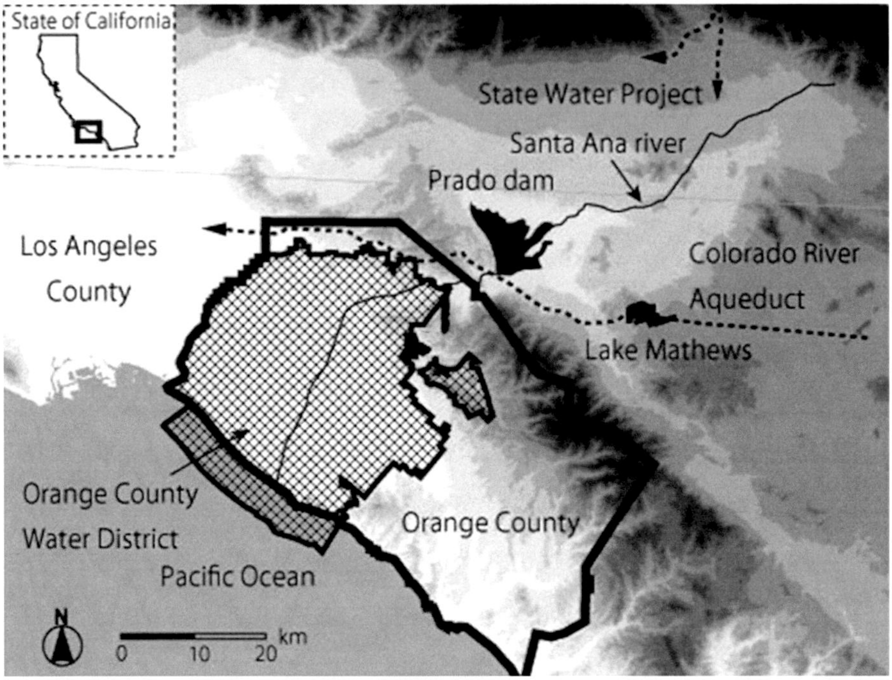

Fig. 12.1 Location of Orange County Water District

OCWD is a special district, a type of local government. When a county or city government provides various public services, it works for a specific purpose such as fire protection, park management or mosquito abatement. A special district that provides a water-related service is generally characterized as a special water district and is often called an "irrigation district" or "municipal water district" in accordance with its purpose. The service area of a special district does not always coincide with that of a county or city government. OCWD's jurisdiction is limited to the northeastern part of Orange County which is endowed with local groundwater (Littleworth and Garner 1995; Senate Local Government Committee 2010) (Fig. 12.1).

OCWD is given various authorities by the state legislature. These include powers to ① sell or store imported water, ② conserve and recharge local waters, ③ bring lawsuits against outsiders to secure a stable water supply for local users, ④ require a tax on groundwater pumping. Although OCWD's authority is wide-ranging, there are limitations. OCWD is not given authority to directly control groundwater pumping. All OCWD can do is control that pumping indirectly by managing the rate of the pump tax. In addition, it does not have power over groundwater quality. Therefore, if there is groundwater contamination within the OCWD service area, it would deal with the polluter using persuasion or lawsuits instead of directly ordering

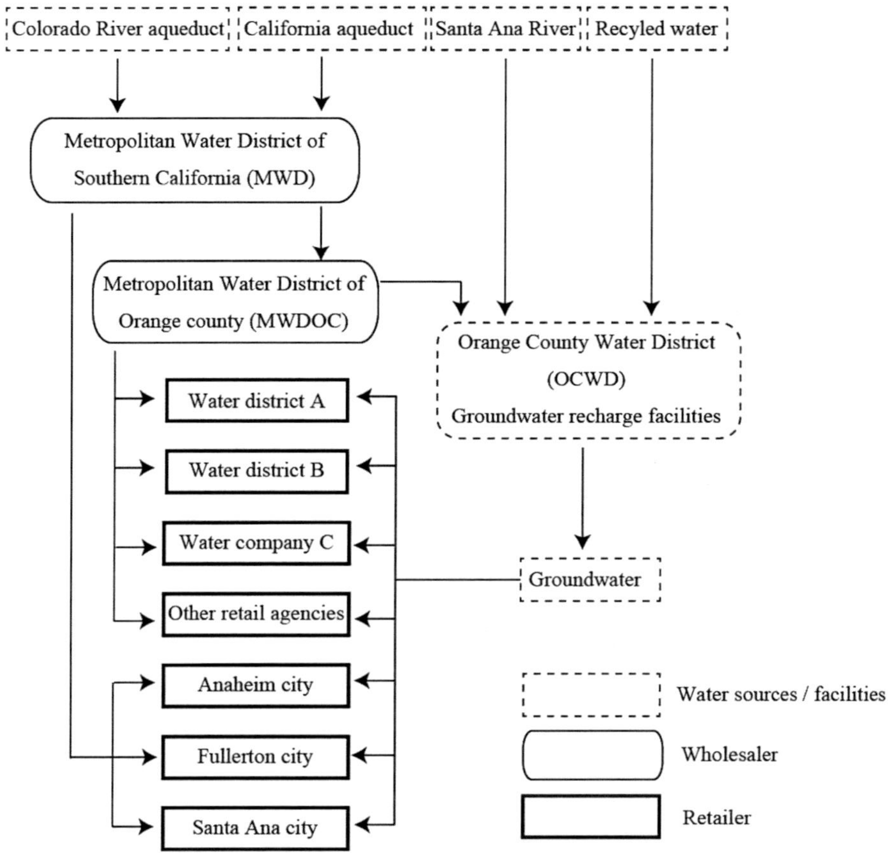

Fig. 12.2 Water supply system in OCWD

the polluter to cease and clean up the contamination (Crooke 1963; Weschler 1968; Blomquist 1992; OCWD 2015).

12.2.3 Water Supply System in OCWD

The OCWD service area is 243,968 acres and provides water to about 2.4 million people (OCWD 2015). Figure 12.2 shows the water supply system in OCWD. The sources of water supply are classified into surface water and groundwater. The former includes not only local water (e.g., the Santa Ana River) but also imported water, which is delivered from outside through long-distance aqueducts such as the California and Colorado River aqueducts.

The main local surface water is from the Santa Ana river. The river's water is rarely diverted for direct (potable) use now. Instead, it is stored in underground

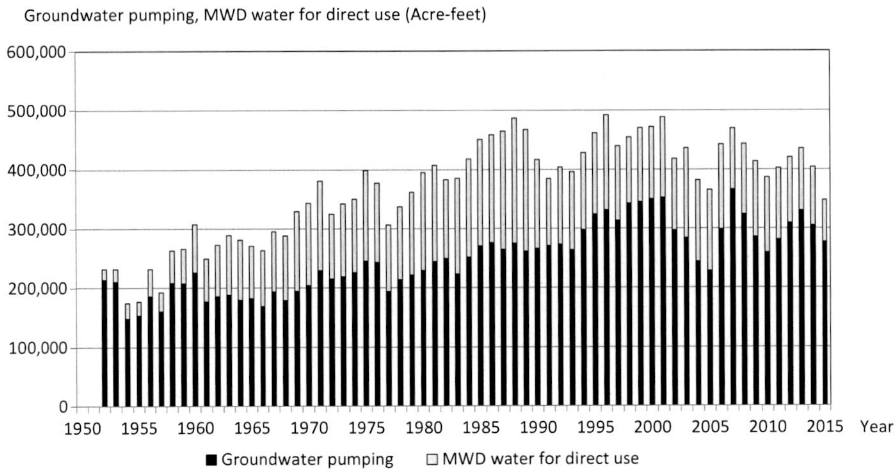

Fig. 12.3 Changes in volume of groundwater pumping and MWD water for direct use

space and then pumped out as groundwater. In other words, the river is used as a huge recharge facility. This approach is also applied to another local water source, recycled water. OCWD is famous for developing elaborate systems of water recycling. This highly advanced treatment changes local sewage into purified recycled water, which is used for groundwater recharge and water barriers against seawater intrusion in coastal areas.

Imported water is for direct use and serves as a recharge source. It is delivered to end users via a few steps. First, it is delivered to Orange County and other Southern California areas by a large wholesaler called the Metropolitan Water District of Southern California (hereafter referred to as MWD). Therefore, the imported water is often called MWD water. This water reaches end users in the OCWD service area mainly via two routes. First, MWD transfers water to a secondary wholesaler called the Municipal Water District of Orange County (hereafter referred to as MWDOC). The MWDOC service area covers all of Orange County and provides water to end users via subordinate retailers (e.g., water districts). Along the second route, as retailers, the cities of Anaheim, Fullerton and Santa Ana receive water directly from MWD and supply its water to citizens.

Groundwater is generally for urban uses. It is unevenly distributed in northwestern Orange County. Therefore, although there are many water districts across the county, only about 20 can access groundwater or imported water. They include the cities of Anaheim, Fullerton, Santa Ana and MWDOC's subordinate water districts, which are rich in local groundwater. OCWD consists of these districts. To put it another way, OCWD does not pump and deliver groundwater to end users directly. The main task is to aid member district groundwater pumping via groundwater recharge and other technical support (Herndon 2013).

Figure 12.3 shows changes in volume of imported water (for direct use) and groundwater pumping. Although imported water became an important supplemental supply after the 1960s, groundwater has been the main water supply. That means the supply has not fully switched from groundwater to surface water.

12.3 OCWD Groundwater Management

12.3.1 Groundwater Management in the State of California

Institutions of groundwater management are important in determining how scarce groundwater is allocated to competing demands (Tarlock 1985). In California, this allocation is determined traditionally by courts, not by administrative permission. The guiding principle is called the correlative right doctrine. First, it classifies groundwater users into two groups, those who use groundwater within or outside the basin. Members of the former group are called overlying users who are usually farmers. Members of the latter are non-overlying users, typically urban users. Then, the doctrine determines allocation priorities considering three cases: between overlying users, between non-overlying users, and between overlying users and non-overlying users. However, this rule does not promote efficient use of groundwater. This is because it plays a role only after disputes and has no preventative function (Sandino 2005).

Given such institutional failures in the region, groundwater is managed locally in diverse ways, reflecting its specific surroundings. The main approaches to groundwater management include formation of a special district, adjudication, and local groundwater management plans (Lipson 1978; Littleworth and Garner 1995). OCWD is considered a representative example of groundwater management by formation of a special district (De Lambert 1984). Nevertheless, this does not mean that any of the three approaches covers California entirely. There have been many locations called groundwater management vacuums, where there is no institution for groundwater (Governor's Commission to Review California Water Rights Law 1978). In 2014, the Sustainable Groundwater Management Act was legislated and many areas were required to establish groundwater sustainability agencies. This legislation was an attempt to fill the vacuums. OCWD has already been admitted as the exclusive local agency to manage groundwater within the district's statutory boundaries, with powers to comply with provisions of the Sustainable Groundwater Management Act (OCWD 2015).

12.3.2 Artificial Recharge

Excessive groundwater pumping can be counteracted either by recharge augmentation, pumping restrictions, or a combination of both. As mentioned in detail later, OCWD imports water from outside areas using revenue from the pump tax and

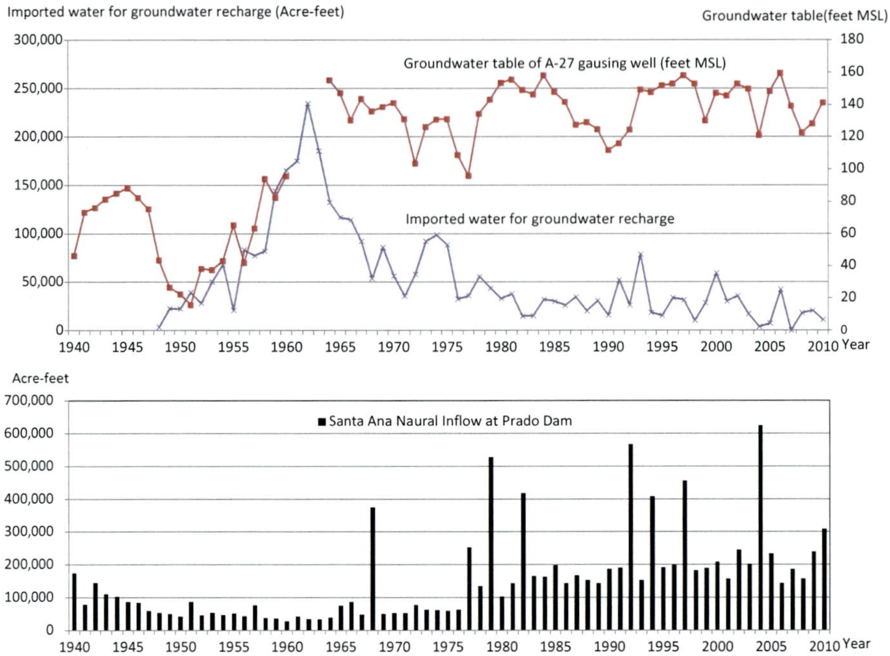

Fig. 12.4 Changes in the volume of imported water, groundwater table, and flow of the Santa Ana River (inflow at Prado Dam)

stores it in local basins to stabilize the groundwater supply. This concept is different from adjudication, which focuses on pump restriction. Rather, OCWD aims to avoid groundwater shortages by expanding groundwater recharge (Blomquist 1992).

The most downstream users of the Santa Ana River formed OCWD to engage in lawsuits with water users who diverted water in the upstream area. Although they won the lawsuit, they could not secure enough water. In addition, years with low precipitation have continued since 1942. This increased groundwater pumping and resulted in a decline of the groundwater table and expansion of areas suffering from seawater intrusion. To cope with this situation, OCWD began to recharge local groundwater using imported MWD water in the late 1940s (Weschler 1968).

Figure 12.4 shows changes in the volume of imported water, the groundwater table, and flow of the Santa Ana River (inflow at Prado Dam). From the late 1940s to 1964, the groundwater table (at A-27 gauging well) recovered rapidly. Flow in the Santa Ana River decreased during that period. Therefore, it may be assumed that artificial recharge by imported water supported the recovery. This supply-oriented policy was called "fill-the-basin". However, this strategy was modified after 1964. This was because there was a great difference between regions in the rise of the groundwater table. Groundwater levels in some of the coastal areas did not recover as expected. Further, it was feared that some parts of the upper basin might revert to swamp because of a high groundwater table (Weschler 1968). The groundwater table recovered even after 1964 when the fill-the-basin policy ended. Figure 12.4 shows that the recovery corresponded with flow in the Santa Ana River. Therefore, it might be that the increase in natural groundwater recharge from the river had a role in maintaining the groundwater level after 1964.

12.3.3 Pump Tax

OCWD uses economic tools to manage groundwater. The district collected property tax for lawsuits against the upstream users in the 1930s and used the revenue to buy MWD water for artificial recharge. However, this revenue was insufficient to continue artificial recharge. Thus, another fiscal base became necessary. In 1953, the state legislature admitted a new authority to impose a pump tax (Crooke 1961). OCWD was the pioneering public water district to introduce a pump tax system in California (Weschler 1968).

The pump tax is levied on all groundwater users, whether the well is for irrigation or urban uses. However, tax rates per pumped acre-foot vary. The rate for irrigation wells has usually been set lower than that for non-irrigation wells. As mentioned above, almost all groundwater pumping is currently for urban uses. The tax rate for such uses was $322 per acre-foot in the 2015–16 water year (July 1, 2015 through June 30, 2016) (OCWD 2017).

Groundwater users are also required to install a meter to gauge pumped volume and report the data to OCWD every 6 months. The amount of tax is determined by multiplying the volume by the tax rate. There is one exception to this rule, i.e., the owners of “small wells” are exempt from the metering requirement. Here, “small” wells include those from which the discharge outlet is not greater than 2 inches (20.4 cm^2) in size and which provide domestic and irrigation water for an area not exceeding 1 acre (Crooke 1961; Weschler 1968).

Basically, the above information-gathering process is based on an honor system. Prevention of cheating is very important under this system. If necessary, OCWD checks the amount of electricity used to pump groundwater and the past pump record to improve their data accuracy. In addition, information sharing enables mutual checks among users to prevent cheating (Ostrom 1990). OCWD promotes mutual checks by publishing the annual pumping volume of major groundwater users (non-irrigation users of >25 acre-feet per year).

OCWD employs engineers who investigate groundwater conditions, especially the amount of overdraft, every year. Overdraft is the estimated quantity by which groundwater pumping exceeds its natural replenishment during a year (OCWD 2017). The engineers check groundwater conditions including the average annual overdraft over the past 10 years and the total accumulated overdraft of the preceding year. Then they consider the water quantity for artificial recharge necessary to mitigate overdraft and calculate the required budget to purchase that amount of water. Last, based on information of anticipated groundwater pumping in the ensuing year, they determine how much money should be levied on 1 acre-foot of pumped

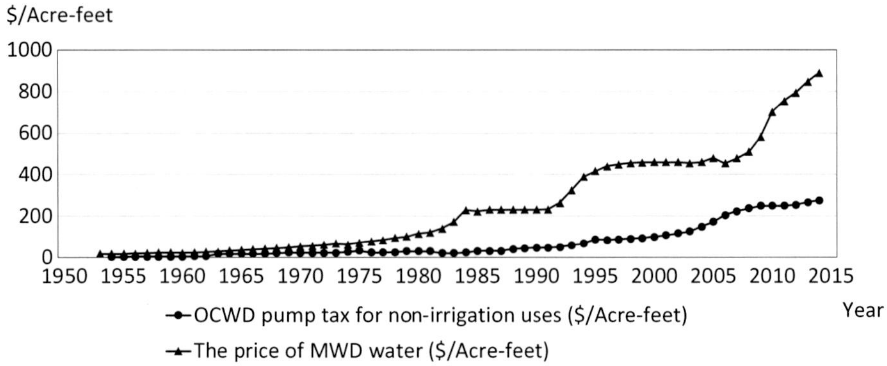

Fig. 12.5 Changes in the pump tax and the price of MWD water

groundwater. This is how the tax rate is determined each year. Therefore, the tax is imposed directly on water, not electricity. As mentioned above, each groundwater user is supposed to pay according to their pumped volume (Crooke 1954).

How OCWD allocates groundwater is different from the correlative right doctrine. Under this doctrine, when there is a dispute between an overlying and non-overlying user, the former is given the superior right to use groundwater. For allocation between non-overlying users, "first in time, first in right" is applied. That is, the party who began to use groundwater earlier is granted the superior right. However, OCWD does not utilize such classifications in terms of time and place of use. Local groundwater users are equal and they can pump groundwater freely as long as they pay the pump tax.

12.3.4 Basin Equity Assessment

The pump tax has been set lower than the price of MWD water (Fig. 12.5). Therefore, retailers in OCWD have had incentive to use groundwater instead of MWD water to satisfy customer water demand, which results in groundwater depletion.

To avoid the above situation, OCWD once asked retailers to use both MWD water and groundwater in a 50–50 ratio. However, this request did not work, simply because it was not mandated. A sharp decline of the groundwater table was observed in part of the OCWD service area.

The aforementioned problem was mitigated by an additional demand control tool called Basin Production Percentage (hereafter referred to as BPP), which was introduced in 1968. Under this system, OCWD established a percentage of water supply originating from local groundwater and supplemental water from outside the basin each year, using data on volume of groundwater storage and available MWD water. BPP is applied to a user who pumps groundwater at more than 25 acre-feet per year. Such users are often retailers like a city or water district, not individual well owners (OCWD 2009; OCWD 2017).

For example, when the BPP for a water retailer is set to 60%, it can satisfy as much as 60% of local water demand through cheap groundwater; the remainder should be met by expensive imported water. If the retailer pumps groundwater at more than the BPP rate, it is required to pay another assessment called a basin equity assessment for the excess. Conversely, when the retailer pumps at less than that rate, it benefits from the basin equity assessment paid by another retailer who pumps more than BPP rate. Thus, BPP is a system that regulates groundwater pumping by controlling the ratio of groundwater and imported water. The basin equity assessment was originally set equal to the difference between the prices of groundwater and imported water. However, it is now calculated so that the cost of groundwater production is higher than purchasing imported potable water. This means that a retailer who pumps groundwater at more than the BPP rate must eventually pay the same or more as it imports water from outside. This prevents retailers from pumping cheap groundwater as much as they want (OCWD 1970; OCWD 2009).

In summary, the pump tax and basin equity assessment are two primary economic tools for groundwater management in OCWD. Generally, the pump tax plays a role in preventing excessive groundwater pumping. However, this function cannot be exerted fully because OCWD has set the pump tax lower than the rate of imported water. This is why OCWD introduced a second tool of basin equity assessment. Simply raising the pump tax rate may prevent local retailers from excessive groundwater pumping but it would also deprive them of access to cheap groundwater, because the tax is imposed on all groundwater users. The aim of the basin equity assessment is to allow for utilization of cheap groundwater with some deterrents against excessive pumping; the additional assessment is only imposed on those who pump groundwater more than stipulated (Herndon 2013). OCWD's groundwater management approach has been considered primarily supply-enhancement (i.e. artificial recharge), but it also controls demand through the basin equity assessment.

12.4 Lessons toward Sustainable Groundwater Management

12.4.1 Sustainable Development and Capital

Various definitions have been proposed for the concept of sustainable development. Economists often define it in terms of capital accumulation. That is, they think sustainable development depends on how a stock of capital and technology should be carried over from one generation to the next. In the classical approach, the concept of capital was limited to produced goods or man-made capital such as factories, but more recent approaches include skill and knowledge embodied in humans (human capital), trust and social bonds between individuals (social capital), and natural resources in general (natural capital) (Pearce and Atkinson 1998).

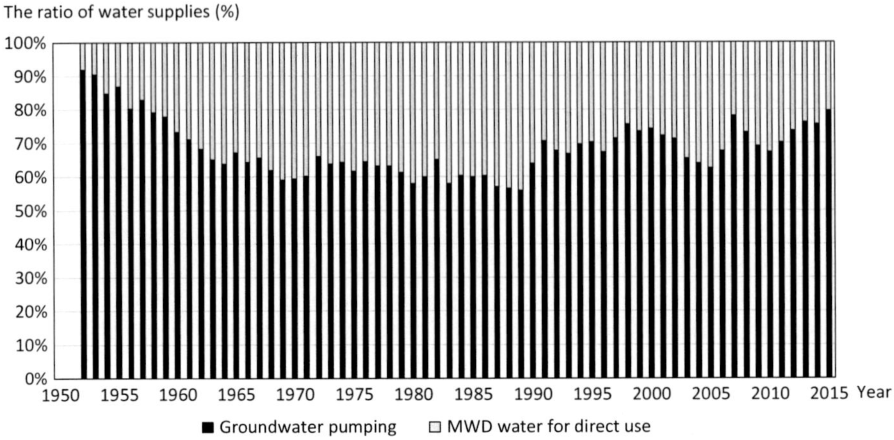

Fig. 12.6 Shares of groundwater and surface water in local water supply

The extent to which artificial and natural capital can be substituted has been controversial. Through debate, two approaches have emerged, weak and strong sustainability. The former assumes that the aforementioned two capitals are interchangeable. It is true that consumption of natural resources (especially non-renewable ones) is associated with costs such as resource depletion. However, if a part of the consumptive benefits can be invested to increase artificial capital which covers the loss of natural resources, the total volume of stocks can be maintained, resulting in sustainable development. However, strong sustainability does not take it for granted that the two capitals can be substituted without limit. Although natural capital is endowed with reproduction capacity, there are limits. Beyond a threshold, such capacity cannot be realized and the stock continues to decrease. In such a situation, artificial capital is no longer useful, so sustainable development cannot be achieved (Daily 1995; Pearce and Atkinson 1998).

12.4.2 Weak/Strong Sustainability in Groundwater Management

The concepts of weak/strong sustainability can be applied to groundwater management in OCWD in the following way. In the former, groundwater is fully consumed to depletion and then imported water is used as a substitute, which is delivered across natural hydrologic boundaries. In the latter, groundwater is used to some extent and the water supply is supplemented with imported water as needed.

Figure 12.6 shows shares of groundwater and surface water in the local water supply. This shows that the share of groundwater decreased from the 1950s to mid-1960s. This trend can be evaluated in terms of weak sustainability. Thereafter, however, OCWD implemented demand control through BPP and prevented further conversion. This can be interpreted as a departure from weak sustainability-oriented policy.

The method of groundwater use in OCWD sharply contrasts with that of Tokyo, Japan. Both areas have faced excessive groundwater pumping, but the countermeasures were very different. In Tokyo, the national Industrial Water Law of 1956 was enacted to cope with excessive groundwater pumping, promoting water supply conversion from groundwater to surface water. As a result, groundwater utilization has been restricted so severely that it has had an unexpected side effect, i.e., a high groundwater level causing floating pressure against underground infrastructure. In the short run, switching the water supply from groundwater to surface water would induce groundwater conservation. However, prohibiting groundwater use may have a completely negative impact on groundwater. This is because such prohibition causes people to potentially neglect groundwater and raises the risk of contamination. Nobody will take care of groundwater unless they use it constantly. The differences between OCWD and Tokyo include the subsidy system, legal definition of groundwater, and local governmental authority. This suggests that legal and administrative institutions influence the sustainable use of groundwater (Endo 2015).

12.4.3 Safe Minimum Standard

The irreversibility problem is often used to justify strong sustainability (Pearce and Atkinson 1998). The concept is that natural capital will continue decreasing to depletion once the stock diminishes beyond a threshold, because of human consumption. However, our knowledge of nature is so limited that it is very difficult to know that threshold precisely. Therefore, the maximum allowable consumption level should not be established by an assumptive threshold, but set within more restrictive levels, considering the uncertainty. This is called the safe minimum standard (Randall and Farmer 1995; Farmer and Randall 1998). However, constructing such a conceptual framework is one thing, while incorporating such an idea into policy is another completely.

The concept of "operating range," which OCWD introduced in 2007, suggests how the safe minimum standard could be applied to groundwater management. The operating range defines the upper limit of groundwater pumping. OCWD determines this limit considering the long-term groundwater budget (the difference between groundwater recharge and pumping), not a yearly one. When groundwater pumping exceeds recharge in a year, groundwater storage diminishes because of the deficit. If this occurs in the subsequent year, the reduction in storage will grow further. From past experience, OCWD engineers assume that seawater intrusion will become out of control once the accumulated reduction reaches 500,000 acre-feet. In reality, OCWD tries to recover groundwater storage using a preventive approach. That is, it takes countermeasures for increasing recharge and raises the pump tax when the accumulated reduction reaches 430,000 acre-feet. This operating range was made possible after the 1980s by a long-term monitoring system (Herndon

2013; OCWD 2017). Although OCWD sets the upper limit, it promotes groundwater pumping within the range. This suggests countermeasures alternative to complete prohibition of groundwater pumping.

12.4.4 Resource Characteristics and Sustainability

The extent to which artificial capital can be an alternative to natural capital is a focal point in the controversy surrounding strong and weak sustainability. Price has been considered an important factor in determining the substitution relationship. Generally, values of natural capital are not shown in the form of price. Therefore, even if natural capital becomes scarce because of depletion, price will not rise sufficiently to promote substitution (Barbier 1990).

However, the case of OCWD suggests that physical resource characteristics also influence the substitution relationship. For example, it takes time for forests to grow. Thus, once a forest is cut and we substitute the water storage function with artificial capital such as dams, it will be extremely difficult to use the forest again after a short period. In contrast, it is easier to switch groundwater with imported water in a short period, because the groundwater volume changes frequently through wet and dry seasons. As Fig. 12.6 shows, the ratio between imported water and groundwater changes every year. The method of switching the water supply between groundwater and surface water depending on precipitation is often called "conjunctive water management" (Blomquist et al. 2004).

Progress in artificial recharge techniques would make such substitution easier. Theoretical studies of sustainable development have not assumed that natural and artificial capital can be substituted frequently. The OCWD case suggests that such flexible uses of natural and artificial capital may be possible, depending on the resource characteristics of reproduction capacity.

12.5 Conclusions

The path to realizing sustainable groundwater management is now a global concern. This paper describes groundwater management in OCWD, with special emphasis on various policy options such as the pump tax and basin equity assessment. The following policy implications were deduced from the OCWD experience, within a framework of weak and strong sustainability:

First, legal and administrative institutions influence the sustainable use of groundwater. Whether local groundwater management is weak sustainability-oriented or not can be evaluated by how groundwater is used. OCWD and Tokyo faced excessive groundwater pumping during the same period, but the countermeasures were very different, resulting in sharply contrasting historical changes in groundwater use. While groundwater is used to some extent and the water supply is supplemented with imported water as needed in OCWD, imported water has almost completely replaced natural groundwater use in Tokyo. The differences between OCWD and Tokyo are partly caused by differences in the subsidy systems, legal definitions of groundwater, and local governmental authority.

Second, the safe minimum standard for strong sustainability can be applied to groundwater management. OCWD introduces the upper limit of groundwater pumping, called the operating range, to prevent seawater intrusion. The limit is established by an assumptive threshold, but set within more restrictive levels, considering the uncertainty. OCWD controls recharge volume and the pump tax rate considering this upper limit. This suggests that an upper limit based on long-term monitoring is necessary to establish organized countermeasures.

Lastly, the case of OCWD suggests that physical resource characteristics should be taken into consideration in the sustainability argument. The extent to which artificial capital can serve as an alternative to natural capital is a focal point in the controversy surrounding strong and weak sustainability. Unlike a forest resource, it is easier to switch natural groundwater with artificially imported water or vice versa in a short period. Theoretical studies of sustainable development have not assumed such a high frequency of substitution. This suggests progress in artificial recharge techniques would make such substitution easier, enabling flexible use of natural and artificial capital.

Comparative analysis of groundwater management in other regions remains to be investigated. It is well known that various experimental groundwater management approaches have been pursued in Southern California where OCWD is located. OCWD does not restrict groundwater pumping and promotes artificial recharge to compensate groundwater depletion. In contrast, Raymond Basin, which is northeast of Los Angeles, has introduced a court-appointed water resource manager, called a watermaster, who sets limits on annual groundwater pumping. Then the limits are divided and allocated to each user as groundwater decrees, which are tradable under supervision of watermaster. Although a few comparative studies have already been done by Lipson (1978) and Blomquist (1992), more should be done to evaluate the effects of each groundwater management framework from a longer-term perspective.

In addition, much attention should be paid to natural resources other than groundwater. As mentioned above, how artificial and natural capital can be substituted depends partly on reproduction capacities of the resources. These capacities vary among natural resources, including groundwater, fish and forest. Although there have been many theoretical studies on the concept of weak and strong sustainability, many case studies should be done to understand how to apply that concept to each natural resource management system. This paper is a starting point for such analysis.

Acknowledgements This research was supported by the research project "Human-environmental security in the Asia-pacific ring of fire" (Director Aiko Endo, Associate Professor, Research Institute for Humanity and Nature) under the Research Institute for Humanity and Nature Project and "Fusion of global water resources assessment and social institution studies" (Director Shinjiro Kanae, Professor, Tokyo Institute of Technology), under the Japan Society for the Promotion of Science KAKENHI (Grant-in-Aid for Scientific Research (B)) grant number 15H04047. Special thanks are due to Mr. Roy L. Herndon, OCWD chief hydro-geologist, for providing the data. The views presented here are those of the author and should be in no way attributed to organization mentioned above. Last, I thank the reviewers for their useful comments. Responsibility for the text rests entirely with the author. We thank Edanz Group (www.edanzediting.com) for editing a draft of this manuscript.

References

Barbier EB (1990) Alternative approaches to economic-environmental interactions. Ecol Econ 2:7–26

Blomquist W (1992) Dividing the waters, governing groundwater in southern California. ICS Press, San Francisco

Blomquist W, Schlager E, Heikkila T (2004) Common waters, diverging streams: linking institutions and water management in Arizona, California and Colorado. Resources for the Future, Washington, DC

Braadbaart O, Braadbaart F (1997) Policing the urban pumping race: industrial groundwater overexploitation in Indonesia. World Dev 25(2):199–210

Bruun E, Givoni M (2015) Six research routes to steer transport policy. Nature 523:29–31

California Department of Water Resources (1959) Santa Ana River investigation (bulletin No.15), Sacramento

Crooke HW (1954) Orange County inaugurates first tax for pumping water from underground basin. Western City Magazine, issue of October

Crooke HW (1961) Ground water replenishment in Orange County, California. J Am Water Works Assoc 53(7):831–837

Crooke HW (1963) Financing a ground water replenishment program and ground water basin management in the area of Orange County Water District. In: proceedings, biennial conference on ground water recharge and ground water management, University of California, Berkeley, California, 26–27 June 1963

Daily HE (1995) On Wilfred Beckerman's critique of sustainable development. Environ Values 4:49–55

Dales JH (1968) Land, water and ownership. Can J Econ 1:791–804

De Lambert DA (1984) District management for California's groundwater. Ecol Law Quart 11:373–400

Eisler MC, Lee MRF (2014) Steps to sustainable livestock. Nature 507:32–34

Endo T (2015) Groundwater management: a search for better policy combinations. Water Policy 17:332–348

Farmer MC, Randall A (1998) The rationality of a safe minimum standard. Land Econ 74(3):287–302

Governor's Commission to Review California Water Rights Law (1978) Final report. Sacramento

Green TR, Taniguchi M, Kooi H et al (2011) Beneath the surface of global change: impacts of climate change on groundwater. J Hydrol 405:532–560

Hardin G (1968) Tragedy of the commons. Science 162:1243–1248

Heal G, Schlenker W (2008) Sustainable fisheries. Nature 455:1044–1045

Herndon RL (2013) Personal interview regarding groundwater management in Orange County Water District, Orange County, California, December 9

Lippincott JB (1925) Report on water conservation and flood control on the Santa Ana river for Orange County, Los Angeles

Lipson AJ (1978) Efficient water use in California: the evolution of groundwater management in southern California. Rand Corporation report no. R-2387/2-CSA/RF, Santa Monica

Littleworth AL, Garner EL (1995) California water. Solano Press Books, California

Lorphensri O, Ladawadee A, Dhammasarn S (2011) Review of groundwater management and land subsidence in Bangkok, Thailand. In: Taniguchi M (ed) Groundwater and subsurface environments: human impacts in Asian coastal cities. Springer, Tokyo, pp 127–142
Orange County Water District (1970) Orange County Water Disrict (Annual report). Orange County Water District, Orange County, California
Orange County Water District (2009) Orange County Water District groundwater management plan update 2009. Orange County Water District
Orange County Water District (2015) Orange County Water District groundwater management plan update 2015. Orange County Water District. Available via https://www.ocwd.com/media/3622/groundwatermanagementplan2015update_20150624.pdf. Accessed 13 July 2017
Orange County Water District (2017) 2015–2016 Engineer's report on groundwater conditions, water supply and basin utilization in the Orange County Water District. Available via https://www.ocwd.com/media/5396/2015-2016-engineers-report.pdf. Accessed 13 July 2017
Ostrom E (1990) Governing the commons, the evolution of institutions for collective action. Cambridge University Press, New York
Ostrom E, Burger J, Field CB et al (1999) Revisiting the commons: local lessons, global challenges. Science 284:278–282
Pearce D, Atkinson G (1998) Concept of sustainable development: an evaluation of its usefulness 10 years after Brundtland. Environ Econ Policy Stud 1:95–111
Randall A, Farmer MC (1995) Benefits, costs, and the safe minimum standard of conservation. In: Bromley DW (ed) The handbook of environmental economics. Blackwell, Oxford, pp 26–44
Ruff LE (1970) The economic common sense of pollution. Public Interest 19:69–85
Sampson GS, Sanchirico JN, Roheim CA et al (2015) Secure sustainable seafood from developing countries. Science 348:504–506
Sandino DA (2005) California's groundwater management since the Governor's commission review: the consolidation of local control. McGeorge Law Rev 36(2):484–486
Schuerhoff M, Weikard HP, Zetland D (2013) The life and death of the Dutch groundwater tax. Water Policy 12:1064–1077
Senate Local Government Committee (2010) What's so special about special districts? A citizen's guide to special districts in California (fourth edition). Senate Local Government Committee, Sacramento
Tarlock AD (1985) An overview of the law of groundwater management. Water Resour Res 21(11):1751–1766
The World Commission on Environment and Development (1987) Our common future. Oxford University Press, Oxford
Wada Y, van Beek LPH, van Kempen CM et al (2010) Global depletion of groundwater resources. Geophys Res Lett 37:L20402. https://doi.org/10.1029/2010GL044571
Weschler L (1968) Water resources management: the Orange County experience. Institute of Governmental Affairs, University of California, Davis

Chapter 13
Utilization of Environmental Water Resources in the Reconstruction of Otsuchi Town After the 2011 Tsunami

Seiichi Mori

Abstract In Otsuchi Town, which suffered the effects of a tremendous tsunami in 2011, civil engineering reconstruction work is now progressing rapidly. In such situations, if the scale of a reconstruction project is excessive and assessments of the condition of the water environment are made too hastily, a burden in excess of that caused by tsunami damage itself might be placed on the aquatic ecosystem. In the Otsuchi case, concern has arisen about the deterioration of natural springs caused by the construction of embankments, landfill and floodgates, as well as the fact that the inland water has become disconnected from the sea due to the formation of a huge tidal breakwater. These aspects could possibly cause the degeneration of the spring water environment, thereby jeopardizing the region's principal "local blessing", namely its high-quality natural water resources. Responding to these challenges is not only a matter of preserving biodiversity, but is also a serious matter for the future development of Otsuchi Town, especially given its long history of enjoying the benefits of its natural water springs and the sea. In this context, there is a danger that any civil engineering reconstruction project might destroy the attractiveness of the region and cause disruption to future town management. Even 2–3 years after the earthquake, the plan for a so-called 'Reconstruction Park' in the disaster risk area was only a sketchy outline with no firm budget. However, since the early summer of 2014, the town administration has devised more concrete plans for the construction of embankments, floodgates, and tidal banks. A full-fledged study on the content of reconstruction work was initiated, based on the results of research carried out by local authorities. Having established a forum for the exchange of views among the local authorities and specialized researchers, I believe that Otsuchi Town's natural resources, its flora and fauna, and its water environment should be regarded as "local blessings". On that basis, I recommend in this chapter some rational and scientific contributions to town planning in the reconstruction of Otsuchi town.

S. Mori (✉)
Biological Laboratory, Gifu-keizai-University, Gifu, Japan
e-mail: smori@gifu-keizai.ac.jp

© Springer Nature Singapore Pte Ltd. 2018

A. Endo, T. Oh (eds.), *The Water-Energy-Food Nexus*, Global Environmental Studies, https://doi.org/10.1007/978-981-10-7383-0_13

Keywords Tsunami · Reconstruction · Spring water · Biodiversity · Local blessing

13.1 Introduction

On March 11, 2011, Otsuchi Town in Iwate Prefecture suffered severe loss of life and damage. Due to a tremendous tsunami, nearly 10% of the population perished, and over 85% of its urban area was devastated. This survey investigates how an important regional resource—spring water—has been used as the central factor in Otsuchi Town's reconstruction project. In addition, I propose a scenario that will contribute to the town's reconstruction by effectively conserving the biodiversity of its spring water environment. The local population has traditionally cultivated the land in the Otsuchi area, located on the Sanriku Rias coast, through the conservation and utilization of spring water. This relationship between spring water and the local residents can form the backbone of the reconstruction project. It is a relationship that should be maintained and handed down to future generations to guarantee a bright future for the area.

In its "Basic Plan for Reconstruction of Otsuchi Town from the Great East Japan Earthquake and Tsunami disaster," revised in March 2014, the Otsuchi Town administration established an important strategy for revitalizing landscapes as regional resources. The town is endowed with a beautiful, natural landscape, surrounded by greenish mountains and the rich sea. In other words, it has abundant spring water and a biodiversity-rich water environment, which is inhabited by the threespine stickleback, (*itoyo* in Japanese) and a myriad of other species. The town also has certain other distinctive properties, such as the coexistence of its natural environment with the activities of its local population. In this report, I define these properties as "local blessings" and examine the rebuilding of the community based on the utilization of these blessings. In association with other reconstruction projects, and based on an investigation of how best to utilize these local blessings which included meetings between local officials and researchers, this chapter summarizes the framework for a the building of a 'Reconstruction Park' that will improve the area while conserving and maintaining the town's 'local blessings'.

Otsuchi Town's 'local blessings': The term 'local blessings' refers here to things and events—local resources—that people regard as treasures within their local environment (Mori 2016). Otsuchi Town's local blessings are the spring water environments that its local residents use in their daily lives and as a production base for water services, agriculture, and as industrial water (Fig. 13.1). Before the disaster, Otsuchi Town had many wells and springs, whose waters flowed through a channel into its urban area. Throughout the year, the water temperature was stable at 10–13 degrees Celsius, and the water was highly valued because of its good quality. Artesian wells and springs also exist in Otsuchi Bay, located in a cove of the Rias coast. In fact, this whole range was known as the 'Otsuchi Burning Well and Spring

Fig. 13.1 Spring bath in a dirt floor kitchen which was found in many houses, Otsuchi Town, Iwate Prefecture (Aug. 30, 2008). A watermelon is being chilled in the water tub

Belt'. The area also has distinctive flora and fauna, such as the threespine stickleback (sympatric inhabiting anadromous and freshwater types) (Fig. 13.2) and the water crowfoot, a symbol of the area, that thrive in the spring water environment. As a result, rich biodiversity is fostered in the brackish waters where the seawater enters.

13.2 Brief History of Efforts in Otsuchi Town to Utilize Spring Water as a 'Local blessing'

Since conducting a hydrological survey with my colleagues in the 1990s, I have continued to pursue an ecological study of the threespine stickleback living in the spring water of Otsuchi Town. This research continued even in the immediate aftermath of the earthquake. My survey activities thus far have become part of the town's 'new history' vis-à-vis its spring water and its rare fish species, the threespine stickleback (Takamura and Mori 2005). Although this history is only about 20 years old, it is contemporary in terms of conservation and the utilization of regional resources, yet also consistent with the town's traditional history. It is also based on firm scientific evidence.

Fig. 13.2 The threespine stickleback group in a newly formed spring pond in the disaster area (Jun. 14, 2014). Photo by Hata Y

The official activities in Otsuchi town began with the "17th National Meeting for the Healthy Ocean" (1997) and a town-planning symposium entitled "Coexisting with Nature" (2002), in which members of the Imperial family also participated. Subsequently, a range of other activities, including research workshops at several universities (1999) have been carried out. In response to these activities, I conducted a survey on the actual situation of the threespine stickleback (freshwater type) of the Gensui River, a tributary of Otsuchi River, and led a number of study sessions, mainly in the Ogakuchi district. At the same time, local people established a volunteer group to help conserve the threespine stickleback (Mori 2002) and carry out daily environmental conservation activities (Fig. 13.3). Subsequently, an exhibition aquarium for threespine stickleback was installed in the entrance lobby of the government office in Otsuchi Town (although this office was destroyed by the 2011 tsunami, an exhibition aquarium has since been re-installed in the lobby of the new office (Fig. 13.4). In 2005, a threespine stickleback observation deck was built along the left bank of the Gensui River by the Public Interest Foundation of Riverfront Improvement and Restoration. In 2007, as an administrative response to the conservation effort, the threespine stickleback was designated as an endangered species and also as the town's 'Natural Monument'. In November 2010—about 4 months before the earthquake—a symposium on water environment and community development was held. In short, regional activities have been implemented continuously

Fig. 13.3 Conservation activities for the threespine stickleback (*Itoyo* in Japanese) and spring environment in the Gensui River, carried out by the residents of Ogakuchi district Otsuchi (Oct. 25, 2006)

Fig. 13.4 Aquarium to observe the threespine stickleback in the entrance lobby of the government office (Nov 27, 2016)

in Otsuchi Town with regard to the conservation of spring water and the threespine stickleback (Mori 2011).

As a reflection of these continuing activities, the "Association to Support the Reconstruction of Otsuchi Town" was established immediately after the earthquake in March 2011 (in which the author was a participant). This association continued to accept donations and material support from the whole country until 2016. In October 2011, the so-called "Roundtable Conference for Thinking about Past, Present, and Future of Otsuchi" was convened; and in April 2012 a tour was conducted to observe spring water (Mori 2012a). In May 2013, a spring water survey was planned with about 200 participants from inside and outside of Iwate Prefecture, and several results were released in various ways. Furthermore, in February and June 2014, two nationwide symposiums were held on the matter of reconstructing towns utilizing 'local blessings'. The author participated in almost all of the above events.

As mentioned previously, the spring water environment and biodiversity in Otsuchi Town including, for instance the situation regarding the threespine stickleback; see below have attracted attention as an object of conservation and utilization, not only among researchers in many fields, but also among the local authorities and residents of other areas. The close collaboration that has occurred among regions, as well as the deep human relationships that have developed, may themselves be regarded as irreplaceable regional resources, in other words, as additional "local blessings".

Some notes on the threespine stickleback: On the Japanese island of Honshu, the threespine stickleback is a fish regarded as a symbol of a clear stream: it can live only in spring water with a low temperature, i.e. under 20 °C. However, most populations are endangered, with habitats that are diminishing year by year (Kitano and Mori 2016). Downstream in the Otsuchi and Kotsuchi Rivers, the anadromous fish (the Japan Sea species) come in from the sea during the spring breeding season, but they do not crossbreed with the freshwater type. The ecological characteristics of the threespine stickleback are as follows: the male, which develops its marriage color during the breeding season, guards its territory, digs a hole at the bottom of the water, collects water grass roots, and makes a tunnel-shaped nest. The male invites females to his nest to lay eggs and then remains around the nest to care for the eggs. The stickleback's lifespan is about one year; many die after breeding, but some individuals occasionally survive and breed again in their second year.

13.3 Prerequisites for the 'Local blessing Utilization Project'

13.3.1 Planning Study Area

In this study, the "local blessing utilization area" is located in the Machikata area (old urban area) affected by the Great East Japan Earthquake, and, under reconstruction planning, it has become part of the disaster risk area (i.e. planning review

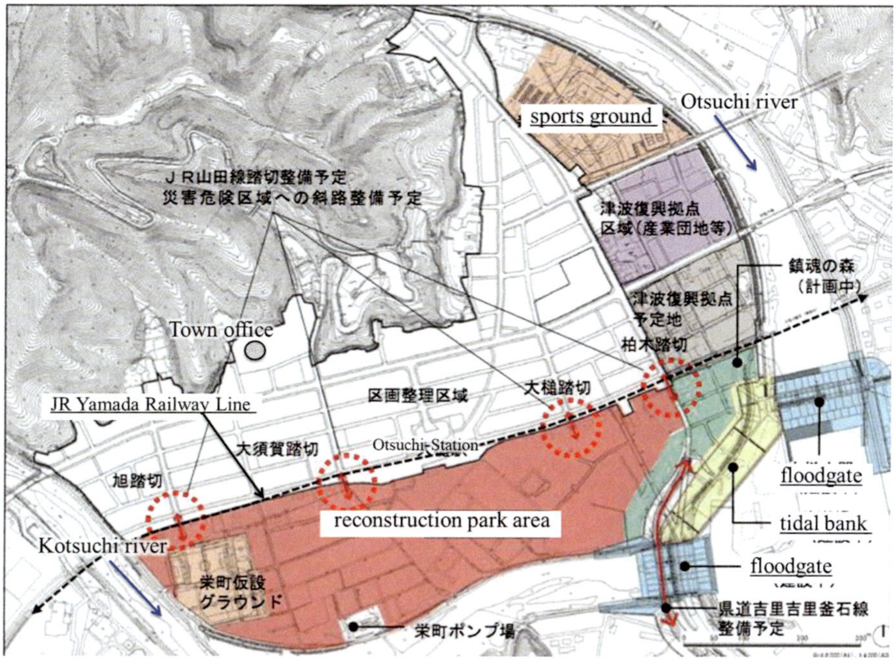

Fig. 13.5 Reconstruction study classification of the central urban area (of which disaster risk area) by town administration. In the Figure, the left side of the Otsuchi River is scheduled for sports grounds, industrial estates, etc., and the upper part of the Kotsuchi River is scheduled to be a planning review area, such as a Reconstruction Park, which is the subject of this research

area) to the south of the JR Yamada Line (Fig. 13.5). This area was densely populated before the earthquake disaster, but after the earthquake, it was designated a disaster risk area, making it impossible to construct houses there. Currently, this site is used as a construction staging area by the prefecture and the town's administration. In addition to the conservation area utilizing local blessings, a plan to use the site, probably for factories, is being considered. The northern side of the JR Yamada Line is an embankment, where, by conducting a land readjustment project, the area could form the site of the new city's center. As further projects related to local blessings, plans to construct a tidal breakwater (14.5 m in height) and water gates at the mouth of the river are being implemented.

13.3.2 *Superior and Related Plan*

As will be explained below, this project has a superior, related plan devised by the Administrative Execution Division of Otsuchi Town, which should be implemented in conjunction with these plans (Otsuchi Town 2011, 2013).

(1) Basic plan for reconstruction of Otsuchi Town (updated in 2014)

The basic aims of reconstruction policy for Otsuchi Town are "to arrange parks and transportation facilities, so as to surround the central city area" and "to make an attractive town through developing an aqueous network and lush greenery in the park."

(2) Otsuchi Town's Urban Master Plan (2014)

While properly securing access to this planned park and evacuation routes on the south of the JR Yamada Line, utilization of industrial sites, farmland, parks, and green spaces will be considered as effective utilization of inundation areas. In addition, other ideas are under consideration for the creation of beautiful landscapes using the town's abundant spring water.

(3) Comprehensive strategy for regional revitalization in Otsuchi Town (2016)

In expanding the interchange population, this strategy rethinks the town's resources and efforts for unique tourism. There are intentions to expand the interchange population, with the concrete goal of expanding it from 37,218 people in 2014, to nearly twice that (70,000) by 2019.

13.4 Present Conditions

13.4.1 A Waterway Running into Saltwater

At present, measurements of the area's ground height are based on a laser survey carried out by the Reconstruction Promotion Division of the Otsuchi Town government. The ground is low on the southeastern side, mostly lowland, less than 0.5 m above sea level (Fig. 13.6, blue-shaded area). Waterways A, C, D, and E run into seawater and become brackish areas (Fig. 13.6). Among them, waterways A and C are connected to the sea at the repaired ditch gate. Waterways D and E are to be constructed with tidal banks and floodgates, and the current plan makes it impossible to get into and out of the sea. Waterway B, which is blocked from connecting to the sea by a new embankment, is a stream of fresh water fed by the spring water supply. Maintaining an environment where the sea and the river interact is the minimum requirement for guaranteeing Otsuchi Town's distinguishing characteristic from the viewpoint of the natural environment and the rich spring water.

The blue-shaded areas in Fig. 13.6 were originally submerged at high tide. Plant communities, such as *Chima-dojotsunagi*, a rare species in the brackish area, grow naturally. Salinity is currently eluted at points W045 and W088, and saltwater moves upstream to these points at high tide. At point W1023, salinity is diluted to 1/20th of the seawater in the peripheral area where spring water emerges. Downstream, there is fresh water close to the surface, and the bottom layer is saltwater; that is, the degree of salinity differs between the waterway's upper and lower parts.

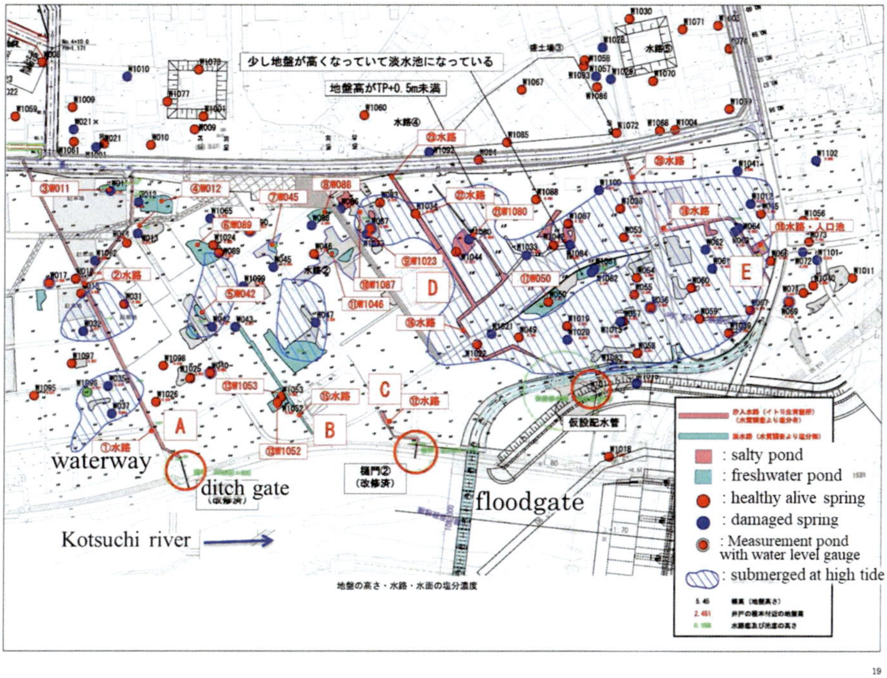

Fig. 13.6 Hydrographic map of a planning review area: spring spots (circles), waterways (A ~ E in square), flood area (blue-shaded area). Red circles indicate healthy alive spring water outlets, and blue circles indicate damaged spring water outlets

13.4.2 Flowing Wells and Spring Ponds that Appeared after the Earthquake

Before the tsunami disaster, many residents used self-injection wells on a daily basis. After many houses were washed away by the tsunami and a huge amount of debris was removed, many ponds were formed, as spring water spouted up from the earth in various places. The presence of approximately 200 artesian wells have been confirmed throughout the whole Machikata district.

The new spring water ponds can be divided into three types: completely independent ponds; flood ponds; and tidal ponds (Otsuchi Town 2014). A completely independent pond is a freshwater pond formed only by spring water, without any effect of saltwater run-up. The flood pond is not influenced by saltwater run-up, but it is a pond that floods during rainstorms and connects with a saltwater pond. The tidal pond is a brackish pond, affected temporarily by saltwater run-up.

Diverse environments have been formed through changes in the degree of mixing of spring water and saltwater due to natural-flowing amounts of spring water, the uphill flow of seawater at high tide, and flooding caused by rainfall. These watersheds form an environment in which the threespine stickleback is able to move freely through the waterways and floodwaters. This environment also causes hybrids of freshwater types and anadromous types, as described below (Mori 2013a).

13.4.3 Habitat of the Otsuchi Threespine Stickleback

Otsuchi Town's threespine sticklebacks (*itoyo*) are not only the freshwater population inhabiting the Gensui River. There are also two ecological populations of freshwater types (living only in freshwater areas) and anadromous types (coming and going between the sea and freshwater) that occur sympatrically; a rare phenomenon. This sympatric habitation is a very rare case, which occurs only in certain areas of Hokkaido in Japan, and has not been confirmed elsewhere in Honshu (Kitano and Mori 2016). Furthermore, Otsuchi Town is located in the southernmost sympatric habitat of the two types. This anadromous type belongs to the Japan Sea Group (*Nihon-itoyo*), which is an independent species. In addition, the presence of a Pacific Group was also once reported, despite the extremely small number of individuals. In some Otsuchi Town waters, genetically and ecologically diverse threespine stickleback groups live, and this biological situation is extremely important in terms of studying the process of biodiversity formation.

Furthermore, after the earthquake, a remarkable environmental situation occurred. The threespine stickleback was newly rediscovered in July 2012 in a pond where a newly formed spring and seawater had mixed in the old urban area (i.e. the disaster hazard area) where the ground had subsided (Mori 2012b). Initially, these fish were assumed to originate from a population of the *Nihon-itoyo* migratory group or a dispersed population from surrounding habitats. Estimation based on statistical analysis using external morphological traits and genetic analysis strongly suggests that this new threespine stickleback population was a hybrid of the anadromous type from the Japan Sea and the freshwater type from the Gensui River. Furthermore, according to analysis of the tsunami moving up the inland river and then pulling toward the sea, this new fish population entered the newly formed spring water pond in the afflicted urban area (Mori 2013a, 2013b). In short, these fish were a source of the Gensui River fish that had moved downstream, pulled along by the tsunami.

Changes in origins, genetic composition, ecology, and life history of this new threespine stickleback group are good subjects for elucidating organisms' adaptation processes in new habitats brought about as a natural phenomenon; in this case, by the tsunami. Furthermore, this situation will enable highly meaningful biodiversity studies in terms of how dramatic environmental changes, caused by the tsunami, have impacted the ecosystem and biodiversity (Mori 2013b). Prior to the tsunami, there was little opportunity for the study of such phenomena.

Fig. 13.7 Drainages from construction site of floodgate in the mouth of Kotsuchi River (Jul. 29, 2017)

13.4.4 Environmental Consideration of Spring Water

At the mouth of the Kotsuchi River, waterway gate construction began using the Super Well Point construction method in 2015. The rich groundwater that emerged during an excavation to a depth of several meters at the construction foundation has been drained into an outside river (Fig. 13.7). Due to that influence, very unfortunately, the disaster hazard area's groundwater level became markedly lower, and through 2016, the majority of the spring water ceased to flow spontaneously until August 2017 (Fig. 13.8). Current spring water does not flow out normally, but, during strong rainfall events, the groundwater level rises temporarily. Also, during several days in mid-August (during the Japanese holiday period known as *Obon*) when use of the Super Well Point method was stopped, the groundwater level rose and the spring water revived. Presumably, these phenomena indicate that when construction work is completed, spring water may flow back again and recover.

To judge from the results of this study, the poor situation, in which spring water became depleted, is improving somewhat. For example, in 10 spring water reservoirs regarded as important for aquatic creatures (including *itoyo*), groundwater is pumped artificially and distributed to the surface (Fig. 13.9), and the spring water environment is somehow maintained (Fig. 13.10). Those reservoirs are also used as a transplant site for rare plants such as *Mizuaoi* (*Monochoria korsakowii*) and *Takonoashi* (*Penthorum chinensis*). However, due to pump failure, two spring

Fig. 13.8 Completely dried-up spring water around the spring spot W053 (Mar. 16, 2016)

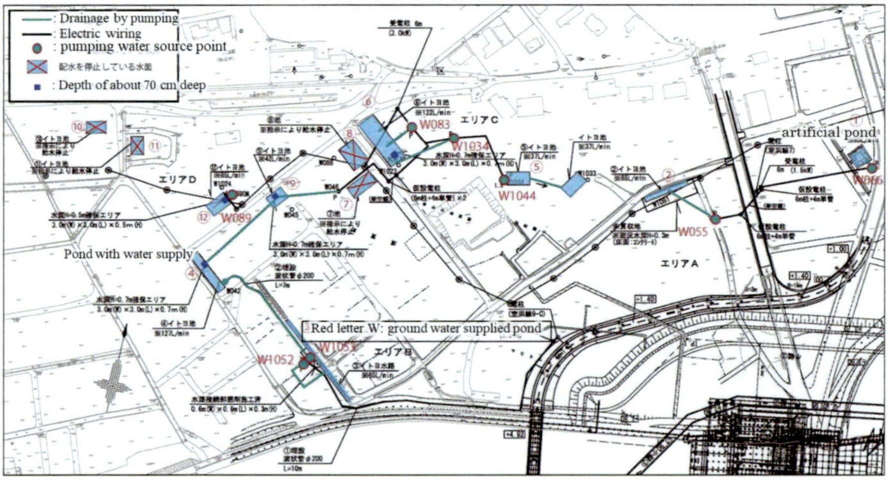

Fig. 13.9 Diagram showing pumped ground water. The ten green circles enclosed in red are the pumping water source points. The blue rectangles are supplied with pumped water

Fig. 13.10 Pond with spring water that was newly artificially created due to the mitigation of aquatic life conservation (Nov. 5, 2015). The blue and orange boxes beyond the pond are the pumping equipment. The depth was also artificially excavated in a rope enclosed

ponds became catastrophically exhausted. Especially in summer, the degree of influence on the spring water environment is great, so we, in collaboration with the prefectural civil engineering department, have been conducting occasional surveys and monitoring activities. After the completion of the foundation construction work in 2018, the local authorities will need to monitor how the spring water environment recovers and consider how to utilize it for the development of the future 'Reconstruction Park'.

Further, to prevent predation, for instance by herons, three spring water ponds, W042, W089, and W045, were dug to a depth of 50 cm to 70 cm in May 2016. This environmental measure has been extremely effective in securing the survival of aquatic animals such as *itoyo*. Although the construction of a tidal bank led to the landfilling of one spring pond in which *itoyo* had been living, a new, alternative spring pond was constructed in March 2016 by Iwate Prefecture Civil Engineering Bureau at a point about 30 m away from the landfill site. At the same time, in collaboration with prefectural officials and civil engineering contractors, we transplanted about 600 *Itoyo* individuals to the new pond (Fig. 13.11).

Fig. 13.11 Since the spring water spot W012 is landfilled, transplant activities (involving, for example, *itoyo*) have been carried out by town officials and civil engineering workers (Jun. 25, 2016). The white building in the top left of the photograph is Otsuchi Town Hall

13.4.5 Plant Conservation

This report focuses on spring water maintenance and the conservation of *itoyo* habitats, but the scope of the survey encompasses the entire spring water ecosystem, including its plant life. This section briefly describes the situation regarding plant conservation.

The situation regarding plant life was the subject of a report by the "Research on Spring Water Environment as a Local Blessing of Otsuchi Town" as a consignment project (Otsuchi Town 2014), and this has been supplemented by other research (Hoshino et al. 2015). These reports confirmed the presence of 266 plant species of 55 families in the whole area surveyed. The area also contains many species growing not only in spring wetlands, but also in sites such as empty lots and roadsides. Marshes, newly emerging in the old urban area, represent a diversity of environmental conditions, specifically: salt marsh; freshwater marsh; marsh through which spring water flows; and flooded marsh.

Due to the influx of seawater, in addition to the presence of spring water, various marshes—ranging from freshwater to brackish water marshes—have emerged, and diverse marsh plant communities have formed in different locations. Plants threatened with extinction have appeared in multiple communities and now grow throughout the marshes. Salt marsh plants in the conservation area of local blessings were not historically found in the main basin of the Otsuchi and Kotsuchi Rivers. After the earthquake, seawater flowed into the disaster area, transforming it into an environment suitable for salt marsh plants.

13.5 Basic Concept of a 'Local blessing Utilization Area'

13.5.1 Establishment of a Local Blessing Utilization Area

In Otsuchi Town, the sources of local blessings are the self-injection wells and the spring water, as well as the environment in which the water lies. To clarify this concept, I have designated the utilization area of local blessings as "the Otsuchi spring zone," and also established "a location to convey the importance of the rich local blessings of Otsuchi" to inform more people about Otsuchi Town's appeal (Fig. 13.12). In the future, it will be necessary to gain the support of local residents, while also widely disseminating information about the characteristics, actual conditions, problems, and utilization of the spring water environment in the Otsuchi Town

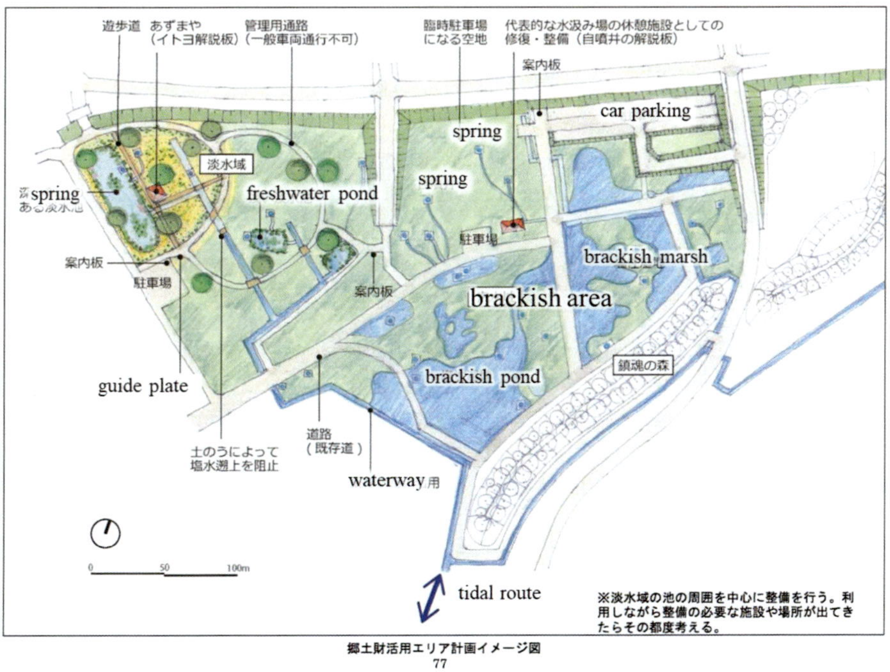

Fig. 13.12 Image drawing of local utilization area for the 'Reconstruction Park'. The water environment, consisting of abundant spring water (freshwater area) and natural flow seawater (brackish area), is a noteworthy feature

by establishing study groups, symposiums, and participatory workshops (Mori 2012a).

The project's target area is the minimum area for satisfying the following conditions, required for the utilization of the area's local blessings:

1) To obtain self-injection wells and spring water.
2) To ensure that brackish water and freshwater environments remain almost as they are.
3) To ensure that water bodies are interconnected in order to facilitate the utilization and management of areas that use local blessings.

Surveys of current ground height show that the ground to the northwest of the planned site is somewhat higher than before. Several freshwater ponds have been secured by pumping groundwater, as mentioned above, and these should henceforth be regarded as 'complete freshwater zones'. In contrast, the planned site's easternmost side remains low-lying and submerged at high tide; therefore, this area should be designated as a 'brackish water zone' (Fig. 13.12).

13.5.2 Inherited Memories of a Livelihood Based on Spring Water

This blessing of new, local spring water results from an environment born of the tsunami disaster. The presence of this spring, emanating from various places and forming ponds, is reminiscent of a time before the earthquake (Mori 2016). This former urban area's residual wells and the foundations of previously existing houses show that the spring water played an important part in the daily lives of many people.

To the extent possible, these places, traditions, and events need to be preserved and regarded as 'iconic' in order for people to understand the significance of where their predecessors lived and how they cultivated the regional culture, and also to remind them of the importance of regional disaster prevention (Terada 1948). At least as a source of basic data, it is very important to record information on geographical locations, ownership, as well as other facts about the remaining wells and springs. Moreover, it is important to preserve records of the water's use in the daily lives of people before the earthquake disaster.

13.5.3 Conservation and Utilization of Local Blessings

(1) Securing both freshwater and salt marshes Various environments which formed after the earthquake—such as freshwater ponds originating from spring water and salty marshes where salt water comes in from the sea—have allowed a characteristic diversity of flora and fauna to inhabit the area. It is desirable to continuously preserve these environments on a scientific basis (Otsuchi Town 2014).

With regard to plans for the future water network, the following points need to be emphasized: (1) water flow from the sea will be secured by utilizing current waterways, (2) the freshwater area and brackish water area will be connected via a network, (3) saltwater run-up will be blocked to secure a freshwater area. Based on the results of the present study, the difference in water level between the complete freshwater area and the saltwater upstream area is about 30 to 40 cm, and, by applying this height difference, the salt water will be blocked.

(2) Utilization of spring water and artesian wells Besides the proposed area for utilizing local blessings, there are many other springs and artesian wells within the town's disaster/hazardous area. We need to consider development that also maximizes the use of these water sources. Furthermore, we need to appeal for the utilization of the 'Otsuchi Spring Zone' for a variety of other purposes, such as sightseeing and ecotourism. Moreover, this spring water will be developed as "disaster prevention water resource" in the future. On this basis, it will also play an important role in the contemporary life of the local community.

Currently, there is a proposal for the construction of a water drawing station, a parking lot and other facilities around the spring spot W053 (see Fig. 13.8), which was often used before the tsunami, with the purpose of making a base for a spring water tour. Moreover, in June 2014, this spring was visited by Fumihito, Prince Akishino and Princess Mako, who rinsed their mouths with the spring water.

13.6 Moving Forward in Otsuchi Town

As the civil engineering reconstruction project has progressed, continuous attention has been paid to Otsuchi's 'local blessings', and tasks and propositions for conserving and maintaining them have been pursued in consultation with local authorities and administrative legal entities. Immediately after the earthquake occurred, we began conducting hydrology and ecological research activities with support from Otsuchi Town. Moreover, since that time, we have held symposiums and study sessions several times a year in order to disseminate the results of our research.

This chapter offers a framework for the future administrative management of the subject area, based on rational grounds and the active collaboration of local residents and administrative organizations (Miyadai 2009). At the same time, it is important for the community to foster individuals, including children, and activity groups, who will focus their activities on Otsuchi Town's 'local blessings'. Under a system managed by the residents themselves, it will be necessary continuously to enlighten the population and disseminate information on activities geared towards the utilization of these local blessings, while also developing activities of benefit for the whole country. As a platform for preserving the local blessings of Otsuchi town, it is sincerely hoped that the measures and solutions advocated in this report will become imbued in the public consciousness, and hence provide a positive example for communities elsewhere the world.

Acknowledgements I am deeply grateful to the people of Otsuchi town, including some acquaintances who passed away due to the tsunami; to K. Sasaki and S. Nishida for supporting the fieldwork and providing with me with a lot of local information; and to T. Akimichi, T. Sumi, J. Kitano and M. Taniguchi for specific discussions during this study. I also thank the staff of Otsuchi Town office and executive officers of the civil engineering department of Iwate Prefecture for the valuable information exchange on some reconstruction projects, and P. Borg for revising the initial manuscript. This work was supported by the Environment Research and Technology Development Fund (ZD-1203), Ministry of Environment Japan.

References

Hoshino Y, Hoshino J, Fukamachi A (2015) Vegetation and plant community of marshes arising in disaster areas of the great east Japan earthquake tsunami. Biostory 23:78–83. SeibundoSinko-sha, Tokyo

Kitano J, Mori S (2016) Toward conservation of genetic and phenotypic diversity in Japanese sticklebacks. Genes Genet Syst 91:77–84

Miyadai S (2009) Difficult task in Japan. Gento-sha shinsho, Tokyo

Mori S (2002) Conservation of the threespine stickleback, Itoyo, in Otsuchi river system. In: Akimichi T (ed) Wild nature and local community. Showa-do Publishing, Kyoto, pp 95–115

Mori S (2011) Conservation of freshwater fishes as utilization of local blessings: a case study of the threespine stickleback, Itoyo, in Otsuchi town. In: Akimichi T (ed) Nature, water and people in Otsuchi. Publishing Plan of Tohoku, Tsuruoka City, pp 272–323

Mori S (2012a) Narrative approach of living thing: nature and people in this country. Biostory 17:69–76. SeibundoSinko-sha, Tokyo

Mori S (ed) (2012b) Biostory as blessings and natural disasters. Publishing Plan of Tohoku, Tsuruoka City

Mori S (2013a) Dispatch from the three-spined stickleback, Itoyo, of Otsuchi spring waterbodies overcoming the earthquake tsunami. Biostory 20:79–85. SeibundoSinko-sha, Tokyo

Mori S (2013b) The Otsuchi three-spined stickleback: status following the 2011 East Japan tsunami. Ichthyol Res 60:177–180

Mori S (2016) Guidelines for rebuilding town development through conservation of spring water fish, Itoyo, as a local blessing. In: Taniguchi M (ed) Ground design for the future from Otsuchi. Showa-do Publishing, Kyoto, pp 71–95

Otsuchi Town (2011) Otsuchi Town reconstruction plan for east Japan great earthquake disaster tsunami- basic plan. Otsuchi town office, 84pp. http://www.town.otsuchi.iwate.jp/docs/2012021500290/. (Refer. 2013–5-23)

Otsuchi Town (2013) Publicity papers Otsuchi, No.560 (5 June 2013). General Policy Division, Otsuchi town office, 16pp. http://www.town.otsuchi.iwate.jp/docs/2013061100020/. (Refer. 2013–6-23)

Otsuchi Town (2014) Study on the spring water environment as a local blessing of Otsuchi Town. The research project of the threespine stickleback and spring in Otsuchi, Iwate Prefecture. Board of Education, Otsuchi town office, 117pp. http://www.town.otsuchi.iwate.jp/gyosei/docs/2014042200012/files/hokoku.pdf

Takamura K, Mori S (2005) Heterozygosity and phylogenetic relationship of Japanese threespine stickleback (*Gasterosteus aculeatus*) populations revealed by microsatellite analysis. Conserv Genet 6:485–494

Terada T (1948) Natural disasters and national defense. In: Torahiko Terada essay writing, vol 5. Iwanami shoten, Tokyo, pp 56–66

Chapter 14
Experience of Disaster and Recognition of Local Re-Sources: A Survey of a Tsunami-Damaged Town in Japan

Takaaki Kato, Ayano Yanagihara, and Aiko Endo

Abstract Otsuchi Town, Japan, suffered significant damage due to the large tsunami in 2011. Rich groundwater and an ecosystem that contains several rare species characterized life in the town before the tsunami. Attracting more visitors to the town is necessary to sustain its economy going forward. This study aimed to identify factors that would increase visitors to the town. A survey of 6262 individuals, residing in nine locations in East Japan and sampled from a survey panel of an Internet research firm, showed that providing learning opportunities with regard to the local ecosystem and disaster recovery activities in Otsuchi would increase individual visitors from those locations by one to six percentage points. The increased publicity of the town after the tsunami is a key driving force in drawing visitors. The provision of information regarding learning opportunities is influential in attracting visitors who have an extensive interest in traveling and learning. These visitors enjoy experiences such as tasting water from natural resources, learning about different ecosystems, and participating in cultural activities. People who are uninterested in traveling or who have a limited range of interest in natural water, ecosystem, or culture are, however, not influenced by such information.

Keywords Disaster recovery · Tsunami · Local ecosystem · Tourism · Otsuchi

T. Kato (✉)
Faculty of Environmental Engineering, The University of Kitakyushu, Kitakyushu, Japan
e-mail: tkato@kitakyu-u.ac.jp

A. Yanagihara
Graduate School of Environmental Engineering, The University of Kitakyushu, Kitakyushu, Japan

Present Affiliation: Nishinippon Plant Engineering and Construction Co., Ltd, Fukuoka, Japan

A. Endo
Research Institute for Humanity and Nature, Kyoto, Japan
e-mail: a.endo@chikyu.ac.jp

© Springer Nature Singapore Pte Ltd. 2018

A. Endo, T. Oh (eds.), *The Water-Energy-Food Nexus*, Global Environmental Studies, https://doi.org/10.1007/978-981-10-7383-0_14

14.1 Introduction

The Great East Japan Earthquake and the subsequent tsunami on March 11, 2011, caused disruption to the coastal town of Otsuchi, Japan. Of Otsuchi's population of 15,000 at the time of the earthquake, 1285 died or were pronounced missing. The tsunami critically damaged 4167 houses, and the central area of Otsuchi almost vanished (Otsuchi Town 2015). The tsunami reached a height of more than 20 meters and this was much larger than the height indicated by early warning information. Lack of a clear lowered tide, which was believed to typically appear before the arrival of a large tsunami, was said to have hampered early evacuation from lowland areas (Kahoku Shinpo 2011). Otsuchi is undergoing the long process of rebuilding its community. Impacts of the disaster are not limited to human society. The local ecosystem of the Otsuchi area suffered great impacts from the tsunami as well. The central area of Otsuchi, which is located between the Otsuchi and Kozuchi rivers, has rich groundwater resources. Clear and cool water from springs is a key element of sustaining the local ecosystem in the area. Among the rare species of the ecosystem is the three-spined stickleback (*Gasterosteus aculeatus*). Major habitats of three-spined sticklebacks were heavily maimed by the tsunami and covered by thick rubble. The marine ecosystem in nearby Otsuchi Bay and Funakoshi Bay were heavily marred as well.

Otsuchi lost many during the disaster, but people who survived and the ecosystem in the area are transitioning in new directions. This transition can provide many learning opportunities for both the people in Otsuchi and visitors to the town. The damage to the habitats of the three-spined stickleback due to the tsunami provides scientific opportunities to understand the habitat conditions and the life of the fish (Kitano and Mori 2016; Mori 2013). Many volunteers from throughout the country rapidly and effectively worked toward the recovery of the tsunami-damaged habitats of the fish. This is a notable example of a successful ecosystem recovery activity after a major disaster. There are many unique learning opportunities in this town after the tsunami disaster.

Otsuchi was not a big tourist attraction before the tsunami, but the wide media coverage of the tsunami damage in 2011 familiarized the name of the town across Japan and outside the country. Thus, tourism could become an important element of the town's economy if the town could create and maintain attractive learning opportunities for visitors. In the short run, however, post-disaster tourism may cause controversy since the main attraction for the tourists would be the destroyed houses and landscape that victims of the disaster would not want to recollect. This type of controversy among tourists, tour operators, residents, and local governments was highlighted in the post-disaster tours of the damaged areas due to Hurricane Katrina in the United States (Robbie 2008; Gould and Lewis 2007). However, disaster sites can be a place of learning as well, as Miller (2008) finds in disaster tourism after Katrina. The objective of our research was to identify factors that would increase visitors to the town. In particular, we investigated the potential of providing learning opportunities on the tsunami-damaged local ecosystem and disaster recovery activities of the local community to attract more visitors to Otsuchi.

14.2 Survey

An Internet-based questionnaire survey of citizens in 14 cities in East Japan was conducted in February 2016. Figure 14.1 shows the locations of Otsuchi and those surveyed cities. Table 14.1 summarizes the collected sample size for each city. The

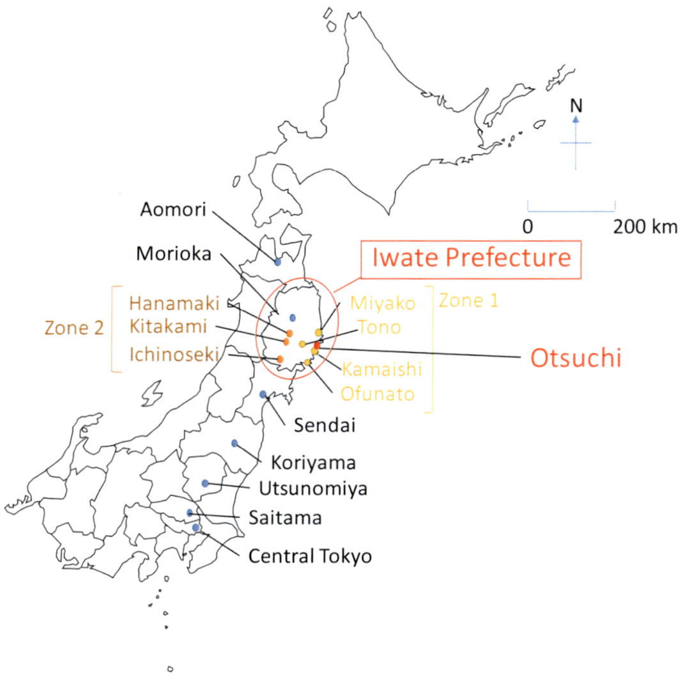

Fig. 14.1 Otsuchi and surveyed cities

Table 14.1 Survey samples

Site	Road distance to Otsuchi	City population (thousand persons)	Targeted sample size (persons)	Sample size (persons)	
				Treatment 1	Treatment 2
Zone 1	10-45	160	130	140	0
Zone 2	85-117	313	150	154	0
Morioka	125	298	500	249	251
Aomori	297	288	500	278	282
Sendai	228	1,082	750	411	413
Koriyama	341	336	500	223	225
Utsunomiya	462	519	750	357	355
Saitama	551	1,264	750	406	419
Central Tokyo	583	9,273	2,000	1,054	1,045
Total	-	13,533	6,030	3,272	2,990

Table 14.2 Gender and age

Age group	Female	Male	Total
18–29	731	426	1157
30–49	1277	1447	2724
50 and above	1035	1346	2381
Total	3043	3219	6262

cities within Iwate Prefecture, except for its capital of Morioka, were grouped into two zones, since sample sizes taken from those cities were small. These samples were collected from the pre-registered panel members of a large Internet survey company, Rakuten Research, Inc. This survey panel was commercially operated by the company and subjected to various surveys including ours. The survey company sent an invitation email to those panel members over 17 years old and collected answers from them in the last week in February 2016. We set up target sample sizes for each location as shown in Table 14.1. The survey company collected answers until they accomplished the target number for each site. For some sites, it was not possible to reach the targets. For other sites, actual sample sizes exceeded the target sizes, since the survey company collected some additional responses when it was possible. The ratios of responses relative to the number of invitation emails sent to the panel members were between 4% and 11% depending on the survey site. These percentages are not comparable with response ratios of traditional social surveys, since Internet surveys intend to collect the required number of responses in a very short period by sending many invitations. Table 14.2 summarizes gender and age distributions of the sample. The shares of respondents aged 50 years old and above among all the respondents were close to the population statistics in zone 1, zone 2 and Aomori. These respondent shares were larger than their corresponding population statistics by 7% to 14% in other survey sites. Thus, our sample overrepresented older generations in the survey sites except for zone 1, zone 2 and Aomori. The samples were divided into two treatment groups. All the respondents within Zones 1 and 2 were assigned to treatment group 1. Those from outside of Zones 1 and 2 were randomly assigned to either treatment group 1 or treatment group 2. Different questionnaires were sent to each treatment group, as we will explain later.

The survey questionnaire consists of questions regarding attitudes toward travel, familiarity with the Otsuchi area, attitudes towards natural water and ecosystem, attitudes toward rural development, personal attributes, and intentions to visit Otsuchi. The visiting intention question had two stages. First, general information on the town, including location, population, damage caused by the tsunami, sea food, and ecosystem, was briefly provided to the respondents. The information on the ecosystem mentioned the existence of the rare three-spined stickleback in the springs and streams in the town, but it also mentioned the lack of opportunities for

Fig. 14.2 Proposed opportunities to learn the local ecosystem

tourists to learn about how the tsunami affected the local ecosystem and the life of three-spined sticklebacks. Immediately after providing this general information, we asked respondents if they would visit Otsuchi within the next 3 years. Then, the second-stage question began by proposing opportunities to learn about the local ecosystem and disaster recovery activities. Treatment group 1 received the information shown in Fig. 14.2. This information included details about ecosystem museums of three-spined sticklebacks and marine creatures in Otsuchi Bay. Opportunities to avail guided ecotours, to learn about the impact of the tsunami on the local ecosystem, were mentioned as well. Treatment group 2 received the disaster recovery activity information shown in Fig. 14.3. This information included learning opportunities on the diversity of recovery needs across neighborhood communities and the recovery of the fishery industry. Opportunities to join guided tours to learn about the culture of Otsuchi, social impacts of the tsunami, and difficulties of decision making in disaster recovery were mentioned as well. After receiving the information, each respondent from both the treatment groups rethought and answered again if he/she would visit Otsuchi within the following 3 years.

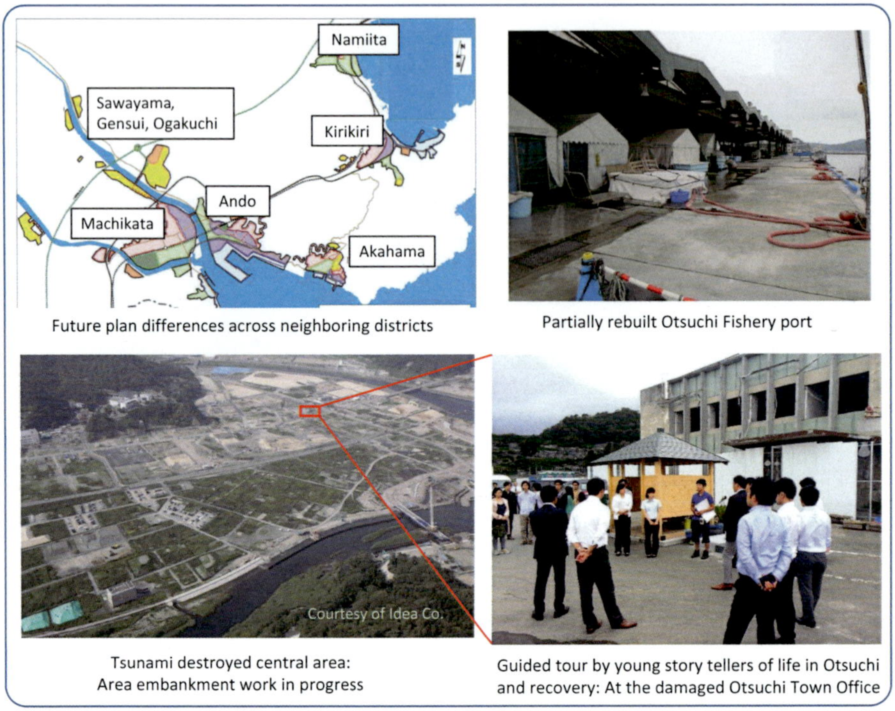

Fig. 14.3 Proposed opportunities to learn about disaster recovery activities

14.3 Characteristics of the Respondents

Figures 14.4 and 14.5 summarize the levels of knowledge about Otsuchi across survey locations. The survey locations on the horizontal axis are listed from the nearest to the farthest from Otsuchi. Figure 14.4 shows that basic knowledge on geography, disaster record, and population declines as the distance from the town increases. Figure 14.5 shows the respondents' knowledge on Otsuchi's ecological and industrial characteristics and Hyoutanjima Island, the town's landmark. Knowledge about these characteristics becomes significantly low outside of Iwate Prefecture.

We then categorized the respondents depending on their preferences regarding traveling, natural water, ecosystem learning, and everyday food selection. We used the k-mean clustering method and divided the respondents into five groups. The k-mean method summarizes respondent attributes and divides respondents into several homogeneous groups. Among various methods of clustering, the k-mean method is commonly used for the analysis of a large sample as in the case of this study. The k-mean method requires the number of clusters to be pre-determined. The objective of this analysis was to make clusters that were clearly different from each other in terms of travel and learning attitudes. Thus, we tried three-, four-, and

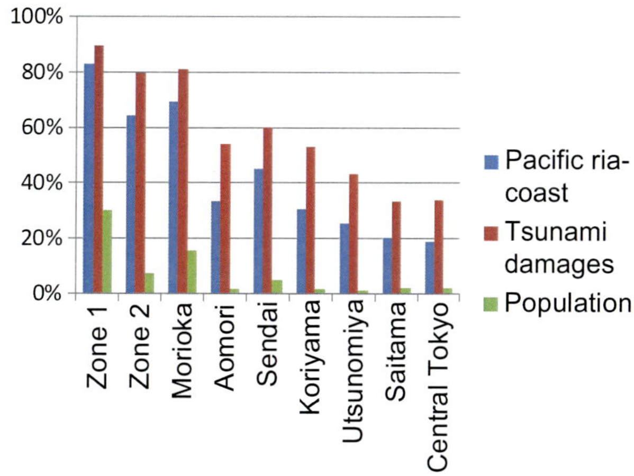

Fig. 14.4 Knowledge about Otsuchi (general information)

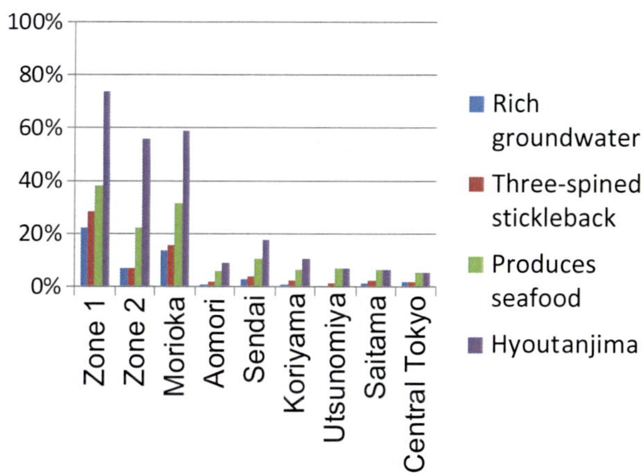

Fig. 14.5 Knowledge about Otsuchi (ecosystem and industry)

Table 14.3 Categorization of respondents

Category	Observation (persons)	Typical personal attributes
Natural water, ecosystem and culture lover	1,191	Suburb
Natural water and ecosystem lover	1,905	Elder
Natural water lover	1,401	Male, middle aged
Natural water avoider	1,597	Female, suburb
Non-interested	1,225	Male, young, urban center

five-cluster cases and found that the five-cluster case created clusters that were the clearest to interpret. Table 14.3 lists these clusters and the number of respondents assigned to each cluster. The k-mean method identified a set of respondent attributes that were typically found in each cluster. For example, the respondents categorized in the first three clusters in Table 14.3 liked to drink and use natural water, while those in the fourth cluster did not like to do so. Those in the fifth cluster did not show a clear preference for or against natural water use. We summarized this kind of information obtained from the k-mean method over a range of attributes and defined the names of the clusters. Preference for natural water plays an important role in creating the clusters. Among the respondents, the natural water lover is one who likes the taste of water from natural springs or wells. The natural water avoider is cautious about drinking such water due to health concerns. The ecosystem lover likes to learn about different ecosystems in field tours, and the culture lover likes to explore local culture in travel destinations. The final category, the non-interested group, does not show clear preferences with regard to traveling, natural water, ecosystem, or culture. The third column of Table 14.3 shows typical personal attributes seen in each category of the respondents. The attributes examined are gender, age, and the characteristics of the place they were brought up in. The characteristics of the place had three categories consisting of nature-rich, suburbs, and urban center, and respondents chose one from these. Many of the natural water, ecosystem, and culture lovers were brought up in suburbs, while the non-interested respondents were brought up in urban centers.

14.4 Intention to Visit Otsuchi

Figure 14.6 shows the percentages of respondents who answered in the affirmative when asked if they would visit Otsuchi within 3 years from the survey in each survey location. The circle markers in the figure show the percentage after providing the baseline information regarding Otsuchi to treatment group 1. The visiting

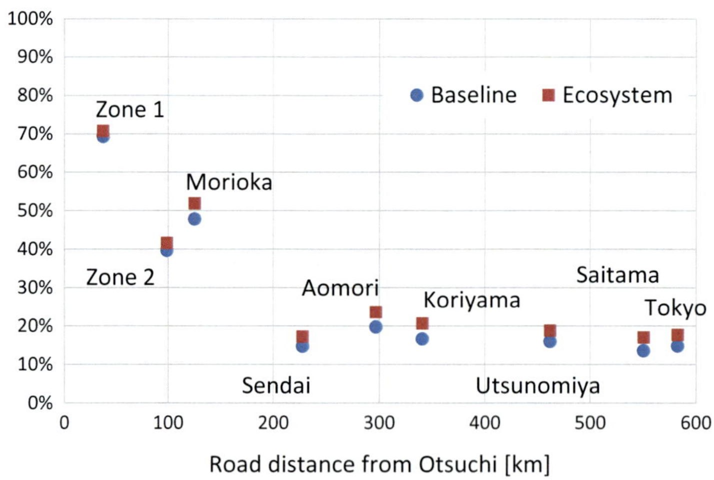

Fig. 14.6 Ecosystem learning and the intention of visiting Otsuchi

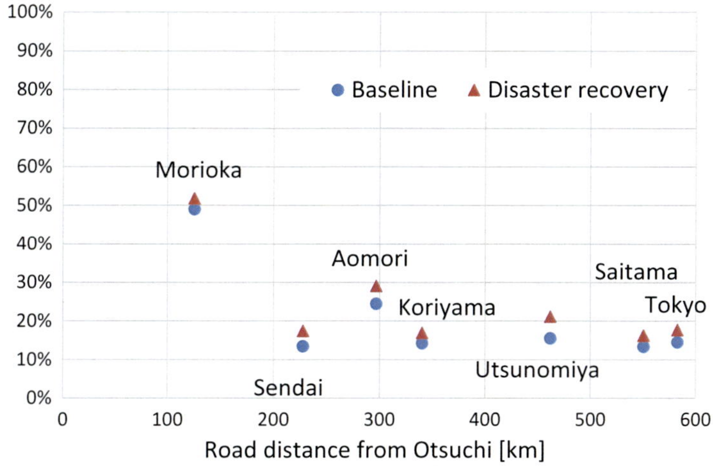

Fig. 14.7 Disaster recovery learning and the intention of visiting Otsuchi

percentages display a downward trend as the distance from Otsuchi increases. This is a natural relationship as the cost of visiting the town increases in accordance with the distance. Morioka's percentage is higher than the trend and Sendai's percentage is lower than that. We will later explain the reason for Morioka's high visiting percentage. Sendai is near the tsunami damaged areas of Pacific ria-coast, which have terrain similar to that seen in Otsuchi. The presence of these alternative destinations might be causing the low visiting percentage from Sendai to Otsuchi. The square markers show the percentages after gaining information on the availability of ecosystem learning opportunities in the town. The visitors' percentages increased by one to four percentage points depending on the survey locations.

In Fig. 14.7, the circle markers show the visitors' percentages after providing the baseline information to treatment group 2. Respondents within Zones 1 and 2 were not subjected to this treatment. Visitors' percentages at the first stage are similar to the ones found in Fig. 14.6 and display a downward trend with respect to the distance from Otsuchi. The triangular markers in the figure show the visitors' percentages after gaining information on the availability of disaster recovery learning opportunities in the town, and the visitors' percentages increased by three to six percentage points.

We then examined the respondents' attributes that would have an impact on their intention to visit Otsuchi. Table 14.4 shows the logistic regression result for visiting intention after the provision of the baseline information in the first-stage question. Both the treatment groups were pooled together for this analysis, since they were provided with the same baseline information. The binary dependent variable is the visiting intention: to visit or not to visit. Independent variables include generalized cost, respondents' relationship with Otsuchi, and respondent types. The generalized cost considered the monetary cost of travel by public transport and the time required

Table 14.4 Baseline information and visiting intention in the next three years

Variable		Coefficient		Odds ratio
Generalized cost (10 thousand yen)		-0.325	**	0.723
Relationship with Otsuchi				
	Past dwelling	1.932	**	6.906
	Past visit	1.785	**	5.959
	Past dwelling by family or acquaintance	0.716	**	2.047
	Family or acquaintances born in Otsuchi	0.818	**	2.265
	Media coverage	0.717	**	2.048
	Purchase product	0.878	**	2.406
	Donation	1.712	**	5.539
Respondent type				
	Natural water, ecosystem and culture lover	1.131	**	3.098
	Natural water avoider	0.339	*	1.404
	Natural water and ecosystem lover	0.686	**	1.986
	Natural water lover	0.687	**	1.987
Constant		-2.016	**	0.133
Sample size		6,262		
Pseudo R squared		0.206		

* and ** = 5% and 1% statistical significance

for the travel from each respondent's house to Otsuchi. Table 14.4 shows the results after statistically insignificant variables at the 5% significance level, were removed from the model. The generalized cost variable has a negative coefficient, meaning that as the generalized cost increases, fewer people plan to come to Otsuchi. Experience of past dwelling and visiting and donations by the respondents largely increase their intention to visit Otsuchi. The odds ratio shows that the odds of visiting Otsuchi among those who had visited the town before is around 6 times larger than those who had not visited there. Having family members and acquaintances in Otsuchi increases the visiting intention. Thus, a strong personal connection with the town is a key factor for visiting the town in this first-stage question without the information on learning opportunities. Existence of strong personal connections with Otsuchi explains the high visiting percentage in Morioka shown in Fig. 14.6. Morioka is the administrative and economic center of Iwate Prefecture and has strong human connections with cities and towns including Otsuchi in the prefecture. For example, 53% of the Morioka respondents had visited Otsuchi before our survey and this percentage is higher than 40% in Zone 2, although distance from Otsuchi is longer to Morioka than to Zone 2. Being familiar with Otsuchi through media coverage or having purchased the town's products increases the visiting intention, as well. Compared with the visiting intention of the "non-interested" group, each of the four other groups displays a stronger intention to visit Otsuchi. Among them, the "natural water, ecosystem, and culture lover" has the strongest intention to visit the town.

Table 14.5 Ecosystem learning opportunity and visiting intention in the next three years

Variable	Coefficient		Odds ratio
Relationship with Otsuchi			
Purchase product	1.130	**	3.472
Respondent type			
Natural water, ecosystem and culture lover	1.348	**	4.162
Natural water and ecosystem lover	0.579	*	1.847
Constant	-3.738	**	0.024
Sample size	2,568		
Pseudo R squared	0.048		

We then investigated the types of respondents who responded after the provision of the additional information on learning opportunities with regard to the local ecosystem and disaster recovery activities in Otsuchi. Table 14.5 shows the logistic regression results of these respondents, who were provided with the information shown in Fig. 14.2. The subjects of this analysis are those from treatment group 1 who did not indicate a desire to visit Otsuchi in the first-stage question with the baseline information. The binary dependent variable is the visiting intention after obtaining the additional information on the ecosystem learning opportunities: to visit or not to visit. We considered independent variables such as generalized cost, relationship with Otsuchi, and respondent types. Table 14.5 lists the variables that achieved the 5% statistical significance level. The generalized cost is no longer an important factor for determining visiting intention in this second stage. The experience of purchasing the town's products increased the intention to visit Otsuchi. The odds ratio shows that the odds of a visit by a respondent who had purchased the town's products was 3.5 times larger than that by a respondent who had not done so. Among the respondent types, "natural water, ecosystem, and culture lover" and "natural water and ecosystem lover" showed a stronger intention to visit the town relative to the "non-interested" category. The odds of a visit by a "natural water, ecosystem, and culture lover" is 4.2 times larger than that by a "non-interested" respondent. The visiting intention of "natural water lover" and "natural water avoider" does not differ from the "non-interested" category. Thus, the additional information on the ecosystem learning opportunities encouraged the groups who have an extensive interest in natural water, ecosystem, and culture, to visit the town.

Table.14.6 shows the logistic regression results of the respondents who were provided with the information on disaster recovery learning opportunities shown in Fig. 14.3. The subjects of this analysis are those from treatment group 2 who did not indicate a desire to visit Otsuchi in the first-stage question. We considered independent variables such as generalized cost, relationship with Otsuchi, and respondent types. Table 14.6 lists the variables that achieved the 5% statistical significance level. Being aware of the town through media coverage and the experience of purchasing the town's products increased the intention to visit the town. The odds of a visit by a respondent who had purchased the town's products

Table 14.6 Disaster recovery learning opportunity and visiting intention in the next three years

Variable		Coefficient		Odds ratio
Relationship with Otsuchi				
	Media coverage	0.525	*	1.691
	Purchase product	1.383	*	3.985
Respondent type				
	Natural water, ecosystem and culture lover	0.691	**	1.996
Constant		-3.308	**	0.037
Sample size		2,433		
Pseudo R squared		0.021		

* and ** = 5% and 1% statistical significance

was around 4 times larger than that by a respondent who had not done so. Product purchasing experience is important for both the ecosystem information case shown in Table 14.5 and this disaster recovery information case. Among the respondent types, only the "natural water, ecosystem, and culture lover" showed a stronger intention to visit the town relative to the "non-interested" respondents.

14.5 Conclusions

We investigated factors for attracting more visitors to the tsunami-damaged town of Otsuchi by providing learning opportunities with regard to the local ecosystem and disaster recovery activities. We focused on individual travelers who decide their destination themselves. Our survey of the residents of nine locations outside Otsuchi revealed that providing such opportunities increases the visiting intention of the respondents by one to six percentage points. Although these percentage gains are relatively small in percentage terms, we can expect many new visitors, given the large population of the survey sites, as shown in Table 14.1. Thus, providing learning opportunities regarding the local ecosystem and disaster recovery activities should be seriously considered as policy options for initiating and sustaining a tourism-based economy in Otsuchi.

The respondents whose visiting intention strengthened because of the additional learning-opportunity information tended to have indirect ties with the town through their knowledge of the town via the media and the experience of purchasing the town's products. Thus, as we noted previously, increased media publicity of the town after the tsunami was a significant factor in attracting more visitors. This may be a race against time, since the media coverage of the town is expected to decrease as time passes. The learning-opportunity information is influential in attracting visitors who have an extensive interest in traveling and learning. These visitors enjoy experiences such as tasting water from natural resources, learning about different ecosystems, and participating in cultural activities. People who are uninterested in traveling or who have a limited interest in natural water, ecosystem, or culture are, however, not influenced by such information. It is noteworthy that preference regarding natural water plays a significant role in segmenting the population. Otsuchi's rich groundwater environment is impressive to them. Making the learning-opportunity information, including information on the town's groundwater, available to the appropriate segments of the population is an important step for Otsuchi Town to gain more visitors.

Another category of visitors that Otsuchi can accommodate is group travelers who are sent to the town for an educational purpose by companies and schools. Since the destinations of these groups are detemined by the decision makers of their organizations, it is important to provide educational programs that are unique and effective. In concluding this chapter, we would like to introduce an important activity seen in Otsuchi. Oraga Otsuchi Yumehiroba, a local non-government organization, provides innovative leadership educational programs. In these programs, citizens in Otsuchi teach the difficult art of collective decision making and the importance of leadership, by referring to the actual cases of resource allocation and decision making that people in the town encountered during the tsunami disaster and by referring to cases they face in the present rebuilding stage of the town (Project Design Editorial 2016). The content of the education evolves as the issues of rebuilding the town change. Thus, they can provide new content to attract not only newcomers but also repeat participants of the education program. Combining this approach of attracting group visitors with the approach of drawing individual visitors would be an effective way to increase the total number of visitors to Otsuchi.

References

Gould KA, Lewis TL (2007) Viewing the wreckage: eco-disaster tourism in the wake of Katrina. Soc Without Borders 2(2):175–197

Kitano J, Mori S (2016) Toward conservation of genetic and phenotypic diversity in Japanese sticklebacks. Genes Genet Syst 91(6):77–84

Miller DS (2008) Disaster tourism and disaster landscape attractions after hurricane Katrina: an auto-ethnographic journey. Int J Cult Tour Hosp Res 2(2):115–131

Mori S (2013) The Otsuchi three-spined stickleback: status following the 2011 East Japan tsunami. Japanese. J Ichthyol 60:177–180. (in Japanese)

Project Design Editorial (2016) Saisei e, Hisaichi o manabino machi ni (reconstruction of community: learning in a disaster damaged town). Proj Des Jan 2016:122–123. (in Japanese)

Robbie D (2008) Touring Katrina: authentic identities and disaster tourism in New Orleans. J Herit Tour 3(4):257–266

Kahoku Sinpo (2011) Goshin Higeki (tragedy caused by misunderstanding), 1 May 2011, Kahoku Shinpo Newspaper (in Japanese)

Otsuchi Town (2015) Higashi-Nihon Daishinsai tsunami: Otsuchicho Hisai Gaiyou (estimation of damage to Otsuchi Town by the great East Japan earthquake tsunami). Otsuchi Town (in Japanese)

Chapter 15
Serious Matters Related to Development of Small-scale Geothermal Power Generation in Beppu-Onsen Hot Spring After 2011

Shinji Ohsawa

Abstract Immediately after the accident at the Fukushima Daiichi Nuclear Power Plant in 2011, the necessity of increasing the power supply using renewable energy was widely discussed in Japan. Small-scale geothermal power generating facilities with ≦2000 kW capacity were assessed as a renewable energy resource. In Beppu-Onsen hot spring, where high temperature hot springs discharge through a wide area, small-scale geothermal power generation started as early as 2013. This report describes difficulties faced by power generating facility operators as well as countermeasures taken by the local governments of Oita Prefecture and Beppu CitTy, and by scientists. The author's ideas related to future conservation of geothermal and hot spring resources in Beppu are also presented.

Keywords Beppu · Geothermal power · Development · Hot spring · Conservation · Countermeasure

15.1 Introduction

Immediately following the Fukushima Daiichi Nuclear Power Plant accident after the Great East Japan Earthquake and its consequent tsunami of March 11, 2011, the necessity of increasing Japan's power supply using renewable energy was widely discussed. As one important resource, geothermal energy attracted considerable attention. With the feed-in tariff scheme for renewable energy proposed by the Ministry of Economy, Trade and Industry of Japan, practical, small-scale geothermal power generation at facilities with capacity of 2000 kW or lower (popularly called "*onsen hatsuden*" or "*onsen netsu hatsuden*" in Japanese) was taken up. At Beppu-Onsen hot spring, where high-temperature hot springs discharge over a wide

S. Ohsawa (✉)
Beppu Geothermal Research Laboratory, Institute for Geothermal Sciences, Graduate School of Science, Kyoto University, Beppu, Oita, Japan
e-mail: ohsawa@bep.vgs.kyoto-u.ac.jp

© Springer Nature Singapore Pte Ltd. 2018

A. Endo, T. Oh (eds.), *The Water-Energy-Food Nexus*, Global Environmental Studies, https://doi.org/10.1007/978-981-10-7383-0_15

area, introduction of small-scale geothermal power was first considered in 2013, and power generation started there in the following year.

As described herein, an overview of Beppu-Onsen hot spring will be presented from the viewpoint of natural science. Then, increased drilling of hot spring wells and its adverse effects on existing hot springs and hydrological environments experienced before 2011 will be described. Third, difficulties encountered in the course of power generation, in addition to countermeasures taken by the local governments of Oita Prefecture and Beppu City and also by scientists will be noted. Finally, the author's ideas regarding the future of conservation of geothermal and hot spring resources in Beppu will be presented.

15.2 Overview of Beppu-Onsen Hot Spring from a Natural Science Perspective

Beppu-Onsen hot spring is a typical volcanic hot spring spreading onto the eastern flanks of Tsurumi volcano, an active volcano located on the northern part of Kyushu Island, Japan (Fig. 15.1). Thermal water of three main types runs beneath the surface of the Beppu-Onsen hot spring area: high-temperature sodium chloride (Na-Cl) type, bicarbonate (HCO_3) type, and acid sulfate ($H\text{-}SO_4$) type (Fig. 15.2). These diverse compositions might be formed from a single parent hydrothermal fluid, which is inferred to be the Na-Cl type of 250–300 °C (Allis and Yusa 1989). Thermal water derived from the parent fluid flow out eastward to the coast. Estimated flow paths of the thermal waters at the southern part of Beppu are depicted in Fig. 15.3.

The history of Beppu-Onsen hot spring can be traced back to the eighth century. Ancient people used natural hydrothermal manifestations ranging from flowing hot springs to fumaroles over much of Japan's history. During the latter half of the nineteenth century, drilling of wells was promoted to enhance the discharge of thermal water and steam from the hydrothermal system. Today, over 2000 active hot spring wells exist in an area spanning 5 km from east to west and 8 km from north to south (Fig. 15.4). The total mass and heat discharge from these wells are, respectively about 50,000 ton/day (600 kg/s) and 350 MW (Yusa and Oishi 1988). Based on field survey data reported by Yusa *et al.* (1975), the ratios of number of wells, mass output, and heat output of steam and boiling wells to those of the total hot spring wells are, respectively 8%, 48% and 79%.

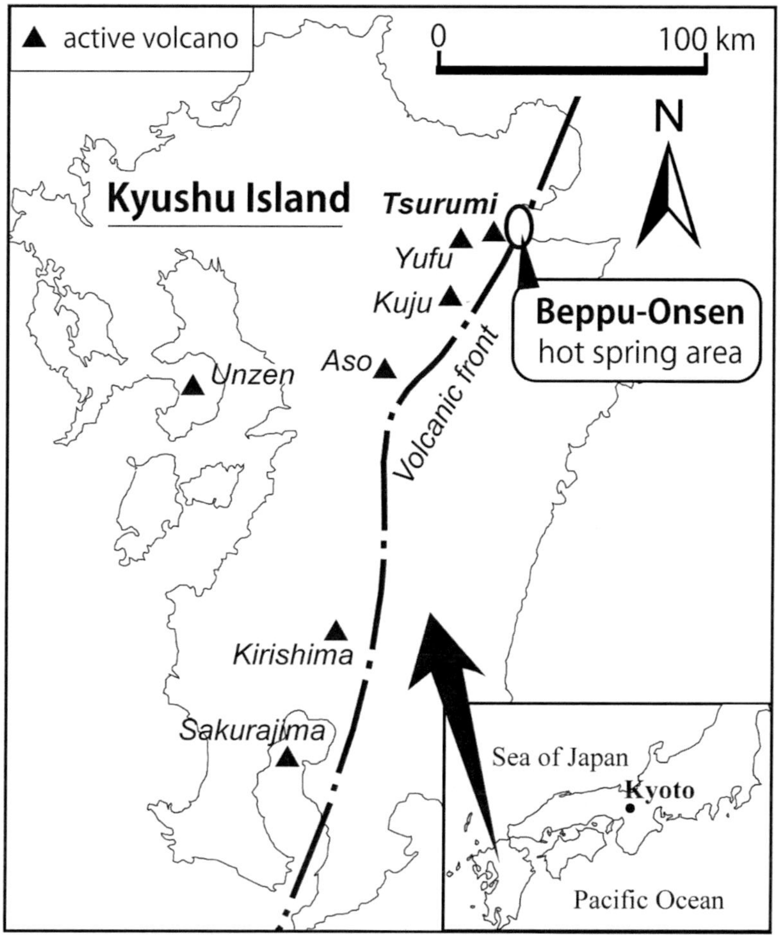

Fig. 15.1 Location map of Beppu-Onsen Hot spring. Closed triangles show active volcanoes

15.3 Increased Drilling of Hot Spring Wells and Adverse Effects on Existing Hot Springs and Hydrological Environments Experienced before 2011

Exploitation of hot springs (drilling of hot spring wells) in the Beppu-Onsen hot spring area started mainly in lowland areas as early as the 1880s. By the 1920s, the number of wells had increased to about 1000. A second flurry of exploitation occurred around the 1960s. By early 1970 over 2500 wells were in use (Fig. 15.5). As the number of pumping wells increased, the number of flowing wells clearly decreased. From 1968, because of restrictions by the local government aimed at preserving the hot spring water resource, the total number of hot spring wells has leveled off.

Fig. 15.2 Conceptual drawing of thermal water formation processes in Beppu-Onsen hot spring

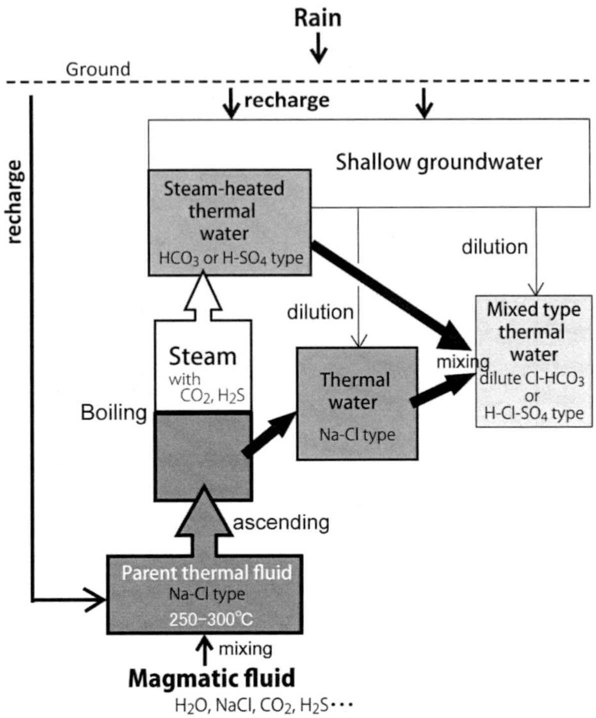

The mass and heat flows had increased mainly because of the discharge of high-temperature sodium chloride thermal waters in highland areas. This caused a decline in the piezometric head of deep chloride water, a decline in the subsurface flow of chloride water towards the lowlands, and intrusion of steam-heated shallow water into the chloride water layer (Yusa et al. 2000). An example of the impact of that exploitation is the chemical variation that has occurred over time in a boiling well of Tenman-cho in the southern part of Beppu (★in Fig. 15.3). Figure 15.6 shows that a four-fold decrease in chloride concentration and a three-fold increase in bicarbonate concentration has occurred since monitoring started in 1968. As little drilling has been done since approximately 1985, the variations in Cl and HCO_3 concentrations have been slight. Thermal water from this boiling well stopped flowing naturally in February 2006. Since then, wells discharging thermal water have been pumped using the air-lift method, in which air bubbles are injected into a hot spring well using an air compressor to raise hot spring water to the ground surface.

The rivers in Beppu are affected strongly by hot spring drainage discharged from boiling hot springs located in unimproved sewage areas (Ohsawa et al. 2008; Ohsawa et al. 2009). The constant drainage of unused hot spring water from the boiling hot springs is suspected of having an influence. For example, arsenic (As) in the Beppu rivers, which exceed the environmental standard value, likely originated from hot spring waters, especially those derived from boiling springs (Fig. 15.7). Annual emissions of As and boron (B), which originated mainly from boiling hot

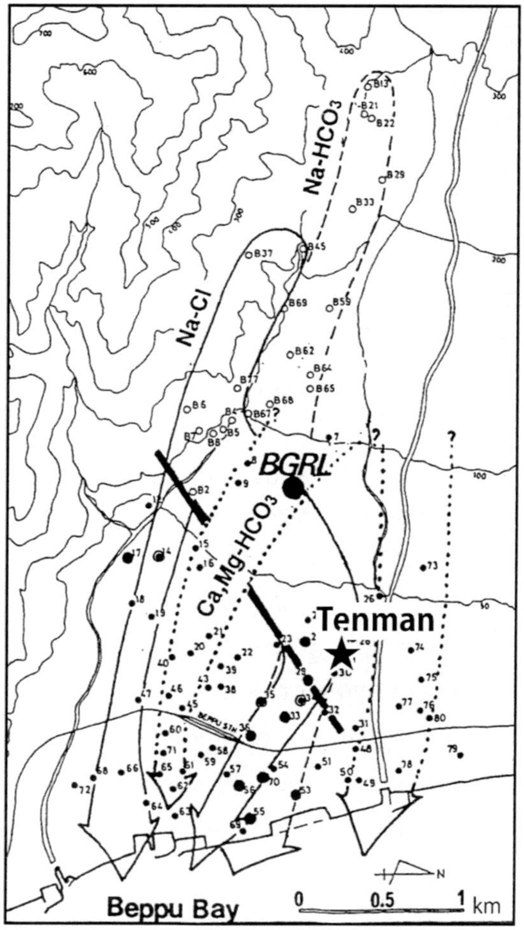

Fig. 15.3 Subsurface flows of thermal waters in the southern Beppu-Onsen hot spring area (Ohsawa et al. 1994): BGRL, Beppu Geothermal Research Laboratory, Kyoto University; Closed asterisk (★), boiling hot spring at Tenman

spring water, are 4.3 and 82 tons, respectively, whereas those of lithium (Li), cesium (Cs) and rubidium (Rb), which are useful metallic elements and which are also mainly derived from boiling hot spring water, are, respectively, 34,0.4 and 5.4 tons. In addition, inflows of hot spring drainage into rivers have increased the amounts of diatoms in several rivers in Beppu. The annual discharge rate of diatoms to the Beppu Bay is estimated as greater than 10 tons annually (Yamada et al. 2017).

Because almost all river sections in the Beppu rivers except for the Hiya River are concrete-lined and also because the Beppu sewage system has not been improved yet in the upstream area, the hot spring drainage from boiling spring wells, which are located mostly in the upstream areas, adversely affects downstream reaches of the rivers and estuary.

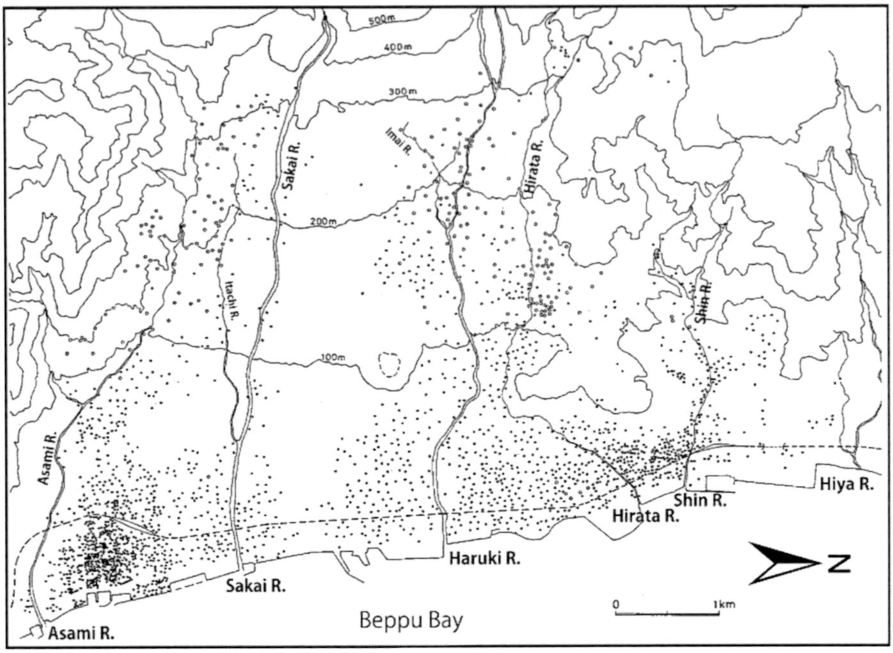

Fig. 15.4 Distribution of drilled wells in the Beppu-Onsen hot spring area in 1985 (Modified from Yusa and Oishi 1988)

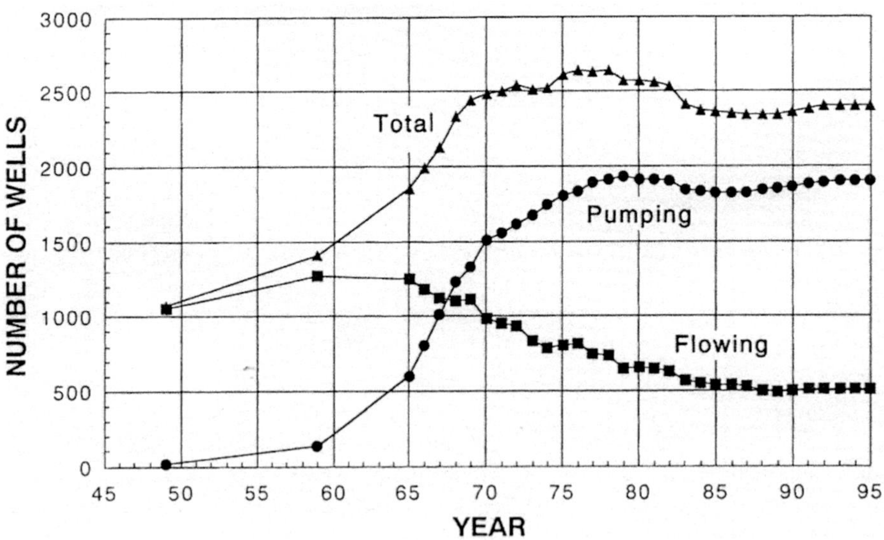

Fig. 15.5 Changes in the number of drilled wells in Beppu-Onsen hot spring area (Yusa et al. 2000). "Flowing" includes wells discharging steam, boiling water, and artesian (sub-boiling) water

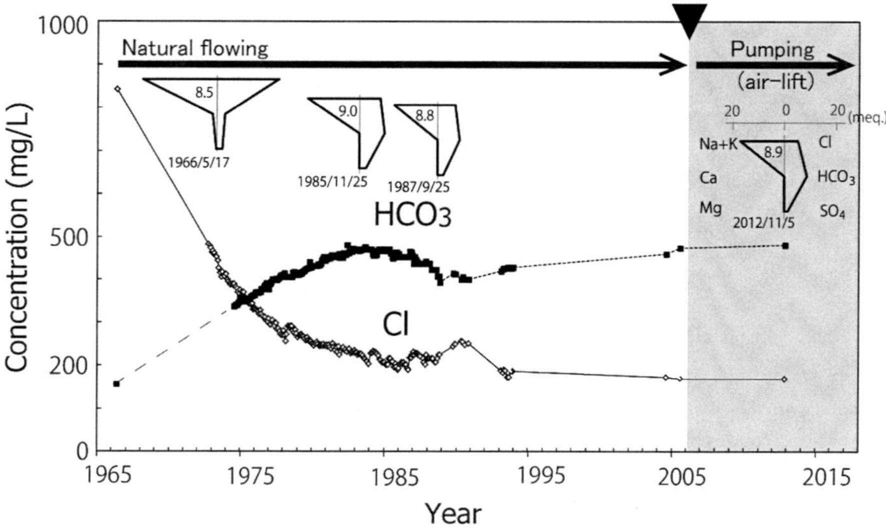

Fig. 15.6 Variations in chloride (Cl) and bicarbonate (HCO_3) concentrations of thermal water discharged from a boiling well at Tenman in Beppu-Onsen hot spring. Data of Cl and concentrations of Cl and HCO_3 before 1995 were referred from Yusa *et al.* (2000). The well location is shown as a closed star (★) in Fig. 15.3

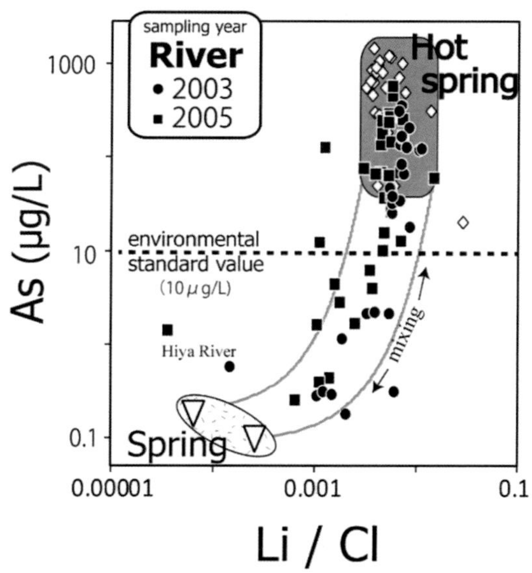

Fig. 15.7 Relation between the Li/Cl ratio and As concentration of river waters running through the Beppu-Onsen hot spring area

Fig. 15.8 Photographs of small-scale geothermal power stations in Beppu-Onsen hot spring: A and B are binary type; C and D are total-flow type ("Yukemuri-Hatsuden")

15.4 Development of Small-scale Geothermal Power Generation (*onsen hatsuden*) After 2011 and Important Development Matters

Geothermal power generation ("*onsen hatsuden*") was anticipated as a renewable energy resource after the Great East Japan Earthquake in 2011 and subsequent shutdown of all nuclear power generation after the Fukushima Daiichi Nuclear Power Plant accident. Beppu soon started to examine introduction of small-scale geothermal power generation facilities. The Japanese government and the Oita prefecture government supported local efforts such as the Feed-in Tariff, deregulation of engineer placement in power generation facilities and partial support for power generation businesses cooperating with local banks and other companies. In January 2013, a small-scale geothermal power plant started operation in Beppu. Photographs of some small-scale geothermal power plants are presented in Fig. 15.8; Fig. 15.9 shows a map of the small-scale geothermal power plant locations as of March 2015, overlaid on flow paths of underground thermal waters. Almost all small-scale geothermal power plants were situated in the upper parts of flow paths of high-temperature sodium-chloride (Na-Cl) thermal water.

The author noticed an important matter that has been passed over. The small-scale geothermal power generation facilities use mixed steam and hot water from artesian boiling wells. The Oita Prefecture government restricts the discharge rate of hot water from pumping wells up to 50 L/min. However, those from artesian wells are exempt from this regulation, as are natural hot springs. The potential discharge

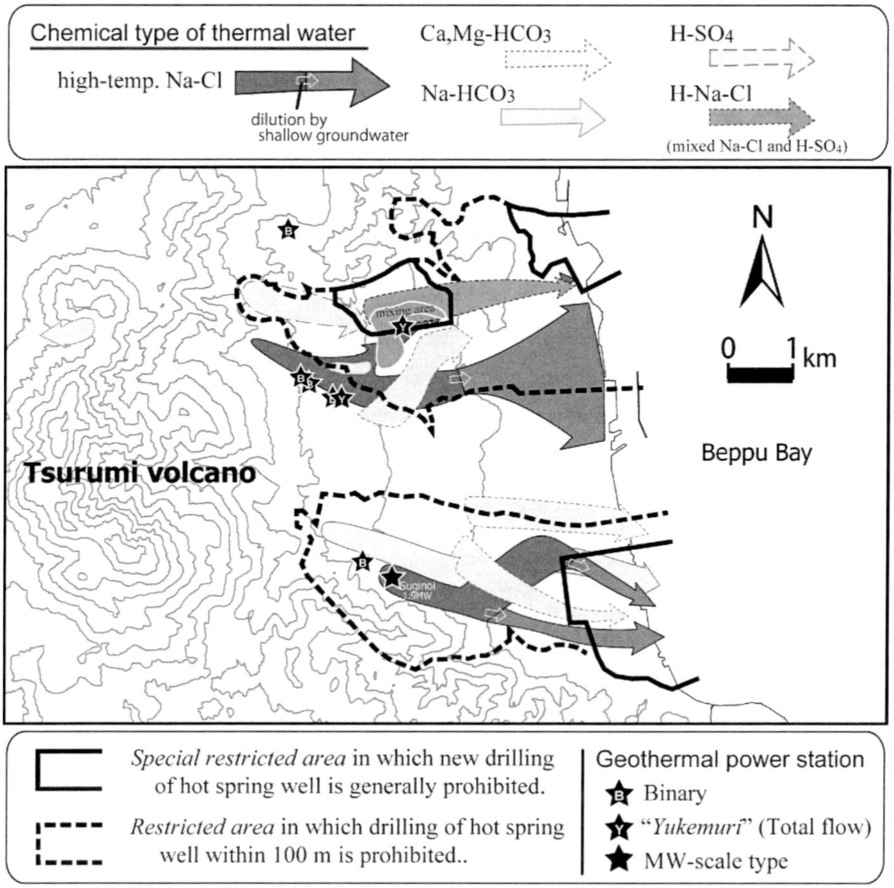

Fig. 15.9 Location map showing geothermal power stations and restricted areas of drilling of hot spring well with flow paths of underground thermal waters in the Beppu-Onsen hot spring area

rate of hot water from an artesian boiling well is reportedly about 1000 L/min. Even if someone extracts all of the steam and hot water from such a well for geothermal power generation, they would not be subjected to the 50 mL/min restriction on pumping wells. Therefore, at the end of 2012, the author expressed concern at meetings of Oita Prefectural government groups, such as the Onsen Group of the Oita Prefectural Environmental Council that some difficulties might occur for traditional users of hot springs if a great amount of underground thermal water were extracted via artesian boiling wells for geothermal power generation.

Unfortunately in this case, new drilling aimed at small-scale geothermal power generation was applied for on July 2013 and was permitted because the application met the prescribed requirements. According to an official report by the Beppu City government, nine plans for small-scale geothermal power station operation were underway as of June 2015: five applications had been approved by prior consulta-

大分合同新聞（朝刊） 2016年（平成28年）12月6日 火曜日

別府市が噴気沸騰泉100カ所調査

温泉資源の現状把握

利用可能な量を判断

基礎資料作成へ

別府市は5日から、温泉資源の現状を把握するため、市内の噴気沸騰泉100カ所を対象にした源泉調査を始めた。温度や湧出量などを測定し、利用可能な資源量を判断するための基礎資料を作成する。同様の調査は旧温泉調査研究会が1986年に実施して以来、30年ぶりとなる。

別府市内の源泉で蒸気の温度などを計測＝5日

初日は2カ所を調査。市内竹の内の源泉では、専用の機器を使って噴出する蒸気や温泉の温度と量、塩化物イオンの濃度などを計測した。源泉を管理する会社の男性役員(69)は「昔に比べ、地下水は減少傾向にあると感じる。自社でも源泉は調べているが、別府全体の温泉資源の動向を把握するためにも市の調査はいい試みだと思う」と話す。

市は本年度中に、熱源となる伽藍岳・鶴見岳に近い標高200㍍以上の地域で抽出した源泉を調査し、発電や熱利用に使える資源量を算定する。新年度に、地熱の適切な利用指針を示す審議会を立ち上げ、調査結果の公表方法などを具体的に検討する。

地熱が新たなエネルギー源として期待される中、別府市は適地として注目を集め、温泉熱を利用したバイナリー発電施設の立地が進む。市環境課によると、既に13カ所が稼働し、今後、45カ所の計画がある。

一方で、急激な開発に不安を感じる市民も少なくない。鶴見岳に近い堀田地区では、ここ2年半ほどの間に3件の地熱利用計画が浮上。住民はいずれも強く反対してきた。後藤上行自治会長は「堀田の泉源は地元だけでなく、市街地に温泉を供給する役割も担う。新たに地下を掘削すれば、どんな影響が出るか分からないし、問題が起きてからでは取り返しがつかない」と懸念を示す。

市環境課の堀英樹環境企画室長は「今回の調査は、いわば温泉の健康診断。資源枯渇を防ぎ、余剰のエネルギーを有効活用していく施策を進めるため、判断材料となるデータをそろえたい」と話している。

（田尻雅彦）

Fig. 15.10 Newspaper clipping describing a survey survey of boiling/steam wells in the Beppu-Onsen hot spring area conducted by the Beppu City government in 2016

tion of the Beppu City government as of the end of September 2016. The number of small-scale geothermal power plants in Beppu has been increasing.

15.5 Countermeasures by Governments and Private Citizens

On October 1, 2014, the Oita Prefecture government partially revised the criteria for permission to drill hot spring wells to support small-scale geothermal power generation at capacities of less than 2000 kW. They drew up the prefectural master plan for hot springs on March 2016 to protect hot spring resources and promote sustainable, safe and secure use of hot springs. The prefectural master plan for hot springs is the basis for hot spring administration in Oita Prefecture. The contents are comprised of a list of hot springs resources and their uses; the current status of and issues in public hot spring administration; specific measures to achieve the basic goals of sustainable, safe, and secure hot spring use; as well as the roles of the citizens, business persons and administrative entities. Later, the Beppu City

試練の再生エネ

先進県・大分は今

▶▶4

『地熱』開発と保護 どう両立

ルール策定 行政模索

目立つ新規の掘削

湧き出る温泉の泉質に変化はないか—。

別府市内の温泉旅館にある源泉井戸。京都大学地球熱学研究施設(同市)の大沢信二教授(55)は3年近く前から月1回、独自のモニタリング調査を続けている。

日本一の源泉数、湧出量を誇る泉都では、温泉熱発電の開発が全国に先駆けて増えている。市によると8施設が既に稼働し、他に37件もの計画がある。

温泉熱発電は当初、既存の源泉を使い、浴用以外の捨てられているエネルギーを活用できると期待された。だが、実際には発電目的の新規掘削や掘り替えが目立つ。「温泉の使用量が増え続ければ、資源の枯渇につながる恐れがある」。そんな懸念から、大沢教授は4カ所で調査に取り組む。

市内では住宅地に温泉熱発電施設が幾つもでき、住民から発電に伴う騒音や激しい湯煙に苦情が出ている地域もある。

モラル信じるしか

温泉熱を含む地熱資源の開発と保護の両立、住環境との調和をどう図るか。市は地域活性化に向けて温泉を最大限に活用する方針の下、方法を模索している。

昨年、地熱開発をする事業者に対し、地元説明会の開催、モニタリング調査(新規掘削のみ)などを求める要綱を策定した。ただ、罰則はなく、事業者のモラルを信じるしかない。

地熱開発にビジネス優先の投資マネーが流入している現状も考慮し、現在、さらに踏み込んだ条例の策定を進めており、ルールが厳守される環境を整備したい考えだ。

市環境企画室の堀英樹室長(48)は「温泉資源の有効活用は望ましいが、乱開発は防がねばならない。アクセルとブレーキを同時に踏むような課題」と説明。うまく機能する条例作りの難しさを感じている。

温泉の発電利用で先を行く別府が直面した課題には、同様の事態が起こり得る他の温泉地や国も注目する。別府の事例を踏まえ、資源エネルギー庁資源・燃料部政策課は「地熱開発を推進する上で方向性を少し見直す必要がある」とし、国としても何らかの対応策を講じる姿勢だ。

県内の温泉掘削申請件数

※■は温泉熱を含む地熱発電目的

(件)

年度	2007	08	09	10	11	12	13	14
申請件数	76	62	62	55	50	49	63	88
■	2	10	0	0	0	5	11	29

源泉井戸から採取した湯の泉質をチェックする京都大学の大沢信二教授（後方）ら＝別府市内

メモ　地熱開発の活発化を背景に、県全体でも温泉の掘削申請は急増している。地熱資源の利用促進と保護を目的に、県の環境審議会温泉部会は昨秋、内規を改めた。小規模な地熱発電（出力2メガワット程度）開発の場合、使える管の最大口径を80ミリから150ミリに拡大した。規制を緩和する一方、事業者には源泉の継続的なモニタリング調査、地域との合意形成などを義務付けた。

Fig. 15.11 Newspaper clipping describing introduction of our scientific monitoring of hot springs in the Beppu-Onsen hot spring area

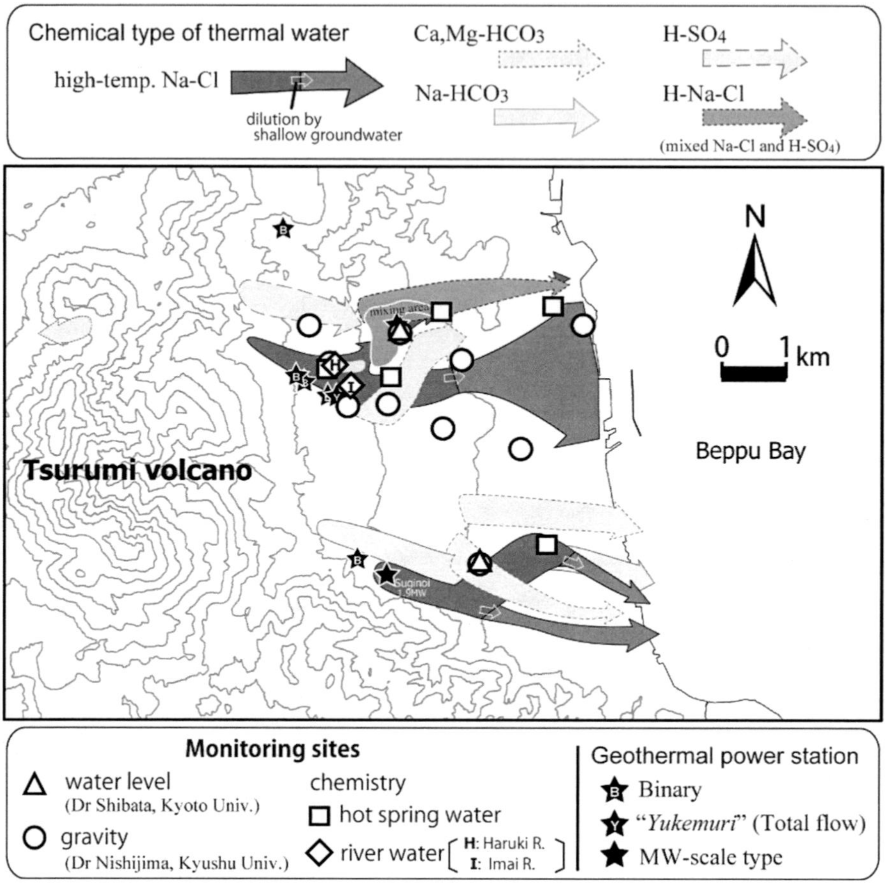

Fig. 15.12 Scientific monitoring site locations provided by the author and private citizens with geothermal power stations and flow paths of underground thermal waters in the Beppu-Onsen hot spring area

government enacted a municipal ordinance for community involvement related to small-scale geothermal power generation on May 1, 2016, and formulated the Beppu City Countermeasures Council for small-scale geothermal power generation in April 2017. At the national level, the Advisory Committee for Geothermal Resource Development was established by the Japan Oil, Gas and Metals National Corp. (JOGMEC) in June 2016.

Regarding practical measures, the Oita Prefecture government provided six monitoring sites for temperature, pressure, and other characteristics of hot spring water in Beppu city in 2015 and 2016. In addition, the Beppu City government conducted surveys of the temperature and amount of discharge of hot spring water at nearly 100 boiling/steam wells in Beppu from December 2016 to January 2017 (Fig. 15.10).

Table 15.1 Geochemical data of rivers at two monitoring sites at Beppu-Onsen hot spring shown in Fig. 15.12 (open triangles labeled with the symbols of H and I): *W.T* water temperature; *E.C.* electric conductivity

Haruki River site near Teruyu hot spring (H)										
Sampling date	W.T. (°C)	E.C. (mS/m)	pH	Na (mg/L.)	K (mg/L)	Mg (mg/L)	Ca (mg/L)	Cl (mg/L)	SO_4 (mg/L)	HCO_3 (mg/L)
Sep. 1,2003	26.6	38	8.2	28	5	10	26	9	85	90
Jun. 21 2005	24.0	46	7.9	41	6	11	28	13	98	122
Aug. 20, 2009	28.0	35	8.1	23	4	12	30	4	62	153
Apr. 21,2015	21.1	34	8.2	22	4	11	27	7	62	114
May 26, 2015	26.0	34	8.8	23	4	11	27	7	65	114
Jul.28,2015	34.4	165	8.3	249	37	9	27	375	65	102
Aug. 6,2015	33.6	127	8.2	182	28	11	29	266	63	107
Sep. 29,2015	25.7	35	7.9	23	5	11	27	13	65	97
Nov. 1,2015	26.1	129	8.2	193	27	10	28	282	61	107
Nov. 25.2015	22.9	136	8.1	194	27	10	28	283	62	105
Feb. 9,2016	13.6	110	8.8	155	24	10	25	220	62	112
Imai River before junction with Haruki River (I)										
Sampling date	W.T. (°C)	E.C. (mS/m)	pH	Na (mg/L)	K (mg/L)	Mg (mg/L)	Ca (mg/L)	Cl (mg/L)	SO_4 (mg/L)	HCO_3 (mg/L)
Sep. 1,2003	38.7	100	8	143	20	6	18	206	43	90
Jun. 21,2005	28.8	103	8	157	22	6	17	223	53	101
Aug. 20, 2009	36.4	107	8	153	22	8	23	182	57	112
Apr. 21,2015	36.6	282	8	475	70	5	18	766	58	102

Before full-fledged implementation of small-scale geothermal power generation from January 2014, the author's research team started geochemical monitoring of hot spring water at four hot spring wells in January 2013 (Fig. 15.11). Continuous monitoring of water levels of hot spring wells and repeated measurements of gravity change were begun in April 2014 by Dr. Tomo Shibata (Kyoto University) and Dr. Jun Nishijima (Kyushu University), respectively. All monitoring sites including hot springs and rivers near the area of intensive development of geothermal resources added later are presented in Fig. 15.12. In order to better understand the current state of geothermal activity in the Beppu-Onsen hot spring area, a new approach was developed to investigate water temperature and water quality. Data was collected at

65 hot springs on November 13, 2016 in cooperation with citizens and related organizations in the region (Yusa and Yamada 2017).

Monitoring data are being collected. They will be reported in a separate publication, but a change of river water chemistry will be described here. Concentrations of major chemical components in river waters along with water temperature (W.T.), electric conductivity (E.C.), and pH are represented in Table 15.1. In both rivers, there are no major changes in water temperature, pH and concentrations of Mg, Ca, SO_4 and HCO_3, although concentrations of Na, K and Cl increased markedly since the start of development of small-scale geothermal power generation in 2013. Since, Na, K and Cl are the main chemical components of sodium chloride type water discharged from boiling wells, the observed increase in Na, K and Cl concentrations of the river waters should be attributed to the effects of drainage after the water is used for geothermal power generation. Thus, changes in river water quality and the river ecosystem noted before in mid-stream and downstream reaches of the Beppu rivers will begin to extend upstream.

15.6 Concluding Remarks

The author's ideas about the future of geothermal and hot spring resource conservation in Beppu-Onsen hot spring are described below.

1. Figure 15.9 shows that small-scale geothermal power generation is concentrated outside the area in which drilling of the hot spring wells. Therefore, the restricted area must be changed. Development was concentrated in a non-restricted area because understanding of the underground flow paths of thermal waters may have been insufficient before the Oita Prefecture government set the restricted areas in 1968.
2. Although it is widely said that the thermal water used for geothermal power generation is best injected underground after its use, in reality, determining which area and depth to return it in the Beppu-Onsen hot spring is not easy. In the case of the Beppu-Onsen hot spring area, production wells for small-scale geothermal power generation have been drilled toward the underground flow passages of the high temperature thermal waters that already have existing hot spring wells for traditional use; thus inadvertent injection might adversely affect other hot springs by decreasing the water temperature.
3. In Oita Prefecture, including the Beppu-Onsen hot spring area, although pumping of the hot spring water is limited to 50 L per minute, developers of geothermal power generation are likely aware that discharged hot spring water can be used from artesian boiling wells with almost no limitation. Development plans for the use of even greater amounts of high-temperature hot spring water (e.g. 1000 L/min.) must be examined carefully and critically.

Acknowledgements I would like to express my appreciation for support received from the RIHN project: Human-Environmental Security in Asia-Pacific Ring of Fire: Water-Energy-Food Nexus and the project members. I received valuable technical support from Mr. Taketoshi Mishima (Kyoto University). Discussions with Prof. Makoto Taniguchi (Research Institute for Humanity and Nature), Dr. Masahiko Fujii (Hokkaido University), and Prof. Keiji Takemura (Kyoto University) have been meaningful and useful.

References

Allis RG, Yusa Y (1989) Fluid flow processes in the Beppu geothermal system, Japan. Geothermics 19:743–759

Ohsawa S, Watanabe K, Takamatsu N, Kato N (2009) Basic research and investigation of untapped hot spring resources (II): the amount of useful metal elements flowing into rivers. Oita Prefect Hot Spring Res Soc 59:13–19. (in Japanese)

Ohsawa S, Yamasaki H, Takamatsu N, Yamada M, Amita K, Kato N (2008) The inflow of useful metal elements into rivers: basic research and investigation of untapped hot spring resources. Oita Prefect Hot Spring Res Soc 58:21–30. (in Japanese)

Ohsawa S, Yusa Y, Kitaoka K (1994) Flow paths of thermal Groundwaters in the southern part of Beppu, Japan. J Balneological Soc Japan 44:199–208. (in Japanese with English abstract)

Yamada M, Shoji J, Ohsawa S, Mishima T, Hata M, Honda H, Fujii M, Taniguchi M (2017) Hot spring drainage impact on fish communities around temperate estuaries in southwestern Japan. Journal of Hydrology: Regional Studies 11:69–83

Yusa Y, Noda T, Kitaoka K (1975) Discharge of water, heat and chemical substances through a hydrothermal system –Beppu hydrothermal system. Journal of the Society of Engineers for Mineral Springs, Japan 10:94–108. (in Japanese with English abstract)

Yusa Y, Ohsawa S, Kitaoka K (2000) Long-term changes associated with exploitation of the Beppu hydrothermal system, Japan. Geothermics 29:609–625

Yusa Y, Oishi I (1988) Statistics of mass and heat outputs from wells in the Beppu geothermal system during 1985 to 1987. Reports of Oita prefecture hot spring research. Society 39:1–6. (in Japanese)

Yusa Y, Yamada M (2017) Simultaneous survey of discharge temperature and water quality of hot spring waters in Beppu-Onsen on November 13, 2016. Reports of Oita prefecture hot spring research. Society 68:1–7. (in Japanese)

Part V
Understanding Socio-economic dimension of Resource System

Chapter 16
Social Acceptability of Micro Hydropower in Laguna, Philippines

Karen Ann B. Jago-on, Mary Jaselle M. Din, Christine S. Bellen, and Fernando P. Siringan

Abstract The use of micro hydropower as a source of renewable energy can help augment electricity demand in communities. However, it is necessary to know how the community near the water source perceives the technology and its possible impacts on the community. This study seeks to determine the social acceptability of establishing a micro hydropower project in Calamba City and the Municipality of Los Baños in Laguna, Philippines. A survey was conducted among 400 households in Barangays Mabato and Canlubang in Calamba and Barangays Anos and Bambang in Los Baños to understand their awareness and knowledge on micro hydropower and their perception on possible impacts of the technology on the community. Results of the survey reveal that there is no perceived conflict on water uses as the river and creek waters are hardly used for domestic, agriculture or any other purposes. The results also emphasize three major considerations by the community in accepting the project: reduction of future electricity costs without investment cost on their part; increase economic and social benefits; and community participation. If the project can address these factors, the communities are highly likely to accept the project. The respondents also express their willingness to participate in stakeholder consultations, information campaigns and in the actual construction of the project.

Keywords Micro hydropower · Social acceptability · Water use · Energy · Laguna province

---

K. A. B. Jago-on (✉) · M. J. M. Din
School of Urban and Regional Planning, University of the Philippines-Diliman, Quezon City, Philippines

C. S. Bellen
University of Canterbury, Christchurch, New Zealand

F. P. Siringan
Marine Science Institute, University of the Philippines-Diliman, Quezon City, Philippines

© Springer Nature Singapore Pte Ltd. 2018

A. Endo, T. Oh (eds.), *The Water-Energy-Food Nexus*, Global Environmental Studies, https://doi.org/10.1007/978-981-10-7383-0_16

16.1 Introduction

In 2015, 45% of the power generated in the Philippines is coal-based (DOE 2015). In order to increase the use of renewable energy the Philippines enacted in 2008 Republic Act No. 9513, also known as the Renewable Energy Act of 2008, which seeks to promote the development, utilization, and commercialization of renewable energy resources. Figure 16.1 shows the different sources of power generation in the Philippines in 2015. Part of the efforts of the government toward the use renewable energy is the promotion of micro hydropower in the country. The Department of Energy (DOE) has identified potential areas for power generation which can be generated through the use of micro hydropower. In Region IV, where the present study areas are located, a total of 3.40 MW of potential power generation is identified (DOE 2012). In line with the program to use micro hydropower, this present research complements with the technical studies on micro hydropower by providing a social perspective on project-affected communities and a framework in assessing social acceptability of infrastructure projects.

Following the study on the potential for micro hydropower in Laguna by Bellen and Siringan (2015), this present study is conducted to determine the acceptability of micro hydropower technology among the communities in the study area. The study of Bellen and Siringan (2015) used mixed methods approach to identify potential sites for micro hydropower, while this study particularly examines the acceptability of the project among the residents of Calamba City and the Municipality of Los Baños in Laguna, through surveys and key-informant interviews. Figure 16.2 shows the potential sites identified in the previous study.

According to the preliminary investigation conducted in Laguna for the potential micro hydropower, there is an increasing need for rural electrification to provide illumination at night and to support livelihood projects. An impact evaluation by the Impact Evaluation Group-World Bank (2008) cited domestic benefits, particularly

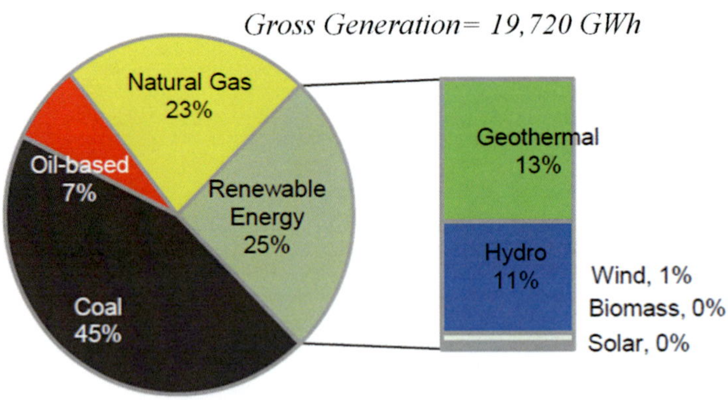

Fig. 16.1 Sources of Power Generation (2015) (Source: DOE Philippine Power Situation Report (2015))

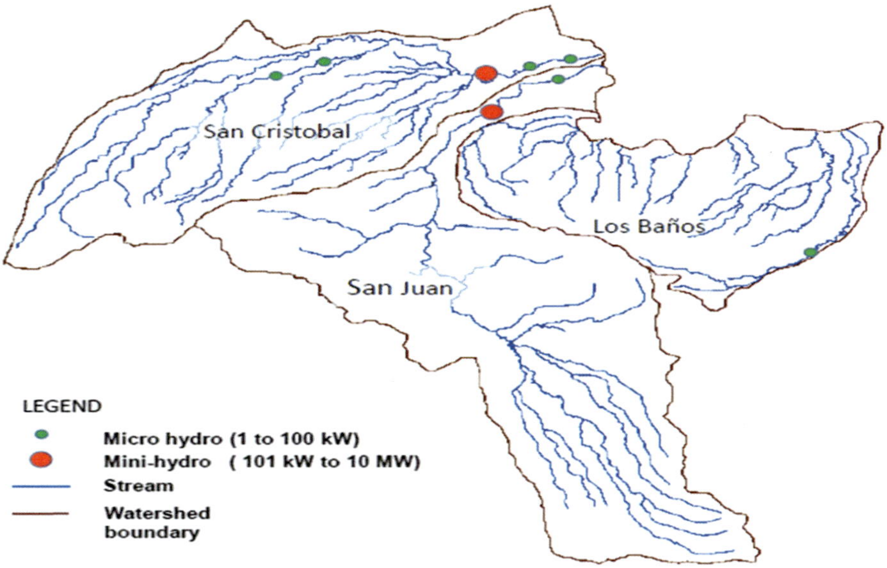

Fig. 16.2 Potential sites for micro hydropower (Source: Bellen and Siringan 2015)

lighting and TV, improved health benefits, time use, education benefits and productive uses as the main welfare impacts of rural electrification. It supports food security and economic productivity. However, extending electricity grids incur high costs. Thus, micro hydropower can be utilized in isolated and off-grid areas for decentralized electrification. Micro hydropower has been proven to have high efficiency (70–90%), high capacity factor (typically >50%) compared with 10% for solar and 30% for wind, high level of predictability, varying with annual rainfall patterns, and relatively low environmental impact compared to large-scale hydropower (Bellen and Siringan 2015). The results of the previous study indicated that small-scale hydropower can be generated, ranging from 101 kW to as high as 2 MW along San Juan River and micro hydropower can be generated from 1 to 100 kW along Cauacauang River in the San Cristobal subwatershed and those around Mt. Makiling.

While the technical and economic feasibility can be established through various scientific and economic studies, a need arises to look into the social context of any intervention, be it government or private sector driven. This provides a more comprehensive approach in examining human-environmental security. The use of existing water systems to provide electrification, while it may increase food and economic productivity, may have externalities to the community. This present study aims to determine the readiness and willingness of the community to accept micro hydropower technology based on their knowledge and perception of impacts and roles.

A study by Moula et al. (2013) examined the social acceptability of renewable energy technologies in Finland by designing a survey which looked into the respondents' background information, awareness of renewable energy technologies

(RETs), and willingness to invest in RETs. Simon (2015) also considered the economic perspective of small-scale hydropower plants as a factor of acceptability. Another factor that determined the plant's acceptability and was considered in the same study is the availability of information about the plant and the working procedures. In Portugal, Ribeiro et al. (2013) also conducted a survey among Portuguese people to look into the public opinion on renewable energy technologies, including hydropower. The survey considered the level of acknowledgement of the technologies by the Portuguese, their position towards energy projects and their perception of sustainable development issues. Their willingness to pay for renewable technologies was also factored in (Ribeiro et al. 2013). Tabi and Wustenhagen (2015) also explored the characteristics that make hydropower projects socially acceptable, particularly, by ensuring that justice, both procedural and distributional, are pursued and impacts are kept minimal and ownership remains local.

Another factor that affects the social acceptability of infrastructure projects is the resettlement and compensation arrangement for project-affected persons, as exemplified in a study by Ha-Duong et al. (2015) in Vietnam. A technical report by the International Energy Agency (IEA 2000) also examined the environmental and social impacts of hydropower development (IEA 2000). The report focused on the physical and biological impacts during and after the implementation of hydropower projects (e.g., changes in biota habitat, water quality and resource use). Drawing from existing literature, the present study utilizes the different factors and considerations put forth that affect social acceptability of a micro hydropower project in Laguna, particularly in Calamba and Los Baños.

16.2 Description of Study Sites

16.2.1 Province of Laguna

Located southeast of Metro Manila, the Province of Laguna is bordered by Laguna de Bay, the country's largest lake, to the north and Mt. Makiling and Mt. Banahaw to the south. Laguna has a total land area of 1759.73 km^2 and a population of 3.04 million people in 2015 (PSA 2015). It is a part of the larger region of CALABARZON (consisting of the provinces of **Ca**vite, **La**guna, **Ba**tangas, **Ri**zal and Que**zon**) in Luzon. Laguna consists of four (4) Congressional districts and six (6) cities, 23 municipalities and 674 barangays. The economic activities in the province vary from agricultural, commercial and tourism to industrial or manufacturing. Calamba City and the Municipality of Los Baños are both part of the 2nd district of Laguna (RDC, 2011). Figure 16.3 shows the location of Calamba and Los Baños in the province of Laguna.

Both Calamba and Los Baños rely on groundwater as their main source of water. While the Calamba Water District and Laguna Water District mainly provide and maintain the piped water systems to Calamba and Los Baños, respectively, other

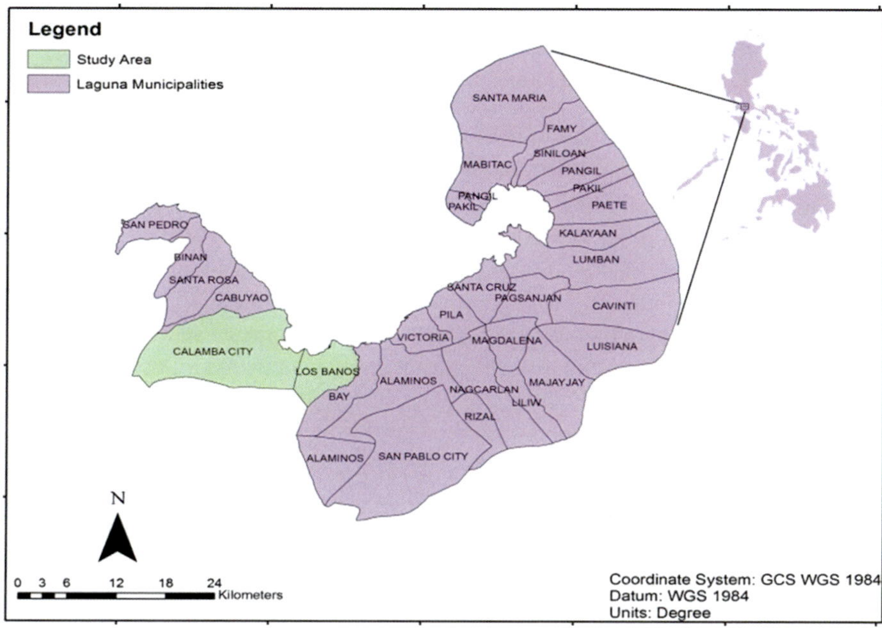

Fig. 16.3 Location of Calamba and Los Banos in Laguna Province

households and commercial establishments install their own pumps without permits. Both Calamba and Los Baños are known for their hot spring resorts, which lead to a huge amount of consumption of groundwater without proper regulation. This causes the decrease of groundwater quantity due to over-extraction. This may cause water resource problems in the future if left unregulated (Jago-on et al. 2017). There are also surface waters in both areas, such as rivers and streams, which could be tapped as resources for energy generation.

16.2.2 Calamba City

Calamba City has a total land area of 149.5 km2 (12.66% of the land area of the province) with a population of 454,486 in 2015, making it the fourth most populous city in the CALABARZON Region. Calamba has 54 barangays (PSA 2016). Aside from being named as the "Hot Spring Capital of the Philippines" Calamba is also considered as a major economic growth center in the region, particularly as an industrial hub, because of the presence of various industrial parks and manufacturing plants (City Government of Calamba 2015).

In Calamba, two barangays were chosen as part of the study sites; Barangays Mabato and Canlubang. A "barangay" is the smallest political unit in the Philippines and is the Filipino term for a village or ward. These barangays were previously identified as potential sites for micro hydropower (Bellen and Siringan 2015). Barangay or Brgy. Mabato has a land area of 273.1 hectares with a population of 687 in 2015 (PSA 2015). It is classified as rural and zoned as upland conservation zone. Due to the barangay's relatively smaller land area and population, Brgy. Mabato has one (1) day care center, one (1) elementary school and one national high school (City Government of Calamba 2015). Only small enterprises, such as sari-sari stores, can be found in the barangay. Being a rural area, the residents usually work downtown or in other villages. It is approximately 12 kilometers or 60 minutes from the center of the city. Located southwest of Metro Manila, Brgy. Canlubang, on other hand, is a major industrial zone in Laguna with seven (7) industrial parks located in the barangay. The Silangan Industrial Park and Carmelray Industrial Parks occupy around 400 hectares in the barangay. According to an interview with barangay officials, majority of the income of the barangay come from the industrial parks. Commercial and trading establishments, such as malls and other shopping centers, are also present in the barangay. It has a land area of 3912 hectares and a population of 54,943 inhabitants (PSA 2015). While it is predominantly known for its industries, Brgy. Canlubang is also considered the most populated barangay in the city due to its large land area. The demand for electricity is high because of the industrial, commercial and domestic activities in the area.

16.2.3 Municipality of Los Baños

The Municipality of Los Baños has a total land area of 56.5 km2 (3.21% of the land area of the province) with 40% of the land area reserved for forest conservation. Declared as the "Special Science and Nature City of the Philippines" through Presidential Proclamation No. 349 in 2007, the municipality is mainly known for its role in agricultural and environmental preservation research through various research institutions such as the University of the Philippines Los Banos (UPLB), International Rice Research Institute (IRRI), ASEAN Center for Biodiversity, Philippine Rice Research Institute, Southeast Asian Ministers of Education Organization (SEAMEO)—Southeast Asian Regional Center for Graduate Study and Research in Agriculture (SEARCA), and the Philippine Council for Agriculture, Aquatic and Natural Resources Research and Development (PCAARRD). While it is also known for its hot spring resorts, the economy of the municipality relies primarily on agricultural production, mostly coconut and banana (Municipality of Los Banos 2015).

In Los Baños, two barangays were chosen as study sites; Barangays Anos and Bambang. Brgy. Anos is classified as an urban barangay with commercial establishments and schools. It has a population of 8039 in 2015 (PSA, 2015). One of the creeks in Los Baños flows through Barangay Anos and feeds the lake in the neighboring barangay. Barangay Bambang extends from the lakeshore area to the Makiling Forest Reserve in Laguna. Brgy. Bambang is classified as an urban barangay primarily because of the presence of hot spring resorts in the lakeshore area. It has a population of 7228 in 2015.

16.3 Framework and Methodology of the Study

16.3.1 Framework of the Study

Based on existing research on social acceptability of hydropower projects, the present study examines acceptability using three factors, namely, (1) awareness of the technology (micro hydropower); (2) perception of impacts; and (3) understanding of roles and responsibilities of stakeholders. The acceptability of a project is dependent on the abovementioned factors and affects the willingness of the community to contribute to the development of the project. This study also considers the need to identify the community's sources of water and its uses, either domestic, commerce or agricultural. The uses of river or stream waters are also determined, either domestic, business, fishing and livelihood, agricultural or navigation. This is important to see if the use of water for micro hydropower can affect other usage of water. Figure 16.4 illustrates the variables used in this study.

16.3.1.1 Awareness of Energy Situation and Micro Hydropower

This component aims to determine the knowledge and awareness of the community on micro hydropower as a source of energy. This will also determine if the people know about the current source and state of electricity provision in the country, the city/ municipality, sources of renewable energy and its benefits to the barangay. This awareness affects the initial acceptability of the project as a community is unlikely to receive an unfamiliar project. This will also help assess the level of information campaign needed should a micro hydropower project is introduced.

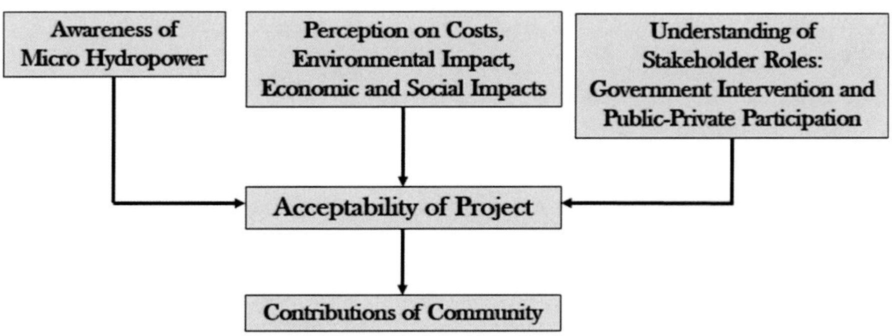

Fig. 16.4 Framework on social acceptability

16.3.1.2 Perception on Costs and Environmental, Economic and Social Impacts

This component entails determining community's perception on the current costs of electricity, costs of construction of micro hydropower and its impact on future electricity costs, which can affect the willingness of the community to use micro hydropower. The section also looks at how the community thinks of the possible impacts of the technology on water use, water quality, land or soil resources and vegetation and agriculture. Aside from environmental impacts, the study looks into the perception on economic impacts such as on livelihood and other economic endeavors, including agriculture, industrial and service activities. Perceptions on the micro hydropower's impacts on the delivery of social services such as education, health, recreational and protective services are also examined in this section.

16.3.1.3 Participation of Public, Private and Community

The community's understanding of the roles of government, private sector and other stakeholders in the project can help increase social acceptability. This component aims to determine activities which the community thinks should be undertaken by the government, private sector and community in project development, planning and implementation.

16.3.1.4 Acceptability of Micro Hydropower Project and Contributions of the Community

Aside from directly asking the willingness to have a micro hydropower project in the area, the study also determines the contributions of the community in project planning and implementation. These include willingness to contribute to the costs of the project, to actively participate in stakeholder consultations in planning and implementation process, to provide useful feedback and suggestions for the project, to help in the construction of the project, to mobilize other residents to participate in community consultations and to organize information campaigns for the project.

16.3.2 Methodology of the Study

16.3.2.1 Survey

Sample respondents from the four (4) barangays have been chosen for the survey. These are Barangays Mabato and Canlubang in Calamba, and Barangays Anos and Bambang in Los Baños. These barangays are identified as potential sites for micro

Table 16.1 Percentage distribution of sample respondents

City/ Municipality	Barangay (village)	2015 Population	% to total population	Number of samples
Calamba	Mabato	687	0.97	4
	Canlubang	54,943	77.50	310
Los Baños	Bambang	7228	10.20	41
	Anos	8039	11.34	45
	Total	70,897	100.00	400

Fig. 16.5 Study areas for the survey on social acceptability

hydropower. The number of sample respondents is determined based on the 2015 Census on population using Slovin's method to estimate sample size. The sample respondents have been selected through convenience sampling, a non-probability sampling technique wherein people who were available and accessible during the time of survey were selected as respondents. Table 16.1 shows the percentage distribution of samples in each barangay (Fig. 16.5).

The respondents were given survey questionnaires which focused on the components of social acceptability as identified in the framework. Statements describing awareness of micro hydropower and perception on social, economic and environmental impacts were formulated and the respondents were requested to indicate their agreement or disagreement to these statements. The survey was conducted from October–December 2016.

16.4 Research Results and Discussion

16.4.1 Characteristics of Respondents

Majority of the respondents come from Barangay Canlubang, about 77% of total respondents, being the barangay with the largest population, followed by Mabato (11%), Bambang (10%) and Barangays Anos (1%). There is almost an equal distribution of male and female respondents at 54% and 56% respectively, and the age ranges from 15–75 years old. Majority of the respondents are engaged in household work and about 23% earn an average monthly income below PhP5,000 (US$ 100). About 36% have income of about PhP10,000 (US$200), and only 21% are within the income level of Php15,000-PhP20,000 (US$300–400). Other respondents are engaged in technical jobs, services and small enterprises with almost the same income level. Among the respondents, about 54% finished high school, while 35% obtained a college education.

The average monthly cost of electric consumption in the household is about Php 1032.00 (US$ 20) and the monthly cost of domestic water consumption is about Php 493.00 (US$ 10). Considering the income levels of the residents, a less expensive electricity source is beneficial.

16.4.2 Current Uses of Water

As shown in Table 16.2, 77% of the respondents in Calamba and 74% in Los Baños are connected to the water supply distributed by the Water Districts, which is sourced mainly from groundwater. Only few respondents (17% in Calamba; 6% in Los Baños) are taking water directly from wells or pumps. The piped water is used primarily for domestic purposes. Respondents from Los Baños do not currently use the river or creek waters, while in Calamba very few respondents use water for domestic purposes (i.e. wash dirt from objects). As the creek or river water is hardly used in the community, it can be utilized to generate micro hydropower. There are also no cultural and traditional practices tied to these water systems. This situation contributes to the generally positive response towards the use of rivers and creeks for micro hydropower.

Table 16.2 Sources of water in Calamba and Los Baños

	Calamba		Los Baños	
Source of water	Number of respondents	%	Number of respondents	%
Groundwater (pumps/wells)	52	17	5	6
Piped/metered	243	77	74	87
Both	17	5	6	7

Table 16.3 Awareness on micro hydropower as a source of energy

Do you know about micro hydropower as a source of energy?	Percentage of Respondents (%)					
	Calamba			Los Baños		
	Yes	No	Uncertain	Yes	No	Uncertain
Calamba	37	48	15			
Los Baños				44	52	5

16.4.3 Awareness of Energy Situation and Micro Hydropower

Being a less popular source of electricity, about 48% of respondents from Calamba and 52% from Los Baños do not know about micro hydropower but are willing to know more about the technology. On the other hand, about 37% in Calamba and 44% in Los Baños have less knowledge of the technology. The rest are uncertain on what they know about micro hydropower. Table 16.3 shows the awareness of people on micro hydropower.

Majority of the respondents have electricity all the time, although interruptions occur during typhoons. Despite the seemingly consistent supply of electricity, the respondents believe that micro hydropower is beneficial to their barangays or local communities. This is either to lower the present cost of electricity or to supplement the present supply for economic and social development. A project on micro hydropower is also in line with the government's thrust to utilize renewable energy resources to promote sustainable development. This openness and willingness of the residents to know more about micro hydropower is a vital component in accepting the technology, provided that the effects on the environment and the economy are beneficial to society.

16.4.4 Perception on Costs and Environmental, Economic and Social Impacts

Table 16.4 shows the perception of respondents on the cost of electricity and the construction of micro hydropower. Most respondents from Calamba (86%) and Los Baños (82%) believe that the cost of electricity is expensive. Although majority perceives that the construction of micro hydropower is also expensive, more than 70% of respondents from both areas think that using micro hydropower will result to reduction of electricity cost in the future. In relation to the willingness to contribute to the project costs, about 35% of respondents from Calamba and 54% from Los Baños are willing to pay to use micro hydropower. This means that while there is a perceived positive impact when it comes to future savings, investment for the project may not be sourced from the residents, but will be provided by the government or the private sector. This is generally understandable given the economic and

Table 16.4 Perception on costs

	Calamba			Los Baños		
	Percentage of Respondents (%)					
	Yes	No	Uncertain	Yes	No	Uncertain
Do you think electricity is expensive?	86	5	7	82	7	8
Do you think that constructing a micro hydropower plant in your barangay is expensive?	76	9	15	54	18	27
Do you think that using micro hydropower will reduce electricity costs in the future?	77	9	13	78	6	15
Are you willing to pay extra costs now to use micro hydropower?	35	46	18	54	31	14

Table 16.5 Perception on economic impacts

	Calamba			Los Baños		
	Percentage of Respondents					
	Yes	No	Uncertain	Yes	No	Uncertain
Do you think there will be jobs generated through the micro hydropower plant?	78	4	15	75	15	9
Do you think there will be more agricultural activity due to the additional source of electricity?	66	11	21	61	16	18
Do you think there will be an increase of industrial activities due to the additional source of electricity?	72	7	18	75	15	9
Do you think there will be an increase in service sector activities due to the additional source of electricity?	74	6	18	89	2	8

financial background of the respondents, as well as the nature of the project, which can be categorized under public utility or infrastructure.

Perceptions of respondents on the economic impacts of micro hydropower are given in Table 16.5. Respondents think that the project will generate positive effects, in terms of added employment opportunities. Because the respondents do not use the river/ creek waters in their current livelihood activities, they believe that there will be no significant loss of livelihood if these waters will be used for micro hydropower. But there is a need to undertake a detailed economic impact analysis to include possible effects on other community members. A general perception also exists that the addition of a source of electricity will lead to an increase in agricultural productivity, industrial and service activities.

Perceptions of respondents from Calamba and Los Baños quite differ on the issue of environmental impacts and these are given in Table 16.6. About 38% of Calamba respondents think that micro hydropower will not affect water use, while 62% from Los Baños have the same perception. This might be because the scale of the technology is small and the river systems in Los Baños are not utilized for any other purpose. In Calamba, 38–40% perceives that the technology will not negatively affect land resources, water, vegetation and agriculture, while in Los Baños a higher percentage (58–68%) of respondents have similar thoughts on the issue.

Table 16.6 Perception on environmental impacts

	Calamba			Los Baños		
	Percentage of Respondents					
	Yes	No	Uncertain	Yes	No	Uncertain
Do you think that using a micro hydropower in your barangay to generate energy will affect water use?	38	38	22	26	62	11
Do you think that micro hydropower will have negative impacts on water?	29	38	31	19	68	13
Do you think that it will negatively affect land resources?	27	39	32	26	58	13
Do you think it will negatively affect vegetation and agriculture?	32	40	26	27	61	9

Table 16.7 Perception on social impacts

	Calamba			Los Baños		
	Percentage of Respondents (%)					
	Yes	No	Uncertain	Yes	No	Uncertain
With additional electricity from micro hydropower project, will there be an increase in the provision of educational facilities?	71	6	12	81	20	5
Will the students have more time to study when there is additional electricity through the project?	64	14	2	89	19	8
Will there be more recreational activities available or accessible to families and children with more electricity?	74	5	7	86	19	7
Will health facilities be more accessible to the community through the project?	73	4	9	88	20	1
With the project, will there be an increase in street lighting?	75	2	2	91	21	6
Will there be an increase in protective services through the project?	77	3	5	87	17	6
Will the project increase the general welfare of the community?	75	3	2	91	20	6

More respondents in Calamba are also quite uncertain on their perception of environmental impacts. Further studies have to be made to identify specific impacts on the environment, which can also help enhance the understanding of the community.

Micro hydropower is also perceived to positively affect the social sector, particularly the provision of educational, health and protective services, as shown in Table 16.7. Respondents also believe that this can help increase other forms of recreational activities. Moreover, about 75% of Calamba respondents and 91% of respondents from Los Baños believe that the project will increase the general welfare of the community. Summing up the respondents' perception of the technology, there is an overwhelming openness and positive outlook on the possible outcome of the project.

Table 16.8 Participation of public, private and community

	Calamba			Los Baños		
	Percentage of Respondents (%)					
	A[1]	D[2]	Neutral	A[1]	D[2]	Neutral
The government should initiate the development of a micro hydropower plant.	83	4	9	84	11	5
The government should provide subsidies for the construction of a micro hydropower plant.	83	5	8	84	11	5
The private sector can also undertake and spearhead the project.	77	12	9	41	48	7
The community must be informed of the development of the project through stakeholder consultations or information campaigns.	88	3	7	95	1	2
Information materials on the project should be made available to the community.	90	3	4	95	2	1
The community should be involved in the planning process of the project.	90	3	4	96	1	1
The community should be involved in the implementation process of the project.	88	4	5	92	5	4
The venue for consultations should be accessible to the community.	90	4	3	92	4	2
An efficient feedback mechanism should be established in the community.	94	3	1	96	1	1
All stakeholders must adhere to the principles of transparency and accountability in all stages of project development.	94	3	2	94		1

[1]A- Agree; [2]D- Disagree.

16.4.5 Participation of Public, Private and Community

Generally, the communities expect the government to provide public infrastructure and utilities, such as micro hydropower to avoid additional costs on the people. If the private sector undertakes the project, respondents believe they will be required to pay, while the private sector generates profit from the project. Majority believe that the government should initiate the development of micro hydropower, although there are few hesitations because of corruption issues in the public sector. The community also believes that strong community participation is essential in this project, both in planning and implementation processes. Table 16.8 shows the possible roles and activities of public and private sectors, community and other stakeholders in the project.

16.4.6 Overall Willingness and Contributions to the Project

The communities generally perceive that the introduction of micro hydropower in their area will be beneficial, particularly in improving electricity condition as well as in supporting existing and future economic activities by providing additional source of energy. One major consideration of the communities is the perceived decrease in electricity costs in the long run. In both areas, more than 90% of the respondents are willing to have the project in their locality (Table 16.9).

In relation to community efforts, many respondents from Los Baños are willing to participate in project consultations, give useful feedback for the project, mobilize and inform project-affected persons. They are also willing to participate in the planning and implementation process as well as in the actual construction of the project. However, in Calamba only few respondents are willing to contribute to the project, especially in the cost of construction and operations (Table 16.10). Respondents from Calamba have less knowledge on micro hydropower and this might have affected their willingness to contribute to the project (Table 16.11).

The results of the survey emphasize three major considerations of the community in accepting the project: (1) reduction of future electricity costs without investment cost on their part; (2) increase economic and social benefits; and (3) community participation. If these factors can be considered in the project planning process, the communities are highly likely to accept the project. The results of the survey have

Table 16.9 Willingness to have micro hydropower in the community

	Calamba			Los Baños		
	Yes	No	Uncertain	Yes	No	Uncertain
Do you think a micro hydropower project will help the community?	97%	1%	2%	93%	4%	3%
Are you willing to have micro hydropower project in your area?	90%	4%	6%	91%	5%	4%

Table 16.10 Community's contribution to the successful implementation of the project

	Calamba		Los Baños	
	Number of respondents	%	Number of respondents	%
Contribute to the cost of project construction and operation.	32	10	32	38
Actively participate in stakeholder consultations during project planning and implementation process.	67	21	67	79
Give useful feedback or suggestions for the project.	56	18	56	66
Help in project construction.	51	16	51	60
Mobilize other residents to participate in stakeholder consultation.	61	19	61	72
Organize information campaigns for the project.	69	22	69	81

Table 16.11 Summary of factors affecting social acceptability of micro hydropower project in Laguna

Factors	Summary of discussion
Current uses of groundwater and rivers/creeks	Groundwater is generally used for domestic and commercial purposes. In Brgy. Canlubang, water is also used for industrial and manufacturing processes.
	River/creek waters are not used by surrounding households. Therefore, these are available for other uses such as energy source.
Awareness of technology	Majority of the residents are not familiar with the technology but are willing to learn about it.
Perception on cost	The construction is perceived to be costly but may have positive long-term effect on electricity cost reduction.
	Government is expected to subsidize project costs.
Perception on environmental impacts	In comparison with Calamba, most respondents from Los Baños believe that micro hydropower will not affect water use. The river water in Los Baños is not used and the scale of the technology is small.
Perception on economic impacts	The project is expected to increase agricultural, industrial and service activities in the area. The positive perception contributes to the acceptability of the project.
Perception on social impacts	This will increase the provision of social services, particularly health, education, protection and recreation, and to the general welfare of the community.
Understanding of stakeholder roles	Government is expected to initiate the project.
	If private sector implements the project, electricity costs are perceived to be higher because of the profit-making nature of the private sector.
	The community has to be involved in all stages of project planning and implementation.
	Stakeholder consultations and information campaigns should be undertaken.

been discussed and validated during the stakeholder consultation meeting with officials of Barangay Anos and representatives from the local government of Los Baños and UPLB. The main concerns raised during the meeting include the cost, source of funds, schedule of project construction and level of participation of the barangay and the community in the project. These concerns can be discussed and recommended to the local government should a micro hydropower project be proposed in the area.

16.5 Summary and Conclusions

Having previously established the technical feasibility of introducing micro hydropower in some communities in Laguna, this study determines the social acceptability of such a project by examining the current, historical and cultural uses of water, their awareness and knowledge of the technology and the perceived economic, environmental and social impacts that will be caused by the project. Different roles of stakeholders are also examined as these can affect community's willingness to accept the technology.

The present study finds that river/creek waters are not generally used by surrounding households. The communities use groundwater for their domestic and commercial needs. Therefore, there is no perceived conflict if the river/creek waters will be used for other purposes, such as for the generation of electricity through micro hydropower. In Calamba, 38–40% of respondents thinks that the technology will not negatively affect land resources, water, vegetation and agriculture, while in Los Baños a higher percentage (58–68%) of respondents have similar thoughts on the matter. Majority of respondents believe that micro hydropower has minimal negative impacts on the environment because of the size of the project, which is relatively smaller than mini and large hydropower plants. However, further studies are necessary to identify specific impacts on the environment, which will also help enhance understanding of the project.

Generally, communities in both study sites perceive that the introduction of micro hydropower in their area will be beneficial, particularly in providing additional source of energy. Although the construction may be costly, the project may help reduce electric costs in the future. However, the communities are generally not amenable to pay for project costs. The project is also expected to increase economic and social services and promote the general welfare of the community. Majority of the respondents suggest that the project should be initiated by the government, while some respondents are hesitant due to corruption issues in the public sector. Although the private sector has enough resources and technical capacity to implement this type of project, some are not willing to let the private sector handle the project. This is related to the perception that electricity costs will increase because of the profit-making nature of the private sector.

In relation to community involvement, many respondents from Los Baños are willing to participate in project consultations, give useful feedback, mobilize and inform project-affected persons. They are also willing to participate in project planning and implementation. However, in Calamba only few respondents are willing to participate in project activities. Respondents from Calamba have less knowledge on micro hydropower and this might have affected their willingness to contribute to the project.

This study finds that small communities are open to different sources and means to generate electricity given availability of water resources. However, there is still a need to provide more information to the community on the characteristics and operations of a micro hydropower plant, its possible effects on the community and the costs and risks involved. Understanding of these information can enhance social acceptability of the project.

Acknowledgment This research is funded by the Research Institute for Humanity and Nature (RIHN) Project R08-Init Human Environmental Security in Asia Pacific Ring of Fire: Water-Energy-Food Nexus, in Kyoto, Japan.

References

Bellen C and Siringan F (2015) Preliminary investigation on the potential of micro hydropower in Calamba and Los Baños, Laguna: a mixed methods approach. Presentation during the project meeting for RIHN on human environmental security in Asia Pacific ring of fire: water-food-energy nexus at the research Institute for Humanity and Nature, Kyoto, Japan on 26 Oct 2015
CALABARZON Regional development council (2011) CALABARZON regional development plan 2011–2016
City Government of Calamba (2015) City profile
Department of Energy (2012) DOE Philippine hydro power resource map
Department of Energy (2015) DOE Philippine power situation report
Ha-Duong Minh, Anh Nguyen Lan, Tracey Strange and An Ha Truong (2015) Social acceptability of large infrastructure projects in Vietnam. Field Actions Science Reports
IEA (2000) Hydropower and the environment: survey of the environmental and social impacts and the effectiveness of mitigation measures in hydropower development. IEA Technical Report
Jago-on K, Siringan F, Tarriela R, Taniguchi M, Reyes Y, Lloren R, Pena M, Bagalihog E (2017) Hot spring resort development in Laguna Province, Philippines: challenges in water use regulation. J Hydrol Reg Stud 11:96–106
Moula (2013) Researching social acceptability of renewable energy technologies in Finland. Int J Sustain Built Environ
Municipality of Los Banos (2015) The special science and nature city
Philippine Statistics Authority (2015) 2015 Census of population
Ribeiro et al (2013) Public opinion on renewable technologies: the Portuguese case.. Sustainable Electricity Power Planning
Simon D (2015) Economic perspectives and social acceptance of possible increase in the small scale hydropower plants in Finland. Tampere University of Applied Sciences
Tabi A, Wustenhagen R (2015) Keep it local and fish-friendly: social acceptance of hydropower projects in Switzerland. Center for Energy Innovation, Governance and Investment (EGI-HSG)
World Bank (2008) The welfare impact of rural electrification: a reassessment of the costs and benefits.. An IEG Impact Evaluation

Chapter 17
Socio-economic Vulnerability and Benefits to the Community Associated with Floating Fish Cages in the Jatiluhur Reservoir

Deny Hidayati, Syarifah Aini Dalimunthe, and Intan Adhi Perdana Putri

Abstract This paper discusses socio-economic vulnerability and benefits of the Jatiluhur Reservoir in the context of Floating Fish Cages (FFCs). The FFC program originally aimed to provide alternative compensation for the local people, but it eventually became an important business, involving investors from outside the local community who dominated FFC ownership. The activity has grown significantly due to the rapid increase in fish demand and markets, not only for local consumption but also in the Province of West Java, Jakarta, and in some cities on the island of Sumatra, such as Bandar Lampung and Palembang. FFC units have increased in recent years and the reservoir has now surpassed its maximum carrying capacity. This study used qualitative methods with focus group discussions, open interviews and field observation as instruments, as well as secondary data collected through desk reviews.

Keywords Socio-economic · Vulnerability · Benefits · Community · Floating fish cage · Jatiluhur · Reservoir

17.1 Introduction

17.1.1 The Importance of Jatiluhur Reservoir in the Context of Floating Fish Cages

Jatiluhur Reservoir is one of the biggest reservoirs in Indonesia, located in Purwakarta District, West Java Province. Its main purposes are to supply water for wet land rice, or *sawah* irrigation, and to generate hydroelectric power at a plant known as PLTA. The reservoir also serves domestic clean water for Subang,

D. Hidayati (✉) · S. A. Dalimunthe · I. A. P. Putri
Research Center for Population, Indonesian Institute of Sciences (PPK - LIPI), Jakarta, Indonesia

© Springer Nature Singapore Pte Ltd. 2018

A. Endo, T. Oh (eds.), *The Water-Energy-Food Nexus*, Global Environmental Studies, https://doi.org/10.1007/978-981-10-7383-0_17

Karawang, Bekasi, and Jakarta, as well as industrial water, particularly in Purwakarta and Bekasi where major manufacturing and textile industries are located. More recently, the reservoir also serves to prevent and manage floods in the surrounding fertile *sawah* area, especially in the Karawang District. The reservoir was constructed and managed by the national government through its management authority, and has been supervised under the State-Owned Enterprise namely Perum Jasa Tirta II (PJT II) since 1999 (Perum Jasa Tirta II, 2014).

The water from Jatiluhur was originally planned to supply about 240,000 hectares of *sawah* using a technical irrigation system in the Districts of Karawang and Bekasi. The development of this reservoir rapidly gained momentum because the availability of reservoir water increased the frequency of rice harvesting in *sawah*, from only once to twice a year. This change resulted in a substantial increase in the total rice production per year in the West Java Province, particularly in these two districts. The additional growing season allowed production to meet the increasing demand for rice in the Provinces of West Java and DKI Jakarta, where populations have been growing rapidly.

The reservoir was built between 1957 and 1967 and covers about 83 km^2 (1.2 km long and 110 m high). The reservoir was a conversion of wet land *sawah*, other agricultural lands, and housing settlements. Construction of the reservoir was challenging because the government had to resettle around 5002 people from 14 villages to neighboring areas of Purwakarta and Karawang Districts and had to provide compensation for their livelihood. The government also introduced floating fish cages (FFCs) in this reservoir as an alternative livelihood for the local people.

The argument for the introduction of FFCs in Jatiluhur was clear; it is suitable for a broad range of open freshwater ecosystems, especially reservoirs. FFCs were claimed to be an economically efficient method for utilizing water bodies in a way that exploits their natural productivity. Nevertheless, few drawbacks of having FFCs in the reservoir were considered. With more and more joint investment ventures, FFCs occupied space on the surface of water bodies and were poorly positioned. They diminished the scenic value of the reservoir, worsened sedimentation, and poor management led to pollution of the environment with unconsumed feed and fish fecal waste.

Water quality was deteriorating. The pH value in the Jatiluhur is seen as critical for fish production. Masser (1997) highlighted that the limit of PH for which failure of fish production can occur is below 4 and above 11. The research shows that there is a decrease of suitability in pH from 2009–2011. The pH is decreasing within the area zoned for net cage aquaculture. The concentration of nitrate was fairly suitable for FCC aquaculture; however, the concentration of nitrate has been increasing in the last five years. This situation should alarm farmers and stakeholders.

Hermana et al. (2014) carried out an analysis of water quality in the reservoir based on West Java Government Regulation No 39/2000. Measured levels of ammonia and biochemical oxygen demand (BOD) exceeded the recommended rate by more than 50 percent, while sulphide is always higher than the regulated limit. This situation is suspected as the cause of corrosion of power plant equipment.

17.1.2 Floating Fish Cages in the Jatiluhur Reservoir

Aquaculture by means of floating fish cages is one of the main uses of the reservoir. The locals call it *keramba jaring apung* (KJA) or floating fish cages (FFCs). The number of FFCs has been skyrocketing each year and has currently reached an alarming level. The burgeoning fish industry expanded the number of units to about 20,000 by the end of 2013. The fish industry in Jatiluhur supplies the Jakarta and South Sumatra demand for fresh water fish, particularly the common carp and tilapia.

The government introduced FFCs to the local communities, in which family lands were converted to construct the reservoir in 1974. It was a pilot project with the main purpose of providing an alternative income activity for these communities. Prior to the project, locals were not interested in the FFC activity because it required capital and skill. However, their interest grew when they saw the success of the government pilot project, in which the FFC production was high and provided a substantial income stream. The success of and the economic benefit from the FFC program has attracted the local communities to practice this activity.

"Ada gula, ada semut" (when there is sugar, there are ants). This proverb seems appropriate for describing the development of FFCs in the Jatiluhur. Information about the high economic benefit from this activity has also reached fishery people in business outside Jatiluhur and Purwakarta. More and more fishery businessmen have come to this area and started to invest their capital in FFC activities. By 1995, around 2100 FFC units had been established in the Jatiluhur, and the number increased to 2400 units in early 2014. (The District Fishery Office of Purwakarta, 2000 and PJT II, 2014).

17.1.3 Research Objective

This chapter explores the Jatiluhur reservoir, focusing on socio-economic vulnerability and benefits of FFCs to the local communities. It begins with a brief history of the reservoir and the major reason for the establishment of FFCs in the reservoir. It then discusses vulnerability and benefits of the FFCs, with a focus on economic, social, and stakeholder issues. The economic aspect focuses on the over-capacity of the FFCs due to a significant increase in FFC units and fish production. The social aspect analyzes the domination of FFC ownership from outside of the reservoir area and the need to revitalize local wisdom. The last aspect discusses the stakeholders' involvement in FFC management, their roles in regulation and operation, and conflicts of interest among stakeholders.

17.1.4 Method

This study utilizes a combination of primary and secondary data collection. The primary data collection includes focus group discussions (FGDs), open interviews, and field observations, while the secondary data collection largely relies on desk reviews.

Focus group discussions aimed to get a brief understanding of the FFC background and its management, stakeholders involved, main issues, and efforts to deal with those issues. The FGDs were conducted with different groups of stakeholders, including the PJT II as the Reservoir Authority, the FFC owners, and the District Government Offices involved in FFC management. Each FGD highlighted a different topic. The FGD with the PJT II focused more on the background of FFC establishment and the role of the PJT II in its management. FGDs with the FFC owners discussed FFC operational activities, including FFC locations, unit numbers, fish types, and how they grow fish (the amount and type of fish feed needed, production, problems related to economic and social aspects, and how to deal with these problems). FGDs with the district government offices were carried out with the head and staff of the District Office of Fishery and Husbandry, the District Agency for Development Planning (BAPPEDA), and the Purwakarta Government Office (Sekertaris Daerah). The discussions focused on relevant regulations and the role of these institutions in FFC management and fishery policies and programs.

Open interviews aimed to get a better understanding of FFC management issues. The interviews were conducted with representatives of the FFC stakeholders, including the PJT II staff, the FFC owners and workers, the Fish Feed Company staff, the Head of BAPPEDA, the District Head Office of Fishery and Husbandry, and Agriculture, and some local key community members.

Some of the data was also collected using a field observation method. The main objective was to get a more comprehensive understanding of FFC management and related issues. Field observations are important and needed to help validate the information gathered from the FGDs and open interviews. The researchers, with this method, could directly observe these aspects in the field, instead of depending only on information from other persons.

In addition to primary data collection, this study conducted library research and secondary data collection. Data on economic aspects, especially related to the increased number of FFC units, the production of fish, and its demand, are based on secondary data collection, such as from statistics collected and published by some of the District Offices, and previous research results.

17.2 Economic and Social Aspects of FFCs

Fishery activities are becoming an increasingly important secondary use of reservoirs, particularly in developing countries, including the Jatiluhur in Indonesia. Reservoir fisheries tend to provide a relatively affordable source of animal protein to their populations. FAO (2014) estimates that about 4.3 percent of the world's food is supplied by aquaculture. According to Welcomme and Bartley (1998), reservoir fisheries development has the potential to contribute markedly to the global supply of fish, particularly in Asia (De Silva 2001).

One of the main objectives of developing FFCs in the Jatiluhur is to meet fish demand in this area, and in addition, to provide an alternative job opportunity for local community members whose lands were converted into the reservoir. This section will discuss economic vulnerability and benefits, focusing on the potential for exceeding FFC carrying capacity due to increasing fish production, demand, and markets.

17.2.1 Increasing Fish Production and Demand

FFC aquaculture in Jatiluhur Reservoir substantially contributes to Purwakarta District fish production, averaging about 32.4 percent between 2004 and 2014. During this period, the fish production increased significantly, almost six-fold with an average annual increase of about 30.4 percent (see Fig. 17.1). The largest increase occurred in the year 2010 when fish production rose around 73.5 percent. The dominant fish types are common carp and tilapia, which comprise about 31.8–32.0

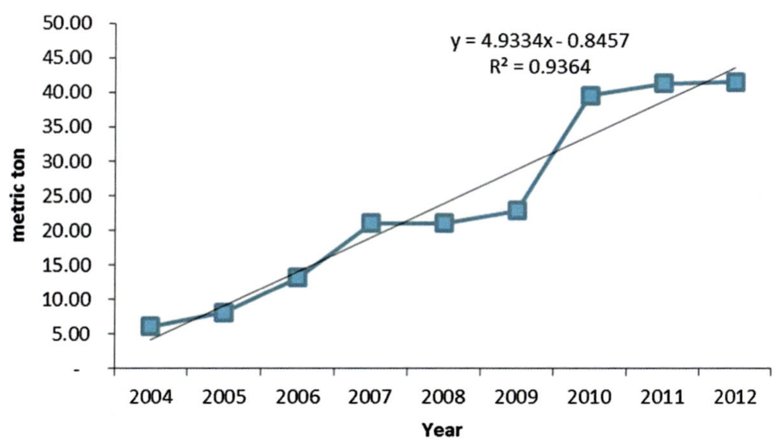

Fig. 17.1 Total Fish Production Using FFCs in the Jatiluhur Reservoir, Year 2004 to 2012. (The Purwakarta District Office of Animal Husbandry and Fishery, 2005; 2008; 2010; 2012; 2013)

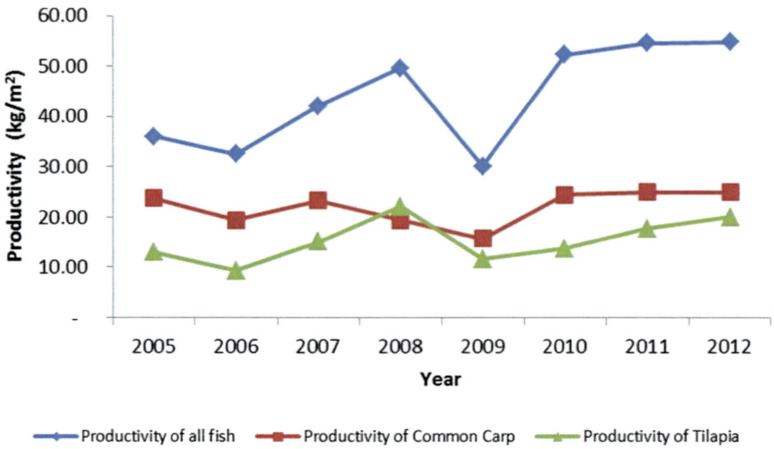

Fig. 17.2 Fish Productivity from FFCs in the Jatiluhur Reservoir. (The Purwakarta District Office of Animal Husbandry and Fishery, 2005; 2008; 2010; 2012; 2013)

percent of fish production in the District of Purwakarta. At the same time, fish productivity in the Jatiluhur has increased, corresponding with the expansion of the FFC area (see Fig. 17.2).

From Fig. 17.2, we can see that FFC productivity is increasing annually in the Jatiluhur for all fish, common carp, and tilapia. Productivity increased by about 52.36 percent between 2005 and 2012. The increase differed sharply among fish types, with the change for tilapia much higher than for common carp. During this period, the total increase in productivity for tilapia was about 54.92 percent, while the total increase for common carp was 5.43 percent. The productivity of these fishes fluctuated over the years. Both fish types experienced reductions in their productivity at certain points in time: common carp in 2005, 2008 and 2009, and tilapia in 2005 and 2009.

The FFCs in Jatiluhur Reservoir produce fish for both local consumption and distribution to consumers outside Purwakarta District (Fig. 17.3). The FFC farmers usually sell their fish to the fish trading collectors that come to their fish cages. These fish collectors then distribute the fish within West Java Province to cities such as Purwakarta, Cimahi, Bandung, and Jakarta. The Jatiluhur fish are also marketed further to Sumatra Island, to areas such as the City of Lampung in Lampung Province and Palembang in South Sumatra Province.

The statistical data from the Animal Husbandry and Fishery Office shows that between the year 2004 and 2012, most of the fish (76 percent) from Jatiluhur FFCs was marketed outside the Purwakarta District, and the rest was consumed locally. The data also indicates that fish demand tended to increase over time, especially from outside Purwakarta.

The growing demand for fish production from the Jatiluhur has driven FFC owners to substantially increase their fish production. As the area available for additional FFC production in the Jatiluhur has become limited, application of fish

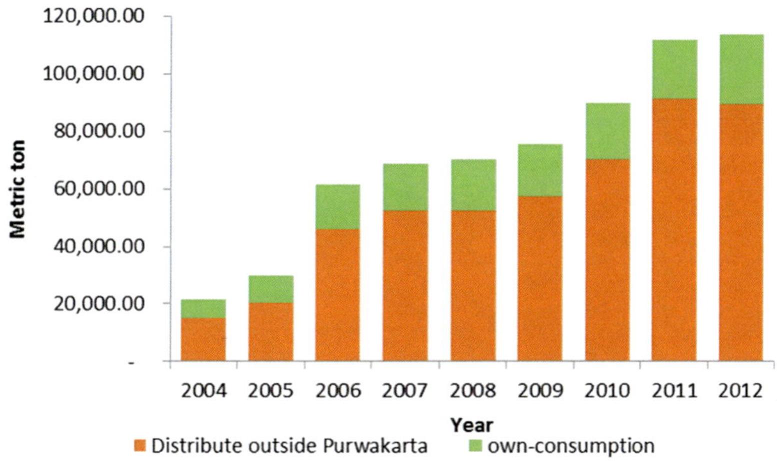

Fig. 17.3 Total Fish Production for Local Consumption versus Distribution Outside Purwakarta, The year 2004–2012 (The Purwakarta District Office of Animal Husbandry and Fishery, 2005, 2006, 2007, 2008, 2009, 2010, 2011, 2012; 2013)

feed (fish pellets) has increased. Currently, according to a staff member of a fish feed company, average utilization of fish pellets for FFC production in the Jatiluhur is approximately 6 metric tons per month.

17.2.2 Exceeding the Carrying Capacity of FFCs

High demand for freshwater fish has increased FFC activity in Jatiluhur reservoir. It is economically attractive to people outside the area to invest more. However, the reservoir has a limit to support FFC aquaculture activities, beyond which self-pollution and negative feedback takes place and reduces fishery productivity (Costa-Pierce 1997). The limit, also known as the carrying capacity, can be reached rapidly if intensively managed units receive a large amount of high-protein feed. According to one of the fish farmers in the Jatiluhur, the first negative feedback occurred in 1997. Over 1560 metric tons of fish (common carp and tilapia) died due to the up welling (turnover) phenomenon Nevertheless, the number of FFC units and the use of fish pellets continued to increase.

The number of FFCs in the Jatiluhur, as well as the number of owners, has increased during the last decade. The FFC area increased from 20.14 hectares in 2004 to 75.62 hectares in 2012 (The Purwakarta District Office of Animal Husbandry and Fishery, 2005, 2008, 2010, 2012, 2013) (see Fig. 17.4). At the same time, the number of FFC owners rose significantly, almost three-fold, from 207 to 572 households (owners). By early 2015, there were around 23,740 fish cages (nets) covering an area of around 116.33 hectares (Susanti 2015).

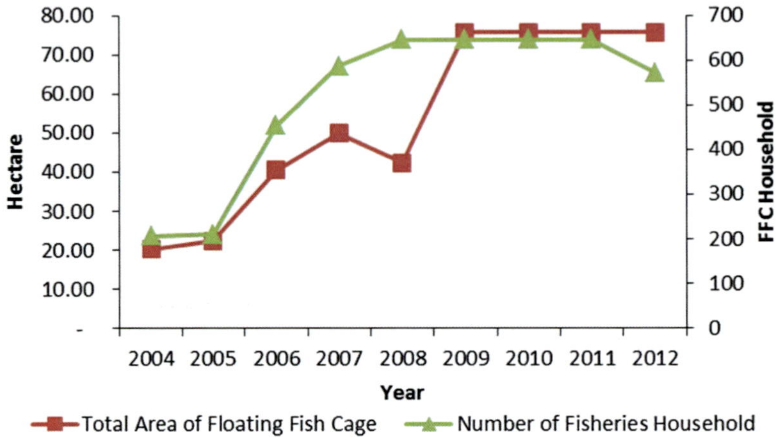

Fig. 17.4 The Total Area and Number of Floating Fish Cage Owners in Jatiluhur Reservoir, 2004 to 2012. (The Purwakarta District Office of Animal Husbandry and Fishery, 2005; 2008; 2010; 2012; 2013)

The number of FFCs in the Jatiluhur, according to the District of Purwakarta Decree No. 6/2000, already surpasses the maximum capacity. The decree states that total number of FFCs that are allowed in this reservoir are about 2100 cages (nets). The existing number of FFCs, therefore, is already ten-fold higher than the number permitted by the decree.

The maximum FFC number determined in the decree, however, is not based on reservoir carrying capacity, particularly as related to pollution load. Machbub (2010) conducted research on pollution load carrying capacity in several lakes and reservoirs in Indonesia, including Jatiluhur Reservoir. His research results showed that the Jatiluhur carrying capacity is approximately 6692 fish cages (nets). The FFC carrying capacity determined by the study is higher than the number of Purwakarta's decree, but it is still below the existing number of FFCs in the reservoir.

Machbub's (2010) research result confirms that the FFC numbers in the Jatiluhur are already over its carrying capacity. Therefore, the number of FFCs in the Jatiluhur, based on this result, should be decreased by approximately 72 percent. Such a decrease in the number of FFCs would be substantial and would have a large impact on the amount of waste, which highly affects water quality and fish production.

Beveridge and Stewart (1998) question the sustainability of intensive FFC aquaculture practices, while considering both environmental and economic aspects. They weigh the pros and cons of FFC culture in the reservoir. The pros include inexpensive operation, as well as ease of construction and management. The cons are mostly related to the fact that the reservoir is not always a simple place for FFC practice, especially as production increases. Our study, based on interviews with FFC owners in the Jatiluhur, attempted to verify Beveridge and Steward's explanation. The FFC owners, during their operation (from the year 1995 to 2014), had to move their fish cages (nets) to other locations at least five times. The main

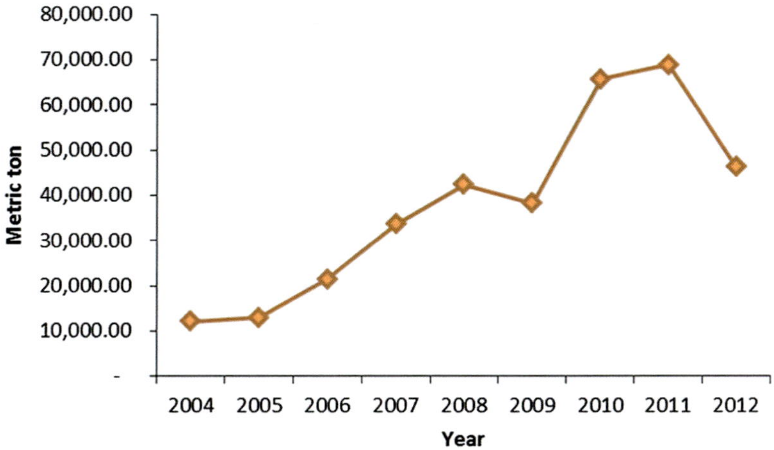

Fig. 17.5 Amount of Fish Feed in the FFC Jatiluhur Reservoir, Year 2004 to 2012

reason for the relocations was to improve fish production because they thought that the water quality of the previous fish net locations had already degraded.

The degradation of water quality was partly due to the significant increase in intensive FFC units. This type of increase affects the use of fish pellets (feed) and can lead to anthropogenic eutrophication problems. Figure 17.5 illustrates that fish feed utilization in the Jatiluhur increased from 12,112.00 metric tons in 2004 to 46,213.46 metric tons in 2012 (the Purwakarta District Office of Animal Husbandry and Fisheries, 2005, 2006, 2007, 2008, 2009, 2010, 2011, 2012, 2013).

17.2.3 The Dilemma of whether to Increase FFC Production

Currently, the Indonesian Ministry of Marine Affairs and Fisheries promotes fish as a major source of food security and nutrition, with a particular emphasis on fish production from aquaculture. One of the main fish production sources in the West Java Province is FFC aquaculture in Jatiluhur Reservoir.

Increasing production in the Jatiluhur is, therefore, viewed by some as an essential strategy going forward. This effort, however, is problematic because it is a two-edged sword. On one hand, it is needed for national food security and nutrition, but on the other it competes with other resource uses (hydro power plant, irrigation, drinking water, and fisheries) and can create social inequity (Beveridge and Stewart 1998). The increase in fish production would also lead to additional reservoir problems related to sedimentation, water quality, and other environmental degradation.

To sustain fish production, we cannot rely only on standard business practices. FFC aquaculture currently uses intensive methods, and farmers are yet to use environmentally friendly fish pellets because the cost is higher than the feed they regularly use.

17.2.4 Social Aspects of FFCs

When reservoirs are built, large communities living in the potential impoundment area can be displaced, depriving them of their traditional livelihoods, often leading to discontent and conflicts with authorities and governments (Roy 1999). The thriving FFC industry in the Jatiluhur is due largely to the government's efforts to boost the displaced community's economy. There are no requirements applied to locals when they start FFC production. However, due to lack of capital and specialized skills, locals often must rely on investors from outside their area of residence.

More than 12,000 FFC occupy the reservoir, most of them owned by investors from areas outside Jatiluhur, such as Jakarta, Bandung, Sukabumi, and Purwakarta. The investor domination is due to the fact that most locals cannot afford to maintain and buy new units, especially in the FFC zoning area. The unit price of a FFC in the zoning areas is very high and its operational cost is also high. Many locals, therefore, can only become laborers for other FFC owners. This condition raises jealousy for some local people because the development of FFCs was originally intended for those whose lands were converted into the reservoir (their parents and/or grandparents). Many of them, therefore, have decided to illegally build and operate FFC units outside the zoning areas.

Buruh (workers) for FFC aquaculture production are largely locals, while the owners are people from outside of the district, such as Jakarta, Bandung, and Sukabumi. There is an internal hierarchy within the buruh itself. There are buruh nelayan (casual workers), juragan nelayan (capital owners), and bandar (middlemen), with each actor having their own role in the FFC production chain.

Buruh nelayan's role is to restock and feed the fish. They also participate in the harvesting process and shipping of the fish to the middlemen. Working hours are from 08:00 am to 04:00 pm. The owners pay a salary ranging from 80 USD to 120 USD/month. When production is complete, workers get another 5 percent of the revenue.

To diversify their livelihood, workers tend to partake in other economic activities, such as *ngampleh* (catching fish using nets instead of FFCs) or setting up their own FFCs outside the official zone. Owning a small piece of farm land (typically less than 0.25 ha) and raising cattle is another alternative. Mostly, the women work on the land while the men are away to tend the FFCs on a daily basis. A good relationship with the FFC owner may create an opportunity for workers to rent the cage for a certain amount of time, particularly in less productive areas. A rental agreement must be settled between the laborer and the owner, usually for a price of around 50 USD to 70 USD/month. This helps the workers to generate some side income and expand their skills. Sundanese, as the majority ethnic group in the reservoir, have their own local wisdom. A well-known Sundanese idiom is *leuweungruksak, cai beak, hirup balangsak*, which means, "no forest, no water, no future". This concept of preventing overexploitation of natural resources in the water catchment area seems to have been forgotten.

The locals also understand how closely linked their lives are to the water stream. This idea is reflected in the geographical names of certain locations where the Sundanese live. For example, the word "ci" in Citarum, the catchment of the reservoir, means water in Sundanese. Protecting environmental sustainability is also part of the local wisdom. Setting up a community-based organization, namely RaksaBumi, or an environmental task force will help to guard and patrol the environment in order to maintain water supply to the downstream areas.

The two examples of local wisdom above need to be translated into reservoir management practices. Adopting those two and involving the locals and reservoir users would lead to better reservoir management, higher water use efficiency, stronger sense of ownership and overall reservoir services satisfaction. Moreover, this integration of traditional and contemporary information would simultaneously preserve local wisdom and generate innovative knowledge.

17.3 Stakeholders and Their Interests

This study identifies three main issues related to FFC management: stakeholders involved in FFC management, their roles in FFC regulation and operation, and conflicts of interest among stakeholders.

17.3.1 The Main Stakeholders

There are many stakeholders involved in Jatiluhur Reservoir FFC aquaculture, but based on their management role, they can be classified into two main groups. The first is responsible for FFC regulation and the second manages operations. The first group includes the PJT II and the Purwakarta District Office of Fishery and Husbandry. The PJT II, as the Reservoir authority, focuses its regulation on the usage of the Reservoir for FFC activity, while the Fishery District Office's main concern is standard FFC operations. The second group consists of the FFC owners and their workers, and fish feed companies. The owners and workers manage their FFCs to achieve desired levels of fish production, and the companies are concerned with the sustainability of fish feed supply and their profit from this activity.

The PJT II oversees three main regulatory processes: designation of the FFC zoning area, the use of zoning licenses, and tax payments. This institution has established a zoning area for FFC operation based on ecological and biophysical criteria to ensure continued functioning of the reservoir. The zoning area covers about 0.1% of the total reservoir size or 10.3 hectares. All FFC units must be operated in this area to allow the PJT II to monitor total unit numbers and avoid reaching the resource's carrying capacity.

The PJT II decreed that any operator of an FFC in the Jatiluhur must obtain permission from the management authority, the PJT II. This institution issues permits for establishing FFC units in the zoning area. The FFC owners receive a license after completing the requirements. The Jatiluhur authority also requires a tax payment from FFC owners of about USD 10 per FFC unit per year.

In addition to the PJT II, the Purwakarta District Office of Fishery and Husbandry has the responsibility of regulating the operation of FFC units. This includes a permit letter for the operation of FFC units, including information such as unit numbers and the size of fishing net per unit. Although FFC owners must apply for a license from the PJT II, they must also get permission for operational activities of their FFC from the District Fishery Office. For this purpose, each owner must pay a one-time fee of about USD 50 to this office.

In order to limit the establishment of FFC units in the reservoir, the Fishery District Office set maximum ownership of FFC units at 20 units per owner/person. The Fishery District Office also provides a standard for fishing net size: 7 m length x 7 m width x 8 m height. This is based on the standard for fishing net size that has been established by the Ministry of Fishery and Marine Affairs at the national level and adopted by the Fishery Office in the Province of West Java.

The FFC owners and workers are responsible for the operation of FFC units. One the proper licenses are obtained, each FFC owner is free to manage his/her units, usually based on his/her capacity, needs, and objectives. Most owners are not local people. They usually are from Jakarta, Bandung, and Purwakarta, and invest their capital in FFCs in the hopes of making a profit from the business. Almost all of them, therefore, hire labor and are dependent on workers for the day-to-day operations of their FFC units.

Meanwhile, the fish feed companies play an important role in FFC operations, especially in relation to providing sustainable fish feed. Many companies supply fish feed (fish pellets) in the Jatiluhur, but only three major companies have been continuously operating: Charoen Phokphand, Sinta Prima Vinyll, and Cargill.

The three main fish feed companies currently supply about 6 thousand tons of fish feed per month. The total volume of supplied feed has increased substantially due to changes in FFC operational activities, especially as related to fish harvesting practices. Fish harvesting frequency has risen from two times a year (every 6 months) to three times a year (every 4 months). In order to increase fish harvesting frequency, the owners have significantly increased their fish feed volume. This change has been received positively by the feed companies. They maintain good relations and provide some incentives to the FFC owners and workers in order to sustain fish feed supply/sales from their companies.

17.3.2 Ineffective Implementation of FFC Regulations

This study suggests that the FFC regulations mentioned above have been ineffective. Regulations require FFC units to be operated in a zoned area, but in recent times, many FFC units, especially those owned by local people, have been established and operated outside the FFC zone. Their justification is that the zoning area has become too limited for new FFC units and thus the cost per FFC unit is very expensive. Most of the local communities, due to their economic situation, cannot afford it. Consequently, many of them set up and operate FFCs outside the zoned area. They feel that they have the right to gain access to FFC activities in the Jatiluhur Reservoir because the area used to be farmed by their families, and the original objective of FFC development was compensation for their family lands.

Issuance of FFC permits is also an ineffective regulation tool. The PJT II issues permits for the use of the FFC-zoned area, but this rule is also not strictly enforced. FFC owners who do not obtain a permit letter from the PJT II can still freely operate their FFC units. The requirement to obtain an FFC operating permit letter from the Purwakarta District Office of Fishery and Husbandry has similarly not been strictly enforced. Almost all of the FFC units have operated long before the required permit letters were obtained.

The ineffectiveness of existing regulations can also be seen from the exceedance of the carrying capacity of FFC units in the Jatiluhur. The total number of units has reached more than 2400 in 2014, whereas according to the District Head of Purwakarta Decree No. 6 in the year 2010, the carrying capacity of FFC in this Reservoir was 2100 units. This problem is related to the ineffective enforcement of the limitation on the total number of FFC units per owner, which is 20 units. Most owners have more than 20 units, and many others, especially investors from outside Purwakarta, have more than 100 units.

The latest regulation being considered is the prohibition of all FFC operations within the Jatiluhur Reservoir, also known as the 'zero FFC' policy. The regulation is planned to be implemented in 2018. According to PJT II (2017), the 'zero FFC' policy will be undertaken because of three issues: (1) decreasing water quality; (2) the weathering of concrete and metal corrosion in some parts of the hydropower facilities; and (3) decreasing fish productivity in the Jatiluhur Reservoir. In order to reduce and control the number of FFCs, PJT will enlist support from the MUSPIDA (Regent, District Military Commander) and the MUSPIKA (sub-district leader, sector police head, Commander of the Rayon Military Command). In 2015–2016 this action succeeded in removing 2731 unlicensed fish cages. This activity was also driven by a concern for national security. Some FFC operations were viewed as being potentially linked to forbidden organizations and terrorism.

The zero FFC policy certainly will affect local peoples, such as FFC owners and laborers, who rely on the dam for their livelihoods. It is estimated that the number of local owners makes up 30 percent of total FFC owners in Jatiluhur Reservoir, and in 2015, PJT II recorded that the number of FFCs is 25,951 units. Consequently, in 2018 when the zero FFC policy is implemented, 633 local owners and 2595 laborers could lose their livelihoods, despite the local government's (Purwakarta Regency) plans for diverting local livelihoods to capture fisheries and tourism. The local government plans to develop a floating resort and traditional floating market, while simultaneously implementing the zero FFC policy in the very near future. This may lead to conflict among the local people, government, and manager of the Jatiuhur Dam (PJT II). Moreover, since Jatiluhur Reservoir supplies freshwater fish to other Provinces such as Lampung, West Java and Jakarta, the ban may cause an increase in fish prices (Koeshendrajana et al. 2011). However, since Jatiluhur Reservoir is a National Vital Asset, the mushrooming of FFC units, besides causing the decrease in water quality that affected hydropower equipment, also increased National Security concerns. In December 2016, four suspected terrorists were arrested in a floating fish cage house.[1]

The ineffective implementation of FFC regulations is due to two main factors. The first is related to limited information of the regulations available to public; most of the FFC owners do not know and are not aware of them. The second is caused by low enforcement for those who do not follow the regulations. The PJT II and the Purwakarta District Office of Fishery and Husbandry for example allow owners to continue operating their FFCs even if they do not have a permit letter.

17.3.3 Conflicts of Interest Among Stakeholders

Each stakeholder has his/her own interests that might differ or conflict with those of other stakeholders in the Jatiluhur. The conflicts of interest are basically related to individual management objectives. Such conflicts could affect FFC management and sustainability of this activity.

The PJT II's main interest, as the reservoir authority, for example, is to optimize the function of the reservoir and its carrying capacity. This institution must maintain water quantity and quality in order to achieve its objective. To do so, the PJT II has to reduce FFC unit numbers and application of fish feed, given that the reservoir is already considered as being both over capacity and over used. This authority then set up a new regulation that prohibits all FFC activities in this reservoir.

The interest of Purwakarta District Office of Fishery and Husbandry differs sharply with the PJT II. The district Office has instructions from the Ministry of Marine Affairs at the national level to increase fish production in this district by about 300 percent. Therefore, it has to increase fish production in an area where the main source of fish production is FFCs in the Jatiluhur.

Similarly, the main interest of FFC owners is to increase fish production, which leads to continual addition of more units. This interest has led to significant increases in the number and density of units in the zoning area. They, especially the locals, even expand FFC production to areas outside the FFC-zone. They also apply higher volumes of fish feed, leaving waste and residual impacts that lead to lower water quality. This condition has a negative impact on the sustainability of their activities.

The increasing demand for fish feed is positively received by the fish feed companies because their main interest is to gain more profit from their production. The companies actually realize that the fish feed generates residual impacts on water quality, but so far, there is no fish feed that is environment-friendly. The government agency in charge of this type of issue, the Ministry of Marine Affairs, has not set up guidance and standard regulations to overcome this problem.

---

[1] https://nasional.tempo.co/read/830456/ini-tanda-pengantin-teroris-akan-bom-waduk-jatiluhur

17.4 Conclusion

This paper illustrates that the local communities only play a small role in the development of FFCs in the Jatiluhur Reservoir. They are unable to compete with the FFC investors that come from outside the reservoir. Many of them, due to their economic condition (low capital and skill), have to become labor workers. Increasing the role of the locals is therefore crucial for achieving the original purpose of developing the FFC industry in this area, which was to provide better livelihoods for the locals, especially those whose family lands were converted into the reservoir.

The FFCs in this reservoir have contributed to a thriving fish business, as indicated by the significant increase in its unit numbers and the expansion of production outside the zoned areas. Demand for fish from the Jatiluhur FFCs has increased substantially, not only from the surrounding area but also other cities. This activity faces many challenges, most of which have socio-economic aspects. The total number of FFC units, according to the reservoir authority and the Fishery District Office, has exceeded the reservoir's carrying capacity. The FFC owners, to meet the demand, apply a high volume of fish feed and thus increase the frequency of harvesting and total fish production. This practice causes a substantial increase in feed waste, which continues over time because regulation and treatment to reduce its negative impact on water quantity and quality are still very limited. As a tradeoff of economic growth, the environment has been degraded, leading to increased risk. Some efforts have been applied, but they are still very limited, especially related to major causes of FFC degradation, supporting regulations and their ineffective implementation.

FFC management, therefore, should be improved by developing standards and regulations based on FFC density and carrying capacity. Making information about standards and the latest regulations readily available to the public, and effective enforcement are crucial, especially to increase awareness and encourage responsible behavior by the FFC owners. It also emphasizes the need to evaluate fish feed regulation to reduce negative impacts on water quality.

The zero FFC policy that primarily focuses on the sustainability of the Jatiluhur environment, but with less attention on the livelihood of the local people, likely cannot be successfully implemented as planned in 2018. More time is needed to introduce this policy to the local people and for preparing their alternative livelihoods. The implementation of this policy in the very near future may cause conflict in this area.

References

Animal Husbandry and Fisheries Agency of Purwakarta District (2005) Animal Husbandry and Fishery Statistics 2004

Animal Husbandry and Fisheries Agency of Purwakarta District (2006) Animal Husbandry and Fishery Statistics 2005

Animal Husbandry and Fisheries Agency of Purwakarta District (2007) Animal Husbandry and Fishery Statistics 2006

Animal Husbandry and Fisheries Agency of Purwakarta District (2008) Animal Husbandry and Fishery Statistics 2007

Animal Husbandry and Fisheries Agency of Purwakarta District (2009) Animal Husbandry and Fishery Statistics 2008

Animal Husbandry and Fisheries Agency of Purwakarta District (2010) Animal Husbandry and Fishery Statistics 2009

Animal Husbandry and Fisheries Agency of Purwakarta District (2011) Animal Husbandry and Fishery Statistics 2010

Animal Husbandry and Fisheries Agency of Purwakarta District (2012) Animal Husbandry and Fishery Statistics 2011

Animal Husbandry and Fisheries Agency of Purwakarta District (2013) Animal Husbandry and Fishery Statistics 2012

Beveridge MCM and JA Stewart (1998) Cage culture: limitations in lakes and reservoirs. Petr T (ed) Inland fishery enhancements. Papers presented at the FAO/DFID expert consultation on inland fishery enhancements. Dhaka, Bangladesh, 7–11 Apr 1997. FAO Fisheries Technical Paper. No. 374. Rome, FAO p 463

Costa-Pierce Barry A (1997) From farmers to fishers: developing reservoir aquaculture for people displaced by dams. 71p World Bank technical paper; no. 369. Fisheries series

De Silva SS (2001) Reservoir fisheries: broad strategies for enhancing yields. In: SS De Silva (ed) Reservoir and culture-based fisheries: biology and management, pp 7–15. ACIAR Proceedings No. 98, Canberra, Australia, ACIAR

FAO (2014) Yearbook of fishery statistics summary tables. Fisheries and aquaculture department. Retrieved 21 Apr 2014, from http://www.fao.org/fishery/statistics/global-aquaculture-production/en

Hermana, Kamiko, Naoyuki, Maryati S (2014) Analysis of water quality status in JatiluhurReservoir: impacts of aquaculture and residential use. Jurnal Perencanaan Wilayah dan Kota 3(3):545–553

Husbandry A, Fisheries Agency of Purwakarta District (2012) Animal Husbandry and Fishery Statistics:2011

Koeshendrajana S, Apriliani T, Firdaus M, Nasution Z (2011) Penebaran Ikan Bandeng di Waduk Jatiluhur: Analisis Dampak dan Kebijakan Pengembangan. Jurnal Kebijakan Sosial Ekonomi Kelautan dan Perikanan 1(1):1–17

Machbub B (2010) Pollution load carrying capacity calculation model on lakes and reservoir (in Indoensian). Jurnal Sumber Daya Air 6(2)

Masser MP (1997) Cage culture site selection and water quality. SARC Publication 161

Roy A (1999) The cost of living. Flamingo, London, 162 pp

Susanti R (2015) January 5. Save Jakarta water, Purwakarta demolished 20.000 cages in Jatiluhur (in Indonesian) Jatiluhur. Kompas.. Retrieved from http://regional.kompas.com

Welcomme RL, Bartley (1998) Current approaches to the enhancement of fisheries. Fish Manag Ecol 5(1):351–382

Chapter 18
Accounting for Externalities in the Water Energy Food Nexus

Kimberly Burnett and Christopher A. Wada

Abstract The WEF nexus can be generally defined as the relationship between food, energy, and water in human-environmental systems. The linkages between each nexus component are often complex, varying across both spatial and temporal dimensions. However, understanding and quantifying those linkages are key to maximizing potential net benefits from the WEF nexus. While there are many ways to categorize the different type of linkages/relationships, we will use the following three distinctions: (i) direct dependencies, (ii) direct competition, (iii) externalities. Research on the third type of nexus relationship is somewhat less prevalent than research on the other two types of linkages, likely because externalities are often difficult to physically quantify, and even more difficult to monetize. Externalities within the WEF nexus are quite common and should not be ignored. In this chapter, we focus on negative externalities and present two case studies from Japan to illustrate how to account for externalities in the WEF nexus.

Keywords Externalities · Water energy food nexus

18.1 Introduction

The water energy food (WEF) nexus has been used to examine burgeoning global issues such as food security, poverty reduction, and environmental sustainability (Biggs et al. 2015, Mirzabaev et al. 2015, Vanham 2016, Ringler et al. 2013), local level issues such as efficient natural resource management and short and long term planning for food and energy production (Hang et al. 2016; Shifflett et al. 2016), as well as household level analyses of these systems (Hussien et al. 2017, Garcia and You 2016).

Overviews of the WEF nexus framework (Al-Saidi and Elagib 2017, Mohtar and Daher 2016, Allan et al. 2015, Bhaduri et al. 2015, Endo et al. 2015, Leck et al.

K. Burnett (✉) · C. A. Wada
University of Hawaii Economic Research Organization, Honolulu, HI, USA
e-mail: kburnett@hawaii.edu; cawada@hawaii.edu

© Springer Nature Singapore Pte Ltd. 2018

A. Endo, T. Oh (eds.), *The Water-Energy-Food Nexus*, Global Environmental Studies, https://doi.org/10.1007/978-981-10-7383-0_18

2015) emphasize that a key component to most WEF nexus analyses is identifying interlinkages and tradeoffs among the water, energy, and food systems (Chang et al. 2016, Fang and Chen 2017, Rasul and Sharma 2016). Hoff (2011) emphasizes that the WEF nexus approach can increase overall resource use efficiency and benefits in production and consumption by addressing externalities across sectors, for example considering the energy intensity of desalination or water demands in renewable energy production.

Case studies are useful in illustrating these concepts (Daher and Mohtar 2015, Smajgl et al. 2016, Walker et al. 2014), but few have deliberately examined the relationships associated with moving within the WEF nexus in order to identify efficient or cost-effective WEF policy and management strategies (Zhang and Vesselinov 2017). Within the WEF nexus, water, energy, and food resources may be inherently dependent on one another, in direct competition, or have positive or negative effects on each other. While externalities have been identified as being significant in the WEF nexus (e.g., Hoff 2011), a typology does not currently exist for the types of relationships and linkages across the WEF nexus. In this chapter we identify three distinct types of WEF linkages/relationships, with a focus on externalities.

18.2 Types of WEF Relationships

The WEF nexus can be generally defined as the relationship between food, energy, and water in human-environmental systems. The linkages between each nexus component are often complex, varying across both spatial and temporal dimensions. However, understanding and quantifying those linkages are key to maximizing potential net benefits from the WEF nexus. While there are many ways to categorize the different type of linkages/relationships, we will use the following distinctions: (i) direct dependencies, (ii) direct competition, (iii) externalities.

18.2.1 Direct Dependencies

Direct dependencies are often the first type of relationship that comes to mind when considering the WEF nexus. Water can be used as an input to energy production (e.g., hydro-electric power generation, power plant cooling) or food production (e.g., irrigation, aquaculture). Energy can be used as an input to water production (e.g., pumping of groundwater, water treatment) or food production (e.g., mechanized harvesting). Food is an input to both water and energy production via labor. Reducing costs of production inputs has the potential to increase social welfare, and the effect grows with the strength of interdependencies between the nexus components. For example, if the current source of energy is coal but technological developments continue to reduce the costs of solar power, then transitioning to solar will eventually reduce the costs of pumping groundwater for municipal use, leading to welfare gains. At the same time, cheaper energy means reduced costs of treating lower quality water for crop irrigation. Consequently, food can be produced at lower costs and potable water that might have otherwise been used for irrigation remains available for household uses.

18.2.2 Direct Competition

Direct competition between nexus components leads to tradeoffs that must be considered in management decisions when resources are scarce. Competition may be between two nexus components over the third or between all three components over a production input coming from outside of the nexus. In either case, economic optimality is achieved when marginal benefits are equal to marginal costs across all uses of the input, as well as over space and time. Consider first the case where a single water source can be used for electricity production, crop irrigation, or both. Because the water source is finite, as scarcity increases over time, allocation between uses becomes increasingly important. If the marginal equality condition does not hold across uses, then there is an opportunity to increase total welfare by reallocating the water; this is sometimes referred to as an arbitrage opportunity. For example, if the marginal benefit of water in crop production exceeds the marginal cost, then an additional unit of water in crop production can increase total surplus. At the same time, if the marginal benefit of water in crop production exceeds the marginal benefit of water in energy production, total surplus can be increased by reallocating water from energy to crop production. The arbitrage exercise can be repeated in this manner until the marginal equality condition holds, at which point further reallocation does not increase total welfare.

The story is similar when the shared input lies outside the nexus. For example, suppose that a large landowner is considering how best to use his/her parcel of land, and options include solar panels or wind turbines for energy generation, agriculture, ranching, and forest restoration for increased groundwater recharge. The simplest approach assumes that each use of the land precludes other uses. A hectare used for cattle grazing cannot be simultaneously used to grow row crops such as lettuce, unless the lettuce is intended to be used as cattle feed. In this case, economic optimization is achieved when the marginal equality condition is satisfied across the landscape. Given the inherent spatial heterogeneity of a large land parcel, some areas may be more desirable for certain uses, e.g., windy areas for turbines and areas with good soil and sufficient rainfall for crop cultivation. The management challenge, therefore, is to determine how best to distribute uses across the landscape to maximize total welfare. A more holistic approach would attempt to identify potential synergies that allow for overlapping uses to achieve multiple objectives simultaneously. Although combining cattle grazing and row crop production in a single hectare is implausible, ranching and wind power generation may not be incompatible and may, in fact, generate additional synergistic benefits. Supposing

Table 18.1 Classification of resource-extraction externalities (p = efficiency price, c = marginal extraction cost, MUC = marginal user cost, MEC = marginal externality cost, t = optimal tax)

		Durability of damages	
		Transient (one-period)	*Dynamic*
Pathway	*Direct*	Flow externality Ex: Air pollution from coal burning p = c + MUC + MEC t = MEC	Stock externality Ex: Greenhouse gases p = c + MUC + MEC t = MEC
	Indirect	Stock-to-flow externality Ex: Landscape amenity values p = c + MUC $t \neq$ MEC	Stock-to-stock externality Ex: Aquifer level effect on nearshore ecology p = c + MUC $t \neq$ MEC

Adapted from Burnett et al. (2015b)

that turbines do not negatively affect the cattle (other than reducing a small fraction of the grazing area), combining the two land uses has the potential to increase total welfare if the cattle reduce turbine maintenance costs by keeping vegetation in check in the surrounding area.

18.2.3 Externalities

Research on the third type of nexus relationship is somewhat less prevalent than research on the other two types, likely because externalities are often difficult to physically quantify, and even more difficult to monetize. Measurement difficulties notwithstanding, externalities within the WEF nexus are quite common and should not be ignored. An externality is defined here as an economic activity that imposes an effect on an unrelated third party. The effect may be positive or negative, but in either case, the effect is often not reflected in the benefits or costs to the producer of the economic activity. Hence, from a social welfare perspective, activities that generate negative externalities are typically overproduced, while activities that generate positive externalities are typically underproduced. Welfare can be theoretically increased (and optimized) if the externality is appropriately internalized. However, the appropriate approach to internalization depends on the type of externality.

In this chapter, we focus on negative externalities (the approach for a positive externality is analogous). An externality can be classified according to its pathway (direct or indirect) and its temporal nature (transient or dynamic), which results in four possible combinations (Table 18.1). Direct externalities result directly from harvest or extraction. For example, extracting coal and using it to produce electricity generates air pollution and acid rain. An indirect externality, on the other hand, is linked to the stock of a resource. For example, the stock of coastal groundwater affects nearshore ecology via submarine groundwater discharge (SGD). The act of extracting the groundwater itself or using it as an input to production does not have

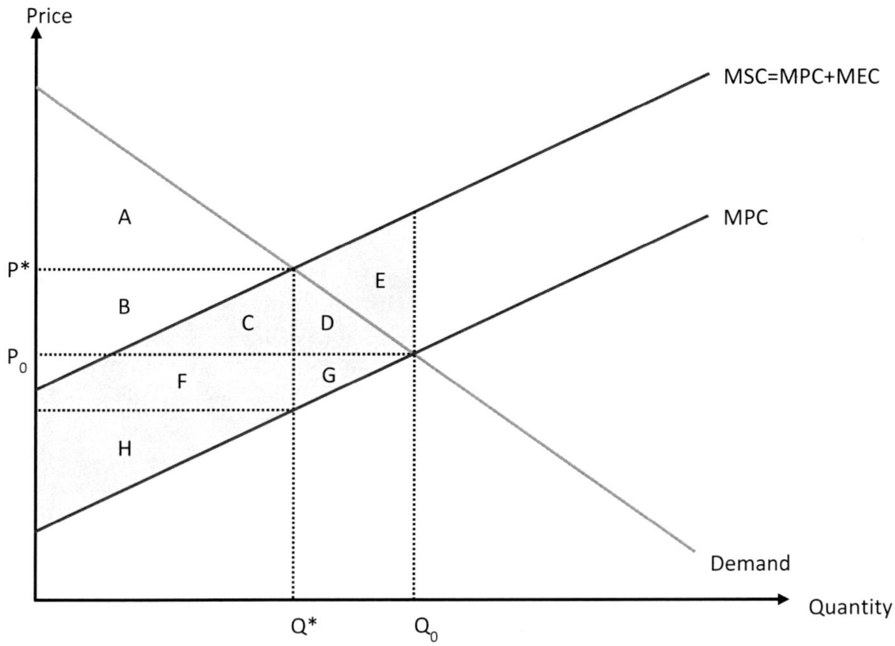

Fig. 18.1 Corrective tax for a transient negative flow externality (MPC = marginal private cost, MEC = marginal externality cost, MSC = marginal social cost). The total externality cost from unregulated production is area CDEFGH. The corrective tax generates a deadweight loss equal to area DG but avoids externality costs equal to area DEG, resulting in net benefit equal to area E > 0

a direct effect on the nearshore environment. However, the cumulative effects of pumping affect the stock of stored groundwater, which is related to the quantity of freshwater discharged at the coast. A reduction in groundwater stock reduces SGD, which in turn can affect the growth rates of important nearshore species that are sensitive to changes in salinity and nutrient levels.

A transient externality lasts for a relatively short amount of time. Going back to the coal burning example, if coal extraction and use cease, soot and acid rain will also stop relatively quickly. A dynamic externality, on the other hand, involves damages that are linked to a stock. In addition to creating acid rain, burning coal adds carbon to the atmosphere, i.e. it increases the stock of greenhouse gases. Because the greenhouse effect is a function of the stock of greenhouse gases, ceasing coal use today will not have an immediate effect on the greenhouse effect.

Burnett et al. (2015a) show that when an externality has a direct pathway, the efficiency price that incentivizes optimal extraction and use of the resource in production is equal to the sum of marginal extraction cost, marginal user cost (MUC), and marginal externality cost (MEC). The MUC captures the scarcity value of the resource, ensuring that the extraction trajectory is optimal over time. The MEC accounts for the damages generated from production. In this case, the socially optimal outcome can theoretically be achieved by levying a tax on resource use equal to the MEC. That is, if the price of coal already accounts for its extraction cost and scarcity value, a tax that reflects the marginal damages incurred will internalize the externality. Figure 18.1 illustrates this idea graphically. For simplicity, we assume that the marginal private cost (MPC) to the producer includes the marginal extraction cost and MUC. Without regulation, the market equilibrium (P_0, Q_0) occurs where the demand curve (representing consumer willingness to pay) intersects the MPC curve. Consumer surplus (CS), the difference between what the consumer is willing to pay and the price actually paid, is equal to area ABCD. Producer surplus (PS), the difference between revenue earned and cost of production, is equal to area FGH. Production of quantity Q_0 generates a total externality cost equal to area CDEFGH. A corrective tax equal to MEC internalizes the externality; the producer now faces a marginal production cost equal to the MSC, resulting in a new equilibrium (P*, Q*). At the new equilibrium, CS is equal to area A, PS is equal to area H, and government tax revenue is equal to area BCF. Consequently, the deadweight loss of the tax is equal to area DG, but the avoided externality cost is equal to area DEG, which means that the net benefit of the tax is equal to area E.

When the pathway is indirect, the formula for the optimal tax is more complex because it needs to account for its own effect on future prices. The tax, therefore, is not exactly equal to MUC or MEC. Moreover, the tax may be changing (imagine the MSC curve in Fig. 18.1 shifting up or down), depending on how the stock of the externality-generating resource is increasing or decreasing over time. Estimation of the optimal tax generally requires information about how the upstream (externality generating) and downstream (externality receiving) resources interact in terms of extraction and stocks, as well as how those interactions are translated to damages. Recalling our groundwater-nearshore ecology example, we need to know (i) how groundwater extraction affects the aquifer stock, (ii) how the aquifer stock affects SGD, (iii) how SGD affects the nearshore ecology, and (iv) how to value the ecological impacts in monetary terms.

Although this chapter does not focus on institutional challenges related to the implementation of externality-correcting policies, it is important to point out that such challenges do often exist.

In particular, complications may arise when the externality is dynamic because the negative impacts may not be realized until long after the externality-generating production has already taken place. Some challenges related specifically to dynamic externalities include assigning responsibility for contamination removal, establishing cost sharing among responsible parties, and applying regulations retroactively (Yoshida 1996; Endo 2016).

In summary, the appropriate approach for internalizing a resource externality depends on its pathway and durability of damages. Flow, stock, stock-to-flow, and stock-to-stock externalities (Table 18.1) often arise in the WEF nexus. As we will see in the two case studies that follow, meeting the information requirements to internalize such externalities are often challenging in complex human-environmental systems. However, ignoring such relationships usually lead to excessive extraction and production, and consequently avoidable environmental damage.

18.3 Case Study 1: Beppu, Japan

Beppu, a city in Oita Prefecture on the island of Kyushu, Japan (Fig. 18.2), is known for onsen (hot springs). Its eight geothermal hotspots attract more than 8 million visitors and 100 million JPY annually (Tourism Association of Beppu; http://www.city.beppu.oita.jp/sangyou/kankou/detail8.html). Although viewed by many as revenue-generating businesses in modern times, onsen in Japan have a rich cultural history dating back hundreds or even thousands of years in some regions. Pressures to modernize notwithstanding, many owners of onsen facilities continue to focus on the historic authenticity of their waters and synergy with nature to appeal to tourists. A recent influx of onsen-energy generation facilities in Beppu has raised concerns with residents and onsen owners about potential negative externalities. An increase in river temperature due to increased drainage water from energy facilities, for example, could potentially affect the distribution of river fish species. Higher river temperatures are more suited for some species than for others. If the species that tend not to thrive in warmer water are valued ecologically, recreationally, as a food source, or otherwise, then the temperature effect is a clear example of a negative externality.

According to our externality classification matrix (Table 18.1), the relationship between onsen-energy discharge and river temperature (the water-energy leg of the nexus) is a stock externality; production of energy directly affects river temperature, and consequently, fish stock. If energy production, river temperature, and fish count data were collected before and after the expansion of onsen-energy generation

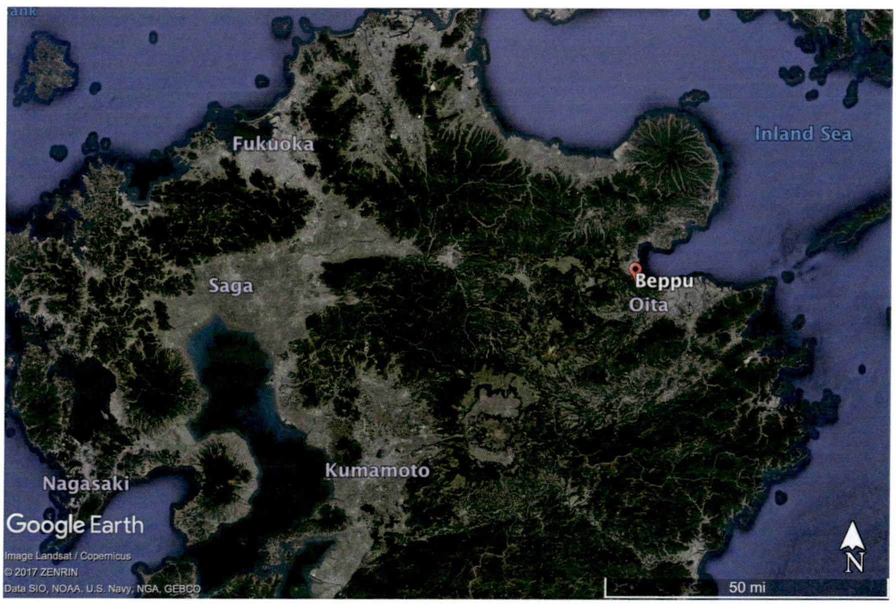

Fig. 18.2 Map of Beppu City in Oita Prefecture on Kyushu Island, Japan

facilities in the area, then the externality effect could be empirically estimated. If instead, however, data was not collected prior to the expansion, then a comparison of post-expansion data for similar rivers (one near and one away from onsen-energy generation facilities) in the region could provide some evidence of the externality, with the caveat that other factors contributing to temperature may not be perfectly controlled for across rivers. Monetizing the externality effect—in this case, damages to the river ecosystem—can be challenging if the affected fish species are not sold in markets, caught for recreation, or consumed for subsistence. However, because the value of a healthy river ecosystem is greater than zero, we can say, at the very least, that energy is being overproduced if our objective is to maximize social welfare. Energy producers who are not incentivized or mandated to account for river ecology effects will generate more energy than those who recognize and internalize social damages generated by their production decisions. Although not first-best optimal, a river temperature constraint (if enforceable) could nudge energy producers toward the socially optimal outcome.

Although some interlinkages are relatively easy to observe, such as the direct effects of onsen-energy production discharge on adjacent river temperatures, other connections may be more difficult to measure precisely. For example, onsen-energy production relies on pumped groundwater, which is discharged to a river after flowing through the generation facility. The drawdown of groundwater eventually reduces SGD, which in turn decreases silica outflow to the nearshore marine environment. Because silica regulates marine diatoms, which maintain dinoflagellates, and consequently prevent algal blooms, this indirect coastal effect may be just as, if not more important than the river effect, especially if nearshore fisheries possess economic, subsistence, and/or cultural importance.

Again, referring to our classification matrix (Table 18.1), the groundwater-nearshore-ecology relationship is a stock-to-stock externality. The cumulative effect of groundwater pumping decisions on the groundwater stock ultimately affects the stock of silica in the nearshore environment. The tax that would incentivize socially optimal groundwater pumping rates (and hence aquifer stock level) is difficult to calculate because the MEC is not separable from the MUC. It requires a clear understanding of the state equations for both stocks (i.e., how they each change over time in response to both natural processes and human decisions), as well as the relationship between them. If the nearshore environment includes a fishery with measurable consumption, then a value can be attached to the stock-to-stock externality damages. It is then possible to calculate the optimal extraction of groundwater for energy production, the optimal stock of groundwater, and consequently the optimal stock of silica in the nearshore environment. If valuation is not straightforward, then a SGD constraint (analogous to the temperature constraint in the river externality example) is one possible management alternative that can partially internalize the externality, although the outcome will not necessarily correspond to the first best social optimum.

This case study showcases several different types of WEF nexus relationships. Onsen-energy production is directly dependent on hot spring water as a production input. The use of groundwater for energy generation reduces SGD along the coast, thus reducing nutrients available for the nearshore environment. Though energy production and nearshore ecology production do not compete directly for the consumptive use of freshwater, water serves as a pathway for nutrient flow, resulting in a negative externality on a potential food source. At the same time, warm water discharge, a by-product of energy production, also generates an externality on the river ecology because river fish productivity is sensitive to temperature.

18.4 Case Study 2: Obama, Japan

Our second case study focuses on Obama, a city in the central Wakasa district, southwest Fukui Prefecture, Japan (Fig. 18.3). Fishing used to be the main industry in Obama, although more recently, the city's economy has shifted largely toward tourism. Groundwater has always been an important resource in the area for consumptive uses, and has also been viewed by residents as possessing cultural and historical significance. Domestic and commercial users currently account for the largest share of groundwater withdrawals in the region, totaling approximately 15,300 m3/day. Another 4000 m3/day is used in winter months for melting snow. However, because Obama City is currently pumping a relatively small fraction of its total groundwater resource, water scarcity in terms of consumptive uses is not an immediate concern (Burnett et al. 2015a; b). More importantly, like in Beppu, the

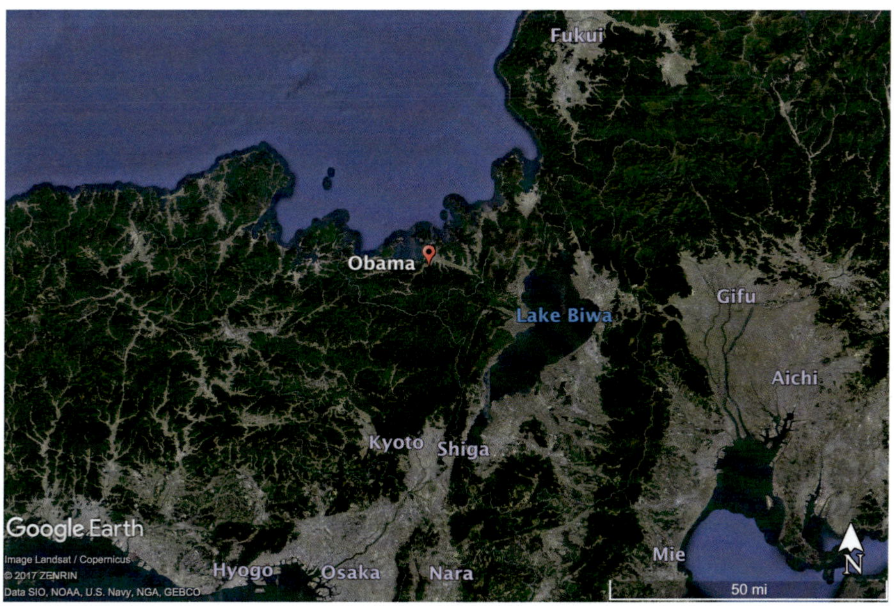

Fig. 18.3 Map of Obama City in Fukui Prefecture, Japan

relationship between the groundwater stock and the nearshore environment is key to characterizing the WEF nexus in the region.

Oyster cultivation in Wakasa started in 1930, and over 20 farms now operate in Obama Bay. Because outputs from the oyster fishery are sold in markets, the value of potential damages from any type of externality (e.g., a stock-to-stock groundwater-oyster externality) affecting oyster production is quantifiable. If available groundwater data is sufficient to characterize the relationship between groundwater pumping, aquifer stock and SGD, then the biggest remaining challenge would be to quantify the relationship between SGD and oyster productivity. In other words, we need to answer the following question: if increased pumping for consumptive uses decreases SGD, how does that, in turn, reduce the quantity and/or quality of oysters being produced in the downstream farms? If all the requisite relationships are quantifiable, then a corrective tax on groundwater use can be calculated.

Although groundwater substitutes for domestic and commercial uses are not readily available, snow-melting can be achieved through other, more energy-intensive means. If the magnitude of the groundwater-oyster externality is large enough, switching from groundwater-based to heat-based snow melting may become efficient. In that case, correcting the water-food externality entails increasing energy use. In other words, protecting or avoiding damages to fishery productivity increases upstream costs in the larger human-environmental system but is economically optimal, as long the aggregate effect on social net benefit is positive.

18.5 Win-Win Solutions and Directions for Further Research

In this chapter, we discussed different ways to look at WEF nexus relationships: direct dependencies, direct competition, and externalities. We also classified externalities into four groups according to their pathways and durability of damages: flow, stock, stock-to-flow, and stock-to-stock. In two Japanese case studies, we focused on externalities within the WEF nexus. One standard approach to internalizing such externalities is a corrective tax on the upstream externality-generating activity. The tax forces the producer to recognize tradeoffs of his/her production activities, e.g. pumping more water for onsen-energy generation reduces groundwater levels, which in turn reduces SGD, and has a negative impact on nearshore fishery resources. In this instance, the corrective action reallocates surplus and increases total social welfare.

In other instances, there may be win-win opportunities to internalize externalities with minimal reallocation. The interdependent nature of the WEF nexus results in many potential conflicts and tradeoffs but simultaneously generates synergistic opportunities. Recall in the Beppu case study that onsen-energy generation results in warm-water discharge that has a negative effect on river ecology. A corrective tax would internalize the externality by incentivizing lower energy production.

However, a secondary use for the water could potentially minimize the externality effect without reducing energy production. For example, suppose that the warm discharged water could be redirected at relatively low cost to provide heat for nearby greenhouses or crop irrigation. Since the secondary user would benefit through this arrangement via reduced energy costs and/or water pumping costs, one can imagine an arrangement where the farmers and onsen-energy producer split the costs of conveyance. This arrangement would allow the energy producer to continue production with a small mitigation cost, avoid damages to the river ecology, and generate cost-savings for secondary users of the groundwater.

Acknowledgements This research was financially supported by the R-08-Init Project, entitled "Human-Environmental Security in Asia-Pacific Ring of Fire: Water-Energy-Food Nexus" Research Institute for Humanity and Nature (RIHN), Kyoto, Japan.

References

Allan T, Keulertz M, Woertz E (2015) The water–food–energy nexus: an introduction to nexus concepts and some conceptual and operational problems. Int J Water Resour Dev 31(3):301–311

Al-Saidi M, Elagib NA (2017) Towards understanding the integrative approach of the water, energy and food nexus. Sci Total Environ 574:1131–1139. https://doi.org/10.1016/j.scitotenv.2016.09.046

Bhaduri A, Ringler C, Dombrowski I, Mohtar R, Scheumann W (2015) Sustainability in the water–energy–food nexus. Water Int 40(5–6):723–732. https://doi.org/10.1080/02508060.2015.1096110

Biggs EM, Bruce E, Boruff B, Duncan JMA, Horsley J, Pauli N, McNeill K, Neef A, Ogtrop FV, Curnow J, Haworth B, Duce S, Imanari Y (2015) Sustainable development and the water–energy–food nexus: a perspective on livelihoods. Environ Sci Pol 54:389–397. ISSN 1462-9011, https://doi.org/10.1016/j.envsci.2015.08.002

Burnett K, Roumasset J, Pongkijvorasin S, Wada CA (2015a) Incentivizing interdependent resource management: watershed, groundwater and coastal ecology. In: Dinar A, Schwabe K (eds) Handbook of water economics. Edward Elgar Publishing, Northampton, pp 150–161

Burnett K, Wada C, Endo A, Taniguchi M (2015b) The economic value of groundwater in Obama. J Hydrol Reg Stud 11:44–52

Chang Y, Li G, Yao Y, Zhang L, Yu C (2016) Quantifying the water-energy-food nexus: current status and trends. Energies 9(2):65

Daher BT, Mohtar RH (2015) Water–energy–food (WEF) nexus tool 2.0: guiding integrative resource planning and decision-making. Water Int 40:748. https://doi.org/10.1080/02508060.2015.1074148

Endo T (2016) An institutional analysis of groundwater quality control: experiences in Hadano, Kanagawa prefecture, Japan. Hydrology 3(2). https://doi.org/10.3390/hydrology3020020

Endo A, Burnett K, Orencio PM, Kumazawa T, Wada CA, Ishii A, Tsurita I, Taniguchi M (2015) Methods of the water-energy-food nexus. Water 7(10):5806–5830

Fang D, Chen B (2017) Linkage analysis for the water–energy nexus of city. Appl Energy 189:770–779. ISSN 0306-2619, https://doi.org/10.1016/j.apenergy.2016.04.020

Garcia DJ, You F (2016) The water-energy-food nexus and process systems engineering: a new focus. Comput Chem Eng 91:49–67

Hang M, Martinez-Hernandez E, Leach M, Yang A (2016) Designing integrated local production systems: a study on the FEW nexus. J Clean Prod 135:1065–1084

Hoff H (2011) Understanding the Nexus. Background paper for the Bonn2011 conference: the water, Energy and Food Security Nexus. Stockholm Environment Institute, Stockholm

Leck H, Conway D, Bradshaw M, Rees J (2015) Tracing the water–energy–food nexus: description, theory and practice. Geogr Compass 9(8):445–460. https://doi.org/10.1111/gec3.12222

Mirzabaev A, Guta D, Goedecke J, Gaur V, Börner J, Virchow D, Denich M, von Braun J (2015) Bioenergy, food security and poverty reduction: trade-offs and synergies along the water–energy–food security nexus. Water Int 40(5–6):772–790

Mohtar RH, Daher B (2016) Water-energy-food nexus framework for facilitating multi-stakeholder dialogue. Water Int 41:1. https://doi.org/10.1080/02508060.2016.1149759

Rasul G, Sharma B (2016) The nexus approach to water-energy-food security: an option for adaptation to climate change. Clim Pol 16(6):682–702. https://doi.org/10.1080/14693062.2015.1029865

Ringler C, Bhaduri A, Lawford R (2013) The nexus across water, energy, land and food (WELF): potential for improved resource use efficiency? Curr Opin Environ Sustain 5(6):617–624. ISSN 1877-3435, https://doi.org/10.1016/j.cosust.2013.11.002

Shifflett SD, Culbreth A, Hazel D, Daniels H, Nichols EG (2016) Coupling aquaculture with Forest plantations for FEW resiliency. Sci Total Environ 571:1262–1270. https://doi.org/10.1016/j.scitotenv.2016.07.161

Smajgl A, Ward J, Pluschke L (2016) The water–food–energy nexus – Realising a new paradigm. J Hydrol 533:533–540. ISSN 0022-1694, https://doi.org/10.1016/j.jhydrol.2015.12.033

Vanham D (2016) Does the water footprint concept provide relevant information to address the water–food–energy–ecosystem nexus? Ecosyst Serv 17:298–307. ISSN 2212-0416, https://doi.org/10.1016/j.ecoser.2015.08.003

Walker RV, Beck MB, Hall JW, Dawson RJ, Heidrich O (2014) The energy-water-food nexus: strategic analysis of technologies for transforming the urban metabolism. J Environ Manag 141:104–115. https://doi.org/10.1016/j.jenvman.2014.01.054

Yoshida F (1996) Geo-pollution and a clean-up system. Econ J Hokkaido Univ 25:61–82

Zhang X, Vesselinov V (2017) Integrated modeling approach for optimal Management of Water, energy and food security nexus. Adv Water Resour 101:1–10. https://doi.org/10.1016/j.advwatres.2016.12.017

Chapter 19
Making Social Networks Visible: Shared Awareness Among Stakeholders on Groundwater Resources

Michinori Kimura, Naoki Masuhara, and Kenshi Baba

Abstract In the sustainable use of groundwater resources, there is a need for smooth governance by stakeholders. Local communities use groundwater resources for a variety of purposes and many different stakeholders are involved, making it difficult to get an overall idea of how groundwater is being used. The purpose of this study is to use social network analysis (SNA) to visualize differences in latent talking point that exist among stakeholders of groundwater resources to structurally understand the potential relationship between the stakeholders of groundwater resources of Obama City, Fukui Prefecture. Results show that stakeholders come to a shared understanding regarding general issues such as "necessity of conservation", "maintaining current usage level", "sanitation", among others. However, the stakeholders do not have a shared understanding regarding specific issues such as "salinization", "temperature", and "natural flow". From a nexus perspective, groundwater is used for various purposes. The fact that concrete issues are not shared between stakeholders with different uses is considered to be a potential conflict.

Keywords Social network analysis · Obama city · Groundwater resources · Stakeholders · Groundwater resource potential problems

---

M. Kimura (✉)
Lake Biwa Environmental Research Institute, Otsu, Shiga, Japan
e-mail: kimura-m@lberi.jp

N. Masuhara
Research Institute of Humanity and Nature, Kyoto, Kyoto, Japan
e-mail: masuhara@chikyu.ac.jp

K. Baba
Tokyo City University, Yokohama, Kanagawa, Japan
e-mail: kbaba@tcu.ac.jp

© Springer Nature Singapore Pte Ltd. 2018

A. Endo, T. Oh (eds.), *The Water-Energy-Food Nexus*, Global Environmental Studies, https://doi.org/10.1007/978-981-10-7383-0_19

19.1 Introduction

Compared to other local resources, the usage of groundwater resources is particularly difficult to ascertain owing to the infeasibility of direct observation. Local communities use groundwater resources for a variety of purposes and many different stakeholders are involved, making it difficult to get an overall idea of how groundwater is being used. The purpose of the use varies depending on the stakeholders, who have different priorities in terms of quantity, quality, and cost performance. The way the current status of the resources is understood, and the way the resources are appraised, will also vary from stakeholder to stakeholder. Thus, local resources that are used for a variety of purposes involving various stakeholders also represent diverse viewpoints.

The differences in understanding among stakeholders of groundwater resources might be a source of conflict when they are confronted with groundwater problems, such as degradation or depletion. For example, actors such as industries, who prioritize quantity and cost performance, will appraise groundwater differently from actors who place more importance on quality and added-value, such as those who use groundwater for food or as a tourism resource. Depending on the condition of the groundwater, these different interests may collide with each other, creating conflict. To promote the sustainable use of groundwater, it is essential to have stable resource governance based on stakeholder consensus, and the stakeholders must therefore properly share information with each other and clarify their talking points. To this end, it is important to ascertain how stakeholders are related to one another before inter-stakeholder conflict erupts. In other words, it is important to identify what potential problems and conflicts are likely to arise among stakeholders.

However, until groundwater problems like degradation or depletion emerge, there is little impetus for stakeholders to communicate with each other, making it difficult to identify the disputes that would potentially arise when the stakeholders start communicating. The latent conflict between stakeholders is thought to arise due to differences in perceptions, values on ways of thinking, such as importance of water quality or quantity and methods of groundwater use.

Therefore, the authors use social network analysis (SNA) to visualize differences in latent talking point that exist among stakeholders of groundwater resources in Obama City, Fukui Prefecture. In this analysis the focus is on stakeholders' shared understanding regarding the talking points on groundwater resources.

19.2 Social Network Analysis (SNA) of Groundwater Resource Stakeholders

Based on network analysis (graph theory), SNA is a process for structurally modeling the relationships between social actors and quantitatively analyzing these relationships based on several metrics. SNA characterizes network structures in nodes (or vertices) and lines (or edges, links, ties, etc.), which indicate the actors and the relationships (or lack of relationships) between them. To model a social network, it is necessary to work out the relationship between each actor and all other actors in the network. There are a number of methods to determine, including direct inquiry (e.g., questionnaire survey) and observing dialogue. However, these methods are usually onerous.

Furthermore, if there is no conflict and stakeholder-driven governance is not yet present, the stakeholders may not be clearly aware of their relationships with other stakeholders. In such cases, using questionnaires or interviews to directly ask stakeholders about their relationships with other stakeholders would not be feasible. Another problem is the multifaceted nature of inter-stakeholder relationships; the relationship between stakeholders might be strong in one aspect, yet weak in another. Due to these problems, it is not feasible to rely exclusively on direct inquiry when investigating the multiplicity of shared understandings that are based on the aggregation of talking points, as doing so would significantly encumber the stakeholders concerned, as well as the researcher.

In view of these concerns, the authors utilize the findings of a study by Baba et al. (2015) which analyzed the interest and concerns among stakeholders who were using groundwater resources directly or indirectly in various contexts. The authors perform SNA on the data to elucidate latent inter-stakeholder relationships, focusing on shared understanding toward the points of concern and contention that Baba et al. extracted. Baba et al. interviewed 38 stakeholders between May and August 2013 as shown in Table 19.1, and they used SNA to extract the 23 talking points shown in Table 19.2. They then arranged these 23 talking points into eight categories.

The relationship between stakeholders and talking points can be represented in a matrix by assigning a score of 1 when a stakeholder is linked to a talking point and a score of 0 when there is no such link. With a matrix that links stakeholders with talking points in this way, one can model a two-mode network graph, as shown in Fig. 19.1, using SNA. A network is called a one-mode network, or unipartite network, if each node can be related to each other node in the same respect. In a two-mode network, or bipartite network, nodes are divided into two sets (such as actors and events) and nodes can only be related to nodes in the other set (de Nooy et al. 2011a).

The 38 stakeholders and 23 talking points in this study yield a two-mode networked structure consisting of two components. This two-mode network can be converted into one-mode networks. As Fig. 19.2 shows, the two modes of the network – talking points and stakeholders – can be converted into two one-mode networks: a talking point network and a stakeholder network.

As shown in Fig. 19.2, in the talking point network, there are three stakeholders who have a stake in both talking points 1 and 2. In the network graph, the relationship between the two talking points is represented by three overlapping lines. A large number of lines between two talking points indicate a large number of stakeholders who have a stake in the two points. Accordingly, line multiplicity can represent the strength of the ties between talking points. As for the stakeholder network,

Table 19.1 Surveyed stakeholders

ID	Affiliation	ID	Affiliation
1	Obama City Government	20	Food
2	Obama City Government	21	Food
3	Obama City Government	22	Food
4	Obama City Government	23	Food
5	Obama City Government	24	Commerce
6	Obama City Government	25	Commerce
7	Obama City Government	26	Commerce
8	Obama Municipal Assembly Member	27	Tourism
9	Fukui Prefecture	28	Consultancy
10	Agriculture	29	Education
11	Agriculture	30	Culture
12	Agriculture	31	Culture
13	Agriculture	32	Local residents
14	Agriculture	33	Local residents
15	Agriculture	34	Citizen group
16	Fishery	35	Citizen group
17	Breeding	36	Citizen group
18	Forestry	37	Citizen group
19	Forestry	38	Citizen group

the strength of the relationships between two stakeholders indicates the amount of talking points in which they both have a stake. When pairs of stakeholder share many talking points, the line multiplicity will be greater, implying a stronger relationship between the two stakeholders.

Converting a two-mode network into one-mode networks generally yields networks with high tie density, making it difficult to visually decipher the relationships therein. To get around this problem, it is necessary to extract the stronger and more central ties from the network. A useful technique for this is the m-slice technique (de Nooy et al. 2011b). The m-slice technique involves removing ties below a certain line multiplicity value (m-value) leaving only the stronger relationships (Fig. 19.3).

When a network is represented as a matrix, a pair of nodes is said to have structural equivalence if their ties with each other and all other nodes are symmetrical such that the two nodes could exchange places in the matrix (Suzuki 2009a). This is the case with stakeholder 1 and stakeholder 3 in the left matrix of Fig. 19.4. Even if these stakeholders exchanged places, there would be no consequences to the network structure. If a pair of nodes has high structural equivalence, this implies that they have a similar role or function in the network. A technique for visualizing structural equivalence is block modeling, which involves color-coding the cells of an adjacency matrix to indicate their relative position in the network (Suzuki 2009b). Simply color-coding the cells in the adjacency matrix will result in a random pattern;

Table 19.2 Points of concern/contention (talking points) among stakeholders of groundwater resources in Obama City, Fukui Prefecture

1. Water quantity	1. Maintaining current usage level
	2. Concerns about excessive use
	3. Concerns about reduction in water quantity
2. Water quality	4. Sanitary concerns
	5. Salinization
3. Purpose of use	6. Domestic use
	7. Tourism / community-building
	8. Industrial
4. Reason for use	9. Fee charging
	10. Underdeveloped waterworks
	11. Temperature
	12. Taste
	13. Cultural tradition
	14. Natural flow
5. Linkage with surrounding environment	15. Impact of soil
	16. Ocean cultivation
	17. Mountain/forest conservation/cultivation
	18. Environment in general
1.1.6. Conservation of groundwater	19. Necessity of conservation
	20. Regulation
	21. Supporting conservation
7. Concerns about residents' viewpoint	22. Concerns about residents' viewpoint
8. Expectations toward scientific survey	23. Expectations toward scientific survey

however, categorizing node pairs with a high structural equivalence can potentially allow the visualization of structural equivalence.

There are several techniques for categorizing nodes with high structural equivalence, such as focusing on node pairs with similar social affiliations. Another statistical technique is cluster analysis. When the two-mode network graph is plotted, this research finds that even stakeholders of the same social affiliation are linked to different talking points, which implies a lack of structural equivalency. Accordingly, the technique employed in this study is hierarchical clustering. When two-mode networks are converted into one-mode networks, the networks are usually weighted by line multiplicities. However, this present study considers line multiplicity as a measure of the strength of relationship. In order to account for clusters of line multiplicity values, Euclidean distance is used for inter-cluster distance.

List of stakeholders' talking points

SH	TP 1	TP 2	TP 3
1	1	1	1
2	0	1	0
3	0	1	1
4	1	1	0
5	1	1	0

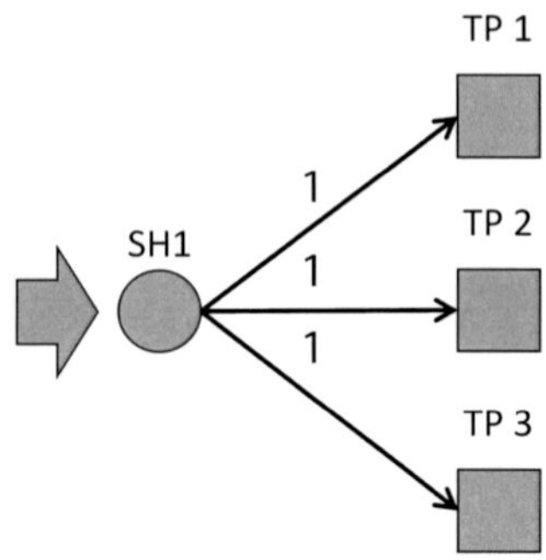

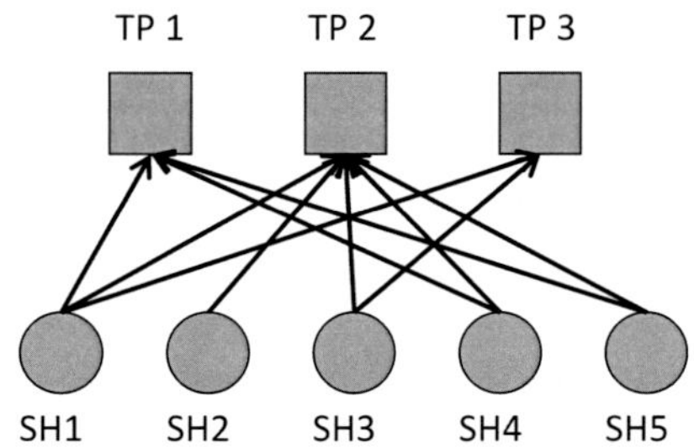

Two-mode network graph of stakeholders / talking points

Fig. 19.1 Overview of two-mode network graph of the relationships between stakeholders and talking points

19.3 Results of Analysis of the Stakeholders of Groundwater Resources in Obama City, Fukui Prefecture

Figure 19.5 shows the two-mode network of shared talking points among the stakeholders of groundwater resources in Obama City, Fukui Prefecture. Square nodes are talking points and round nodes are stakeholders.

This visualization process yields a single coherent network. Some stakeholder pairs only have one shared talking point; for example, the only shared talking point among "4 Obama City Government" and "7 Obama City Government" was "30 culture." However, other stakeholders are linked to multiple talking points, and the number of talking points to which stakeholders are linked varies by stakeholder. Thus, the heterogeneity of the talking points surrounding the groundwater resource has been visually confirmed.

Next, the two-mode network (stakeholders / talking points) is converted into a one-mode network, and this network is represented in a stakeholder network graph as shown in Fig. 19.6.

In the figure, tie thickness indicates line multiplicity value. On the whole, there are no isolated stakeholders; each stakeholder is linked to, and shares talking points

SH	TP1	TP2	TP3
1	1	1	1
2	0	1	0
3	0	1	1
4	1	1	0
5	1	1	0

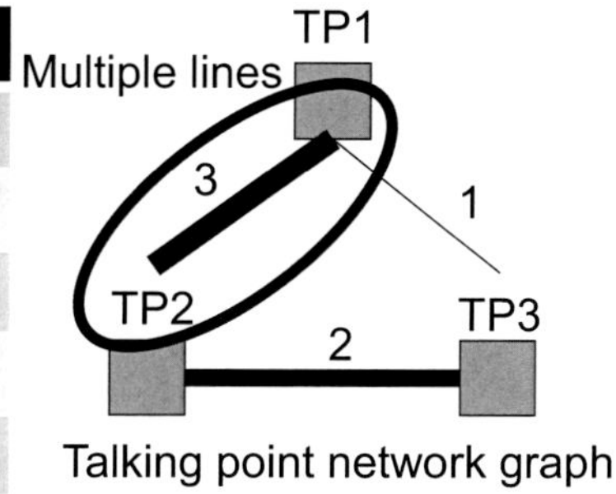

SH	TP1	TP2	TP3
1	1	1	1
2	0	1	0
3	0	1	1
4	1	1	0
5	1	1	0

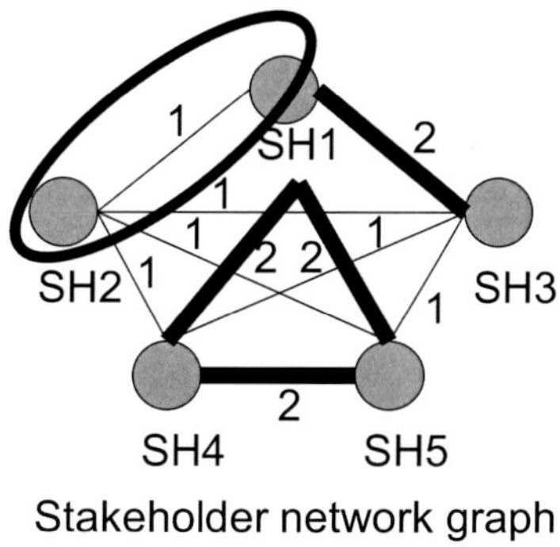

Fig. 19.2 Conversion into one-Mode networks

with multiple other stakeholders. In addition, the network features a myriad of ties, filling up the graph with a mass of color. A complete network, one in which all pairs of nodes are linked, is said to have a maximum density. If maximum density is scaled at 1, then the above network graph has a density of 0.89, which is a high value. Thus, the stakeholders of Obama City's groundwater resources form a cohesive network of dense ties, which are mediated through talking points.

Next, a more central subnetwork structure is extracted from the network. To achieve this, a m-slice technique to extract all ties with an m-value of 10 or more is used, resulting in a subnetwork, as shown in Fig. 19.7.

The resulting network comprises stakeholders of various affiliations. These include the "Obama City Government" and "consultancy" actors, who use groundwater in their business operations; "local residents" and "citizen group" actors, who use groundwater for domestic purposes or engage in water conservation activities; "agriculture" and "fishery" actors, who use or benefit from groundwater; and an "education" actor, who conducts research and education in connection with groundwater. All these actors have a deep connection with groundwater in that they engage in groundwater-related activities on a daily basis, and as such, they have a range of interests in groundwater. It is therefore assumed that they would also share many talking points. The talking point network is shown in Fig. 19.8.

As with the stakeholder network, the resulting talking point network is singular and coherent with no isolated talking points. The network is also very dense. The

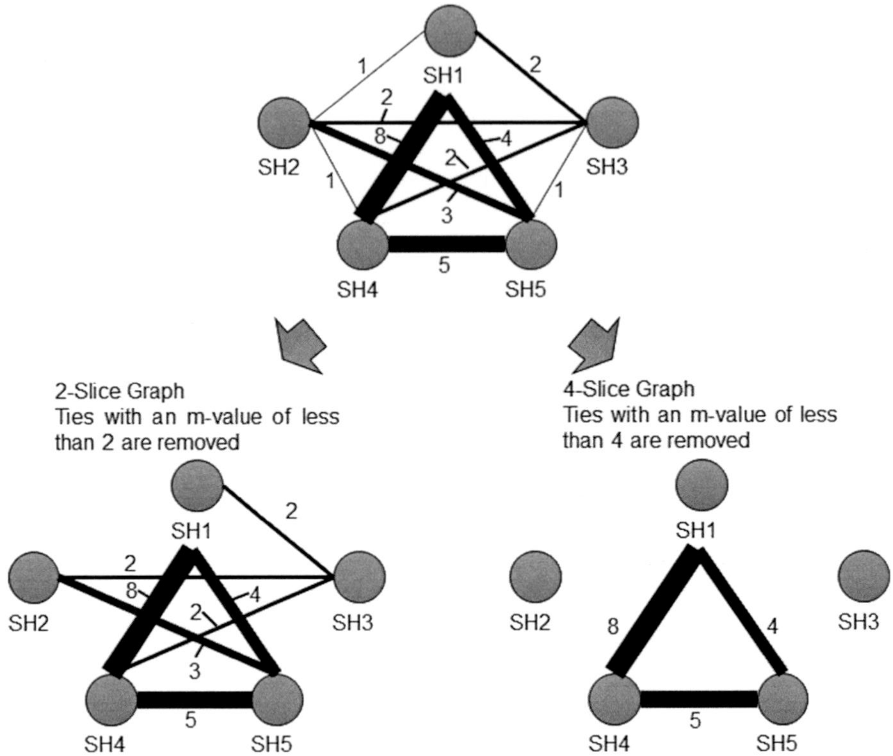

Fig. 19.3 Overview of m-Slice method

Adjacency matrix

	SH1	SH2	SH3
SH1	0	1	5
SH2	1	0	1
SH3	5	1	0

Hierarchical cluster blockmodel

Cluster ID		1	1	2
	SH	SH1	SH3	SH2
1	SH1	0	5	1
1	SH3	5	0	1
2	SH2	1	1	0

Fig. 19.4 Overview of hierarchical cluster block model

density value is almost maximum at 0.96, indicating a nearly complete network. To extract a subnetwork with higher centrality from the network, the m-slice technique is used, extracting lines with an m-value of 12 or more. The results are shown in Fig. 19.9.

The central structure of concerns regarding Obama City groundwater resources consists of concerns about water quantity ("maintaining current usage level" and "concerns about reduction in water quantity"), purpose of use ("industry" and

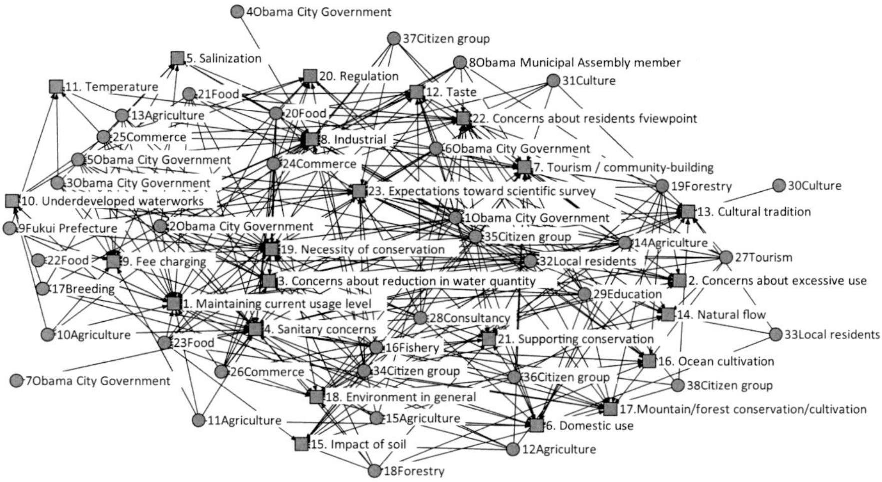

Fig. 19.5 Two-Mode network graph – stakeholders of groundwater resource in Obama city and the talking points

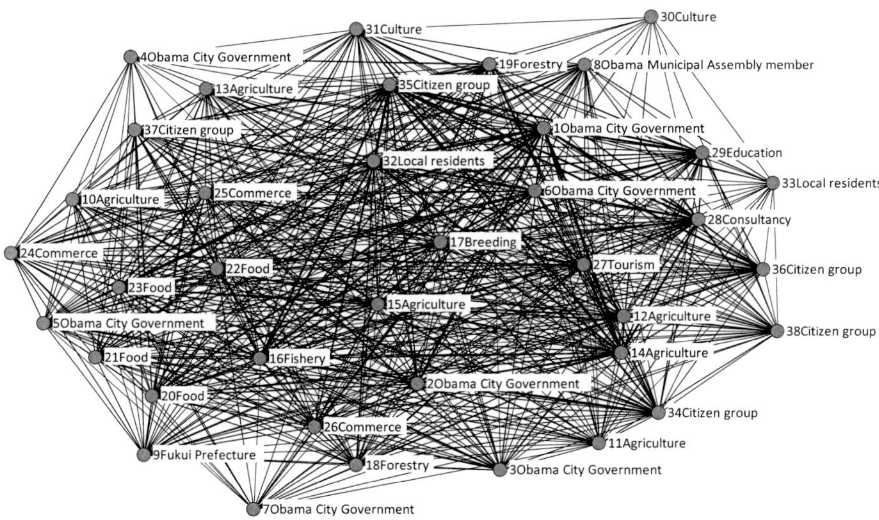

Fig. 19.6 One-Mode network graph – stakeholders of groundwater resource in Obama city

"necessity of conservation"), quality ("sanitary concerns"), and the condition of groundwater ("expectations toward scientific survey"). This structure implies that the central issue regarding Obama City water resources is the propriety of maintaining the current use of water, which is primarily for industrial purposes. It would be necessary to conserve water to prevent a reduction in water quantity or a decline in sanitary standards, which probably explains the expectation toward a scientific survey for the purpose of determining the said necessity.

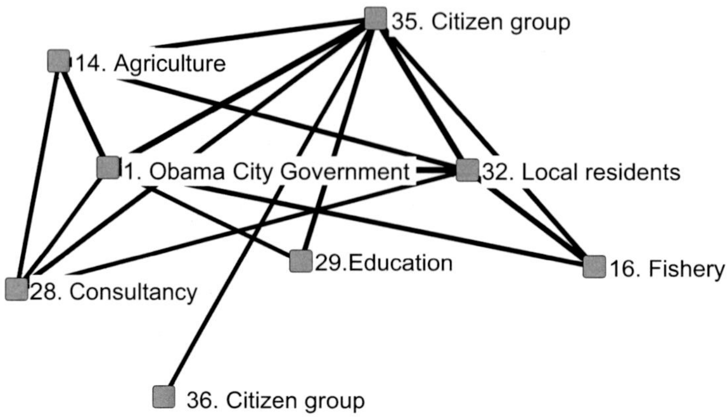

Fig. 19.7 10-Slice stakeholder subnetwork

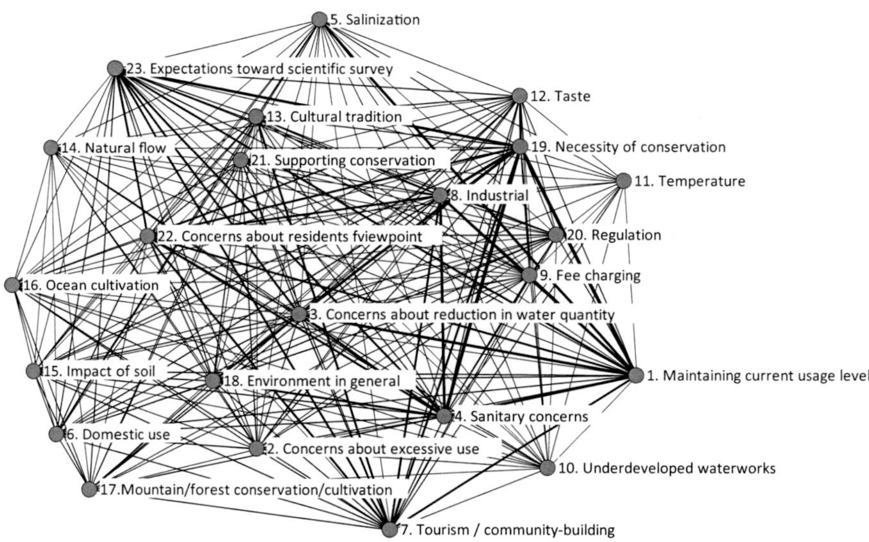

Fig. 19.8 Talking point network graph

Next, the overall structure of the stakeholder network is analyzed by block modeling a 5-slice subnetwork. The results are shown in Table 19.3. A hierarchical cluster analysis yield five stakeholder clusters. The first row and first column of the block model represent the cluster IDs. The second row and second column represent the stakeholder IDs as used in Table 19.1. Cells are shaded according to m-value, with stronger shading indicating larger line multiplicity and thus stronger ties.

Taking aside clusters 4 and 5, cluster 3 forms the central cluster with high line multiplicity as indicated by the strong shading of the cells. Cluster 3 comprises the following stakeholders: "1 Obama City Government," "14 Agriculture," "16

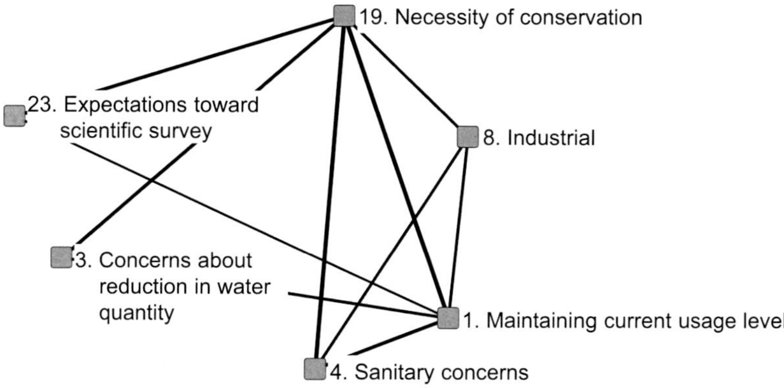

Fig. 19.9 12-Slice talking point subnetwork

Table 19.3 5-Slice stakeholder block model

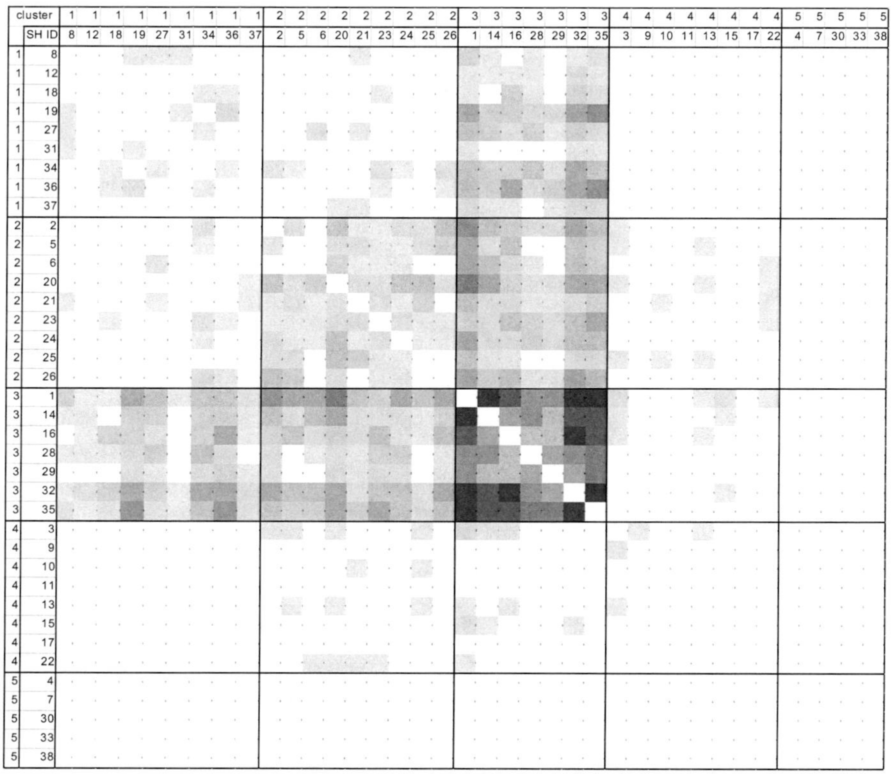

Table 19.4 5-Slice talking point block model

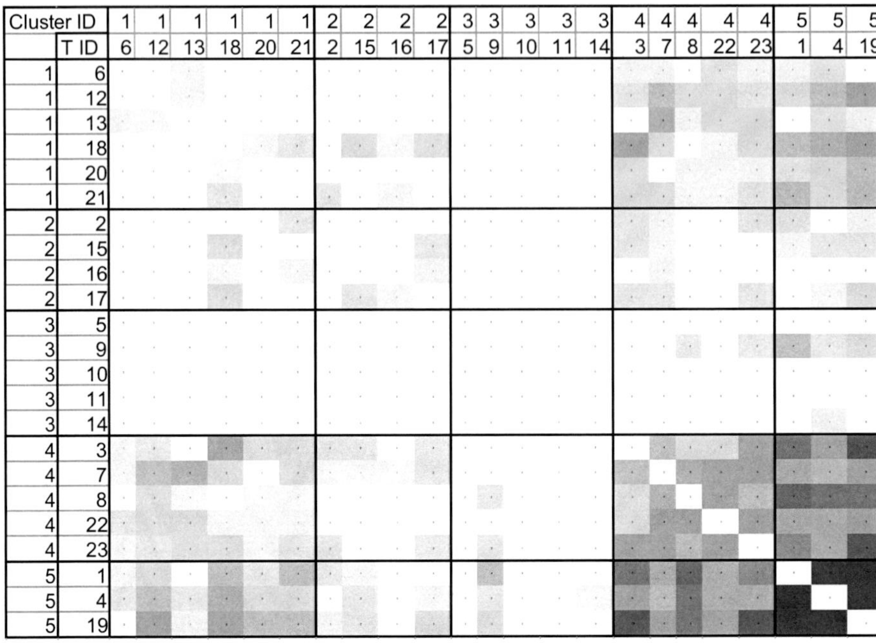

Fishery," "28 Consultancy," "29 Education," "32 Local residents," and "35 Citizen group" These actors have a deep connection with groundwater; they either use it directly or engage in activities related to groundwater.

Cluster 2 has a high centrality next to cluster 3, and the relationship between clusters 2 and cluster 3 is strong. Cluster 2 comprises the following stakeholders: "2 Obama City Government," "5 Obama City Government," "6 Obama City Government," "20 Food," "21 Food," "23 Food," "24 Commerce," "25 Commerce," and "26 Commerce." The actors affiliated with Obama City Government or commerce include public servants and commercial associations who deal with groundwater. The actors affiliated with food are those who use groundwater to produce commodities. Because groundwater plays an important role in their professional lives, these actors occupy a relatively central position.

The most peripheral cluster in the subnetwork is cluster 5. Cluster 5 comprises "4 Obama City Government," "7 Obama City Government," "30 Culture," "33 Local residents," and "38 Citizen group," These actors include public servants in charge of agriculture or food, and actors who use groundwater to a limited extent in their cultural or community-building activities. Compared to those in clusters 3 and 2, these actors have limited ties to groundwater, which explains their peripheral position.

Next, the one-mode network of talking points is analyzed. For this purpose, the m-slice technique to extract a 7-slice subnetwork of talking points is used and the subnetwork is plotted on an adjacency matrix. The authors then block modeled the matrix and performed a hierarchical cluster analysis. The results are shown in Table 19.4. The hierarchical cluster analysis yields five clusters. The first row and first column of the block model represent the cluster IDs. The second row and second column represent the talking point IDs as used in Table 19.2.

Taking aside cluster 3, clusters 4 and 5 form the central clusters with high line multiplicities as indicated by the strong shading of the cells. One difference between clusters 4 and 5 is that cluster 5 × 5 has more strongly shaded cells, indicating that the talking points have stronger ties. Therefore, the talking points in cluster 5 ("1 Maintaining current usage level," "4 Sanitary concerns," and "19 Necessity of conservation") are the more central issues that the stakeholders of Obama City's groundwater resources would argue about.

The talking points in cluster 4 are "3 Concerns about reduction in water quantity," "7 Tourism / community-building," "8 Industrial," "22 Concerns about residents' viewpoint," and "23 Expectations toward scientific survey." When combined with the talking points of cluster 5, these points form a subnetwork with a high degree of cohesion similar to that in the network shown in Fig. 19.6. Thus, these talking points have a central position in the network of talking points.

The edges of clusters 4 and 5 are intersected by cluster 1. The talking points in Cluster 1 are "6 Domestic use," "12 Taste," "13 Cultural tradition," "18 Environment in general," "20 Regulation," and "21 Supporting conservation." Thus, these points would be argued about as a side note to the central talking points. The peripherally positioned cluster 2 comprises "2 Concerns about excessive use," "15 Impact of soil," "16 Ocean cultivation," and "17 Mountain/forest conservation/cultivation." Specific talking points regarding anxieties about groundwater resources or environmental impact thus hold a peripheral position in the network. As for cluster 3, the cells of which are scarcely shaded are the talking points "5 Salinization," "9 Fee charging," "10 Underdeveloped waterworks," "11 Temperature," and "14 Natural flow." The block model suggests that these talking points are isolated, having little connections with the other points.

19.4 Discussion and Conclusion

Based on the results above, this section now summarizes and discusses the networked structure of the talking points surrounding Obama City's groundwater resources and the corresponding stakeholders. The authors initially referred to the results of a previous study which analyzed stakeholders. A social network analysis has been performed on this data to visualize a common understanding on the talking points. This analysis has yielded a single coherent network. The authors then analyze the stakeholder network structure and identify central structures in the stakeholder and talking point networks.

The central structure of the stakeholder network comprises "1 Obama City Government," 14 Agriculture," "16 Fishery," "28 Consultancy," "29 Education," "32 Local residents," and "35 Citizen group." Thus, stakeholders who actively use groundwater in business operations, public duties, or civic activities have a central role. However, some stakeholders in city government, agriculture, fishery, local resident and citizen group had a peripheral position in the network, implying that affiliation is not what characterizes the network structure. Furthermore, the actors whose affiliations include "1 Obama City Government," "28 Consultancy," and "29 Education" deal with groundwater in various contexts, including in public administration, surveys, and educational settings. As such, these actors would have a stake in a wide range of talking points, explaining their central position in the network.

In the talking point network, the central points are "1 Maintaining current usage level," "4 Sanitary concerns," and "19 Necessity of conservation," while the peripheral points include "5 Salinization," "9 Fee charging," "10 Underdeveloped waterworks," "11 Temperature," and "14 Natural flow." This structure suggests that, with a view to conserve the current usage of Obama City's groundwater, the stakeholders come to a shared understanding on the matter of groundwater conservation, and also form a shared understanding about groundwater being a limited resource. However, these issues have few ties with talking points that concern the specific conditions of the groundwater, such as salinization, temperature, and natural flow. This lack of connection might reflect the fact that specific issues have not yet emerged, which means that although the stakeholders have concerns about the groundwater, inter-stakeholder conflict has not yet happened. From a Nexus perspective, groundwater is used for various purposes. The fact that concrete issues are not shared between stakeholders with different uses is considered to be a potential conflict.

Looking forward, to promote discussions about the conservation of Obama City's groundwater, it is necessary to elevate the discussion to a stage where the stakeholders share concerns about the risks to groundwater, such as how groundwater use can lead to salinization or affect its natural flow or temperature. To this end, hopes are pinned on a scientific survey, which can be an essential step to draw the interests of stakeholders.

References

Baba K, Matsuura M, Taniguchi M (2015) Effectiveness and challenges of stakeholder analysis in co-design and co-product of science and society: implications from a case study of identifying potential issues on utilization of groundwater resource in Obama. Environmental. Science 28(4):304–315

de Nooy W, Mrvar A, Batagelj V (2011a) Exploratory social network analysis with pajek second edition. Cambridge University Press, Cambridge, UK, pp 118–119

de Nooy W, Mrvar A, Batagelj V (2011b) Exploratory social network analysis with pajek, second edn. Cambridge University Press, Cambridge, pp 124–125

Suzuki T (2009a) Network analysis. Kyoritsu Shuppan, Tokyo, pp 86–87

Suzuki T (2009b) Network analysis, vol 2. Kyoritsu Shuppan, Tokyo, pp 91–94

Part VI
Inter- and Trans-Disciplinary for Approaching Nexus Issues

Chapter 20
An Interdisciplinary Approach for Water–Energy–Food Nexus

Aiko Endo, Terukazu Kumazawa, Kimberly Burnett, Akira Ishii, Izumi Tsurita, Christopher A. Wada, Takaaki Kato, Makoto Yamada, and Pedcris Orencio

Abstract This chapter summarizes the activities of the interdisciplinary study group under the RIHN Nexus project. The mission of the group was (1) to review the WEF Nexus studies as a means to understand the current state of research on the WEF, (2) to develop integrated methods to address nexus issues, and (3) to design and visualize a WEF nexus system map to understand the complexity of the WEF Nexus system. The future challenges of the interdisciplinary group are (a) to develop the concept of integration in order to understand the concept of WEF nexus, (b) to integrate each method for adopting inter- and transdisciplinary research approaches to address nexus issues, (c) to approach not only disciplines,

---

A. Endo (✉) · T. Kumazawa
Research Institute for Humanity and Nature, Kyoto, Japan
e-mail: a.endo@chikyu.ac.jp; kumazawa@chikyu.ac.jp

K. Burnett · C. A. Wada
University of Hawaii Economic Research Organization, Honolulu, HI, USA
e-mail: kburnett@hawaii.edu; cawada@hawaii.edu

A. Ishii
Yachiyo Engineering Co., Ltd, Tokyo, Japan
e-mail: akri-ishii@yachiyo-eng.co.jp

I. Tsurita
Japan International Cooperation Agency, Tokyo, Japan

T. Kato
Faculty of Environmental Engineering, The University of Kitakyushu, Kitakyushu, Japan
e-mail: tkato@kitakyu-u.ac.jp

M. Yamada
Faculty of Economics, Ryukoku University, Kyoto, Japan
e-mail: myam@econ.ryukoku.ac.jp

P. Orencio
USAID Strengthening Urban Resilience for Growth with Equity (SURGE) Project, Makati City, Philippines
e-mail: porencio@surge.org.ph

© Springer Nature Singapore Pte Ltd. 2018

A. Endo, T. Oh (eds.), *The Water-Energy-Food Nexus*, Global Environmental Studies, https://doi.org/10.1007/978-981-10-7383-0_20

but also sectors using integrated methods focusing on stakeholders to adopt a transdisciplinary approach.

Keywords Water-energy-food (WEF) Nexus · WEF Nexus methods · Interdisciplinary · Transdisciplinary · Qualitative · Quantitative · WEF Nexus system

20.1 Background to Inter- and transdisciplinary Studies of the RIHN Nexus Project

20.1.1 Primary Goals of the RIHN

The RIHN has three primary goals: (1) to promote environmental studies that elucidate the interaction between humanity and nature and critically examine the future potential of human culture, based on the accumulated body of RIHN research and the results of global environmental research in Japan and abroad; (2) to promote solution-oriented global environmental studies involving close collaborations with stakeholders, starting from the research community; and (3) to contribute to problem-solving by applying research results in support of, and participation in, on-site multi-stakeholder arrangements in society. To meet the goals of the institute, the RIHN Nexus project was initiated as a five-year project in 2013, and was requested to adopt interdisciplinary and transdisciplinary research approaches.

20.1.2 The Mission of the Interdisciplinary Study Group of the RIHN Nexus Project

Per the structure of the project explained in Chap. 1, the RIHN Nexus project was designed to fundamentally consist of five groups: (1) the water–energy nexus group, (2) the water–food nexus group, (3) the stakeholder analysis group, (4) the socio-cultural group, and (5) the interdisciplinary study group. This chapter summarizes the activities of the interdisciplinary study group, consisting of experts in the fields of environmental economics, fisheries economics, computer science, water engineering and modeling, and policy studies.

The mission of the interdisciplinary study group was (1) to review the WEF Nexus studies as a means to understand the current state of research on the WEF, (2) to develop integrated methods to address nexus issues, and (3) to design and visualize a WEF nexus system map to understand the complexity of the WEF Nexus system. Interdisciplinary approaches integrate separate disciplinary data, methods, tools, concepts, and theories to create a holistic, systemic view of a complex issue (Keskinen 2010). Thus, we determined to focus on and develop methods to adopt an interdisciplinary approach following this definition of interdisciplinarity.

This chapter describes two peer-reviewed scientific articles in Sects. 20.2 (Endo et al. 2015) and 3 (Endo et al. 2017), and the results of ongoing research in Sect. 20.3.

20.2 Current State of Research on the WEF Nexus

20.2.1 Purposes and Methods

We reviewed 37 nexus case studies to understand the current state of clarification regarding the following: (1) the types of nexus being conducted, such as water–food nexus, water–energy nexus, or water–energy–food nexus; (2) the part of the world in which nexus projects have been conducted; (3) the types of nexus activity that have been conducted; (4) who is leading the nexus projects; (5) who is involved in the nexus projects; (6) who funded the nexus projects and the budget size; (7) the purposes of the various nexus projects; (8) the methodologies used for nexus studies; (9) the project outcomes; and (10) the project's launching year and period of study. We also examined the challenges and outlook for future nexus studies.

We adopted a quantitative approach using secondary data included in publicly available academic publications in journals and on the web in order to: (1) select the target nexus projects; (2) review the documents of the selected projects historically, including a timeline of nexus activities, nexus concepts, and the position of the nexus project in global environmental research; and (3) conduct quadrat analysis from the perspective of nexus type (water–food nexus, water–energy nexus, water–energy–food nexus, and climate-related nexus), nexus region and type, nexus keywords, and stakeholders.

20.2.2 Results of the Review

The review demonstrated that a diverse range of water–energy–food nexus studies and projects have been conducted, based on a variety of actors' interests at different scales worldwide since the first appearance of nexus terminology in 1983, when the United Nations University (UNU) launched a Food–Energy Nexus Programme to acknowledge the important interconnectedness between the issues of food and energy.

20.2.3 Nexus Type

Among the four types of nexus project identified, the number of water–energy nexus projects was highest, including both consumption of energy for agricultural production and wastewater treatment, and consumption of water to produce energy such as hydropower. Many of the water–energy–food nexus projects focused on biofuel production consuming food and water. Another feature identified in each nexus type, i.e., social and governance activities to engage and involve the stakeholders affected by the research results and/or policy decisions through the activities of capacity building and policy planning, was combined with environmental and economic research activities.

20.2.4 Nexus Regions

The regions were divided into Asia, Europe, Oceania, North America, South America, the Middle East, and Africa. We found that North America had a tendency to focus on the water–energy and climate-related nexus types, while Africa mainly focused on climate-related types. The other regions had a relatively balanced interest in each nexus type.

20.2.5 Nexus Keywords

Eighty-four keywords were identified in 37 nexus projects, including the four identified nexus types of water–food (n = 6), water–energy (n = 12), water–energy–food (n = 11), and climate-related (n = 8). The 37 projects were all related to water, and many of the selected keywords (40 out of 84) were also linked with water, mostly focusing on fresh water including river water, rain water, reservoir, groundwater, and seawater and mainly related to terrestrial activities for agriculture production and wastewater treatment.

20.2.6 Nexus Stakeholders

A wide variety of stakeholders were involved in nexus activities such as the United Nations Agency (n = 16), international groups, institutes, and NGOs (n = 28), private companies (n = 7), national governments and agencies, institutes and universities in Europe (n = 19), in North America and Latin America (n = 28), in Asia (n = 28), in Oceania (n = 7), and in Africa (n = 4). We re-categorized the 137 organizations into the eight categories listed by Future Earth, namely research, science-policy interfaces, funders, governments, development organizations, business and industry, civil society (NGOs), and media (Future Earth 2013). The stakeholders with the highest number were research (n = 77) followed by governments (n = 47); the lowest was media (n = 2).

20.2.7 Challenges of WEF Nexus Research

We next identified the challenges of nexus studies, namely: (1) understanding the interrelationships between resources and the subsequent complexity of nexus systems is limited, especially since projects involve the linkage of two resources, e.g., water–food or water–energy; (2) the nexus is likely to be recognized at the research level, but is not fully acknowledged on the ground; (3) the ways to connect local nexus issues within a community to broader national and global environmental issues and themes were often missing from site-specific case studies; (4) the differences and/or changes in tradeoff relationships between different spatial and/or temporal scales were poorly analyzed; (5) the definition of synergy effects in nexus case studies remains unclear; and (6) the academic concept of "WEF nexus" has not been clearly defined due to the above five reasons.

20.3 Methods of the Water–Energy–Food Nexus

The second mission of the interdisciplinary study group was to develop WEF Nexus methods to reduce the tradeoff relationships and produce synergistic effects. We determined the methods, developed existing methods, and/or created new discipline-free methods, based on synthesizing and harmonizing team-based production, collected from individual scientists in different disciplines from each group for interdisciplinary research approaches. We also faced the further challenge of developing these approaches to incorporate non-scientific/non-disciplinary views in the analysis for the transdisciplinary research approach.

We then classified the integrated methods as qualitative or quantitative to contribute to both interdisciplinary and transdisciplinary research studies. The methods for interdisciplinary research can be used to unify a collection of related variables, visualize the research problem, evaluate the issue, and simulate the system of interest. The qualitative methods that we analyzed consisted of questionnaire surveys, ontology engineering (Kumazawa et al. 2017), and integrated mapping. The quantitative methods included physical models, benefit-cost analysis (Burnett et al. 2017, Wada et al. 2016), integrated indices (Orencio et al. 2015), and optimization management models (Burnett et al. 2017).

20.3.1 Pros and Cons of each Method to Address Nexus Issues: Qualitative Methods

As a result of our analyses, we identified the pros and cons of each method along with accompanying case studies from research sites in Japan and the Philippines. Although the case studies are specific to the project research sites, we see these methods as applicable to other areas with appropriate calibration. All methods discussed here are transdisciplinary in that they begin by engaging stakeholders as a means to identify the appropriate research question. They are then used to design the scientific approach to collect appropriate data in order to parameterize and develop models. In turn, this allows researchers to answer the policy or management question of interest.

Regarding the three qualitative methods, we used questionnaire surveys in the case of Laguna de Bay to contribute to a nexus assessment that aimed to address the question of how the population's security is affected when various natural and social hazards disrupt the linkages among the three systems. The questionnaire survey was especially useful in incorporating the local people's general outlook on their level of economic, food, and livelihood security when various shifts occur in terms of the quality and quantity of the water–food nexus. Consequently, this provides the information necessary to make decisions and thus optimally manage local nexus resources. However, we must acknowledge that the quality of the survey instrument always affects the data resulting from this approach.

Ontology Engineering is one of the base technologies in semantic web technology, where the internet is used to create a knowledge base that computers can deal with directly by means of adding metadata (Mizoguchi 2005, 2012). An ontology consists of concepts and relationships that are used to describe the target world. It provides common terms, concepts, and semantics by which users can represent the contents with minimum ambiguity and interpersonal variation of expression. The construction of a well-designed ontology presents an explicit understanding of the system. We used Ontology Engineering in Obama City and Beppu City to display a conceptual map focusing on water. Ontology engineering could be used for (1) designing the project to build a list of common conceptual terms; the linkages of each term among stakeholders included researchers and practitioners, (2) assessing whether the policy/plan would cover all disciplines including natural sciences, social sciences and the humanities, and sectors such as WEF.

An integrated map is an overlay of various single maps, and can be used as a method to support the implementation of synthesized policies between land and sea. In contrast to sectoral management and monodisciplinary research approaches (which often focus on a single ecological system), an integrated map informs policies capable of restoring and maintaining the interdependence between land and sea. The creation of an integrated map brings many benefits. First, it can be used to incorporate individual research results into maps as integrated methods for an interdisciplinary research approach to enhance mutual understanding between members. Second, it can be used to unify the data, information, and knowledge on maps to visualize and disseminate the current status of environment and utilization in river basins and coasts to stakeholders. Third, an integrated map can facilitate the identification of key nexus issues, such as the impact that nutrient flows have on coastal ecosystems. Finally, integrated maps can be used as a transdisciplinary method, engaging stakeholders and policy-makers to discuss through an integrated map how to implement integrated management of land and coastal areas. An example of an integrated map is that created for Beppu Bay. It is possible to create a site-specific integrated map at the local level to visualize the current conditions of water, energy, and food resources, as well as resource users. However, it would be challenging to create an integrated map at the national or global levels. In addition, an integrated map shows a static condition, not future scenarios, which limits the map's ability to demonstrate inter-scale, inter-generational, and inter-area circumstances.

20.3.2 Pros and Cons of each Method to Address Nexus Issues: Quantitative Methods

While we presented the specific per-site results in several forms, the quantitative methods normalized these forms to allow for direct comparison with other results at different project locations in the Asia-Pacific region. This makes it possible to decide on optimal policies regarding the sustainable management of water, energy, and food, not only for project members, but also for stakeholders.

With regard to the integrated physical model, we calibrated the model using data from Obama City. The models (such as those that measure water balance), and hydrological parameters (such as water exchange between rivers and groundwater, and groundwater discharge into the ocean) are useful methods for hydrologists. Material transport (including the transport of nutrients from land to ocean by rivers and groundwater) is important for fisheries. Hydrology, fisheries, and geochemical and biochemical information can be applied to this integrated physical model in an interdisciplinary way.

Integrated physical models can simulate the balance between water, energy, and food production; therefore, simulations based on potential future scenarios can be useful for decision makers. However, the results of integrated model simulation without social and local knowledge may lead people to misconstrue the model's results if the numbers from simulations are unrealistic for political, economic, or other reasons.

Benefit-cost analysis (BCA) entails using benefit-cost models (BCMs), which assess the desirability of a proposed policy or project, either independently or ranked according to highest net benefit if selecting from a range of alternatives. BCMs can be used in the context of evaluating WEF nexus projects to clearly consider the trade-offs in a particular region where one or more of the WEF elements will be utilized. We developed a BCM to analyze the construction of a new dike between the Pacific Ocean and Otsuchi's coastline. The WEF Nexus is inherently about trade-offs. BCA enables researchers to provide decision-makers with information regarding the consequences of these trade-offs and to explicitly examine the net benefits of decisions in order to allocate scarce resources (such as water) toward food or energy. In addition, to improve understanding of the trade-offs, BCA makes the costs and benefits accrued over individual time periods transparent.

The optimization management model provides one possible method with which to examine optimal resource allocation. We used the Optimization Management

Model to study the groundwater allocation problem in Obama City. The model allows researchers to explicitly represent the interaction of natural resources, which is key to understanding the trade-offs inherent in the WEF nexus. Decisions to draw down one resource often affect other resources, as well as the social welfare of the community of interest. For example, the decision to use groundwater for fisheries rather than agriculture depends on the production costs of both fish and agriculture, including energy. Economic optimization allows the researcher to determine how to allocate scarce resources over time, when doing so has consequences for the surrounding ecosystem and society.

20.3.3 WEF Nexus Methods for the Transdisciplinary Approach

To adopt the approach of co-design and co-production through the project process, each method should be developed as a science-policy interface method, despite each method having different uses at various stages. Ontology Engineering would be the most useful method for project design during the initiation stage to build a list of common concepts and terms; the linkages between terms among stakeholders included researchers and practitioners. In addition, Ontology Engineering could be used at the policy planning stage to assess whether the policy/plan would cover all disciplines, including natural sciences, social sciences, and humanities, and sectors such as water, energy, and food (in order to address the key issues that were originally identified during the initiation stage). Questionnaire Surveys would be more useful for collecting information to analyze WEF interlinkages when little data exist; their use would then help to identify the key issues during the initiation stage. Integrated Maps can provide an opportunity to share knowledge showing actual conditions at a spatial scale among stakeholders during the policy planning stage. BCA and an Optimization Management Model would play important roles in clarifying trade-offs during the initiation stage, creating and providing policy options during the policy planning stage. Physical models could be quite essential to understand WEF nexus systems; if it were developed to clarify interlinkages between the physical conditions of water, energy, and food, as well as human activities by working with social scientists, then it could be used to address the key issues more holistically during the policy planning stage. Using an Integrated Index can be a discipline-free method that incorporates and integrates each result with different disciplines, then evaluates trade-offs during the policy planning stage. Simultaneously, interdisciplinary team members could themselves act as interpreters or coordinators for the science-policy interface, using those approaches when they have a commitment to both science and society from the initiating stage.

20.3.4 Challenges of WEF Nexus Methods

From the perspective of spatial and temporal scales, despite covering spatial, physical, and economic dimensions, our approach is somewhat limited in terms of vertical and horizontal elements, as well as on a temporal scale to address the WEF nexus. To address these challenges, it is possible to use global data such as a global model to set our site-specific case studies within a global context on a vertical spatial scale (Guillaume et al. 2015). In addition, the creation of future scenarios that further integrate each integrated method mentioned in this paper remains a challenge; however, doing so will allow the WEF nexus to be analyzed based on a temporal scale (Keskinen et al. 2015). To address the issue of temporal scale, we determined if we could use each method to address the nexus problems during the initial stage, developing stage, and policy planning stage to design future scenarios.

20.4 Design and Visualization of the Water–Energy–Food Nexus System

After reviewing the current condition of nexus research, developing nexus methods to address nexus issues, and identifying the challenges of nexus research and methods, we determined to design and visualize the WEF Nexus system to identify the interrelationships between WEF resources and to understand the subsequent complexity of WEF nexus systems systemically, adopting an interdisciplinary approach.

Object-oriented concepts and ontology engineering methods were applied according to the hypothesis that the chains of changes in linkages between water, energy, and food resources holistically affect the water–energy–food nexus system, including natural and social systems, both temporally and spatially. The water–energy–food nexus system that is developed is significant because it allows us to: (1) visualize linkages between water, energy, and food resources in social and natural systems; (2) identify tradeoffs between these resources; (3) find a way of using resources efficiently or enhancing the synergy between the utilization of different resources; and (4) aid scenario planning using economic tools.

20.4.1 Challenges of Designing a WEF Nexus System

The challenges of designing and visualizing a WEF nexus system include clarifying the linkages between events, identifying: (1) how changes in linkages between water, energy, and food resources affect the WEF nexus system holistically and systemically; (2) what the driving force is; and (3) whether there are hidden or unexpected factors that would impact the WEF nexus system. The second challenge is to identify the tradeoffs between resources, for example, tradeoffs in the use of land resources between energy and food production. The third challenge is to find a way of using resources efficiently or enhancing the synergy between the utilization of different resources, for example, to streamline, recycle, or cascade resources and/or to add higher value to resources. Finally, a fourth challenge is to contribute to scenario planning.

Forecasting scenario planning includes four steps, namely: (1) problem setting; (2) constructing causal networks to represent the targeted system; (3) describing scenario storylines; and (4) describing details, including parameterization and quantification. On the other hand, backcasting scenario planning consists of six steps: (1) determining the purpose of scenario building; (2) specifying goals, constraints, and targets; (3) describing the present system; (4) specifying exogenous valuables; (5) undertaking scenario analysis, including developing scenarios; and (6) undertaking impact analysis, including comparison of scenario results with predetermined goals (Robinson 1990). Both scenarios require designing the target system in the course of scenario planning.

20.5 Future Challenges

We reviewed 37 nexus projects based on academic papers and identified the current situation and challenges of nexus studies. In addition, to address nexus issues, we identified and determined the integrated methods, when and how we could use those methods, and the pros and cons of each. Furthermore, we are now developing a nexus system map.

The future challenges of the interdisciplinary group are (1) to develop the concept of integration in order to understand the concept of WEF nexus, (2) to integrate each method for adopting inter- and transdisciplinary research approaches to address nexus issues, (3) to approach not only disciplines, but also non-academic sectors using integrated methods focusing on stakeholders to adopt a transdisciplinary approach.

References

Burnett B, Wada C, Endo A, Taniguchi M (2017) The economic value of groundwater in Obama. J Hydrol Reg Stud 11:44–52. https://doi.org/10.1016/j.ejrh

Endo A, Burnett K, Orencio P, Kumazawa T, Wada C, Ishii A, Taniguchi M (2015) Methods of the water–energy–food nexus. Water 7:5806–5830. https://doi.org/10.3390/w7105806

Endo A, Tsurita I, Burnett K, Orencio P (2017) A review of the current state of research on the water, energy, and food nexus. J Hydrol Reg Stud 11:20–30. https://doi.org/10.1016/j.ejrh.2015.11.010

Future Earth (2013) Future Earth: research for global sustainability.. http://www.futureearth.org/. Accessed 26 Sep 2015

Guillaume JHA, Kummu M, Eisner S, Varis O (2015) Transferable principles for managing the nexus: lessons from historical global water modelling of central Asia. Water 7:4200–4231

Keskinen M (2010) Bringing back the common sense? Integrated approaches in water management: lessons learnt from the Mekong (Doctoral dissertation). Aalto University, Espoo, Finland

Keskinen M, Someth P, Salmivaara A, Kummu M (2015) Water–energy–food nexus in a transboundary river basin: the case of Tonle sap Lake, Mekong River basin. Water 7:5416–5436

Kumazawa T, Hara K, Endo A, Taniguchi M (2017) Supporting collaboration in interdisciplinary research of water–energy–food nexus by means of ontology engineering. J Hydrol Reg Stud 11:31–43. https://doi.org/10.1016/j.ejrh.2015.11.021

Mizoguchi R (2005) Ontology kougaku (Ontology engineering). Ohmsha, Tokyo. (In Japanese)

Mizoguchi R (2012) Ontology kougaku no riron to jissen (theory and practice of ontology engineering). Ohmsha, Tokyo. (In Japanese)

Orencio P, Endo A, Taniguchi M, Fujii M (2015) Using thresholds of severity to threats to and the resilience of human systems in measuring security. Soc Indic Res 124(2):1–21. https://doi.org/10.1007/s11205-015-1152-x

Robinson JB (1990) Futures under glass: a recipe for people who hate to predict. Futures 22:820–842. https://doi.org/10.1016/0016-3287(90)90018-D

Wada C, Burnett K, Gurdak J (2016) Sustainable agriculture irrigation management: the water–energy–food nexus in Pajaro Valley, California. Sustain Agric Res 5(3):76–83. https://doi.org/10.5539/sar.v5n3p76

Chapter 21
Assessment of Collaboration Process in Interdisciplinary Research of Water-energy-food Nexus by Means of Ontology Engineering

Terukazu Kumazawa, Keishiro Hara, Aiko Endo, and Makoto Taniguchi

Abstract This article discusses an assessment of the collaboration process of interdisciplinary research on the water-energy-food nexus by combining a text mining approach and an ontology engineering approach. For this purpose we first overview a collaboration experiment designed for research development. Second, we identify researchers' concerns in the discussion context of this experiment and represent the communication process focusing on keyword usage. Third, we show the relationships between the researchers' domains and the fixed research issue and its basic direction by means of ontology engineering. Finally, we discuss how the shared concepts among researchers are dealt with in the collaboration process by combining the represented communication process as an assessment of collaboration.

Keywords Collaboration process · Interdisciplinary research · Assessment · Text mining · Ontology engineering

21.1 Introduction

The Committee on Facilitating Interdisciplinary Research (2004) has stated that interdisciplinary research is typically collaborative and involves people from disparate backgrounds because such research is pluralistic in both method and focus. Moreover, Repko (2016) stated that interdisciplinary research consists of heuristic, repetitive, and reflexive processes within the decision-making process.

When we discuss research on the water-energy-food nexus, the discussion inevitably involves interdisciplinary content because the discussion involves interactions among the water, energy, and food domains. In the field of team science, various

T. Kumazawa (✉) · A. Endo · M. Taniguchi
Research Institute for Humanity and Nature, Kyoto, Japan
e-mail: kumazawa@chikyu.ac.jp; a.endo@chikyu.ac.jp; makoto@chikyu.ac.jp

K. Hara
Osaka University, Osaka, Japan

© Springer Nature Singapore Pte Ltd. 2018
A. Endo, T. Oh (eds.), *The Water-Energy-Food Nexus*, Global Environmental Studies, https://doi.org/10.1007/978-981-10-7383-0_21

methods and frameworks to support interdisciplinarity and transdisciplinarity have been proposed. However, as the field of team science matures, in order to assess processes and outcomes of team science, more sophisticated methods and research designs, such as prospective quasi-experimental research designs (Stokols et al. 2010), must be developed.

The approach introduced in the present article introduces a collaborative design approach (Holsapple and Joshi 2002) in the context of the transition management approach in sustainability science (SS; Loorbach 2007; Grin and Rotmans 2010). As Miller (2013) reported, it is crucial to create a space for a more democratic and reflexive research agenda in process-oriented SS. In other words, SS must deal with dynamics in the sense that SS can adaptively react according to the goals/requirements of the user. In Kumazawa et al. (2017), we discussed the potential effectiveness (contribution) of an ontology engineering approach in the process of collaborative research.

Based on this discussion, in the present article, we focus on assessing, in a practical manner, the collaboration process in interdisciplinary research on the water-energy-food nexus by means of ontology engineering. For this purpose, we first overview the collaboration experiment designed for research development. Second, we represent the communication process in the experiment. Third, we indicate, in a practical manner, the relationships between the researchers' domains and the fixed research issue developed through discussion among researchers and its basic direction by means of a map tool based on ontology engineering. A researchers' domain covers researchers' interests and areas of expertise, while a basic direction of research represents the approach to discuss the proposed research issue. Finally, we discuss how the shared concepts among researchers are dealt with in the collaboration process. Linked concepts are concepts shared by multiple researchers, which are defined by the ontology engineering approach.

21.2 How Do we Assess the Collaboration Process in Interdisciplinary Research?

In the present article, we investigate the production process from initial interest to the development of a research question in the collaboration process. But how do we understand this production process? Difficulties in interdisciplinary research involved in developing a research question can be identified through understanding the individual research interests of researchers and their areas of expertise.

Documents used in meetings for setting research issue, individual interviews with researchers, discussions, and meetings during research and development, and structural drawings written on whiteboards and papers are materials that make up the texts used to clarify the context of a proposed research question and the interests of individual researchers. However, the question arises how individual events,

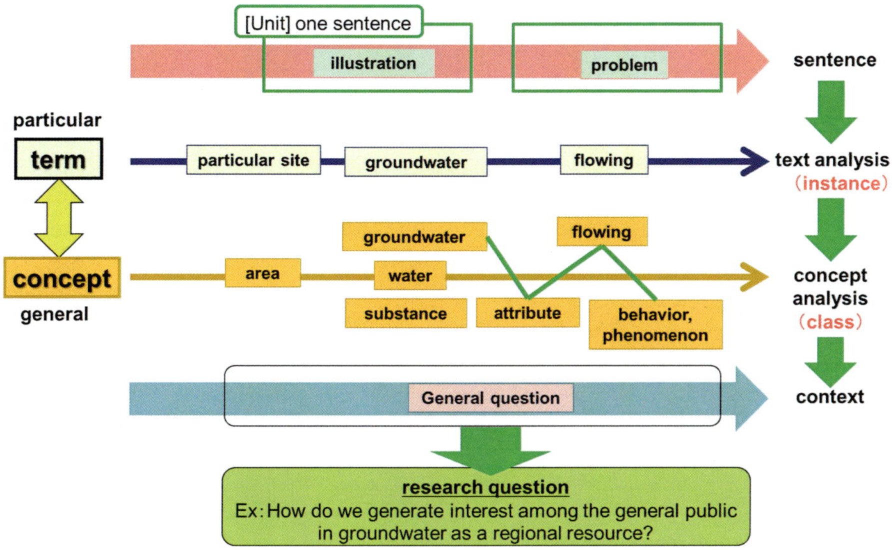

Fig. 21.1 Hierarchical classification of the analysis target and method (Kumazawa et al. 2017)

processes, reasons, questions, emerging problems and impacts, and goals are interconnected and should be considered in developing the research question.

As shown in Fig. 21.1, we can find research questions from the context underlying discussions. We then must consider what connects text and context. The authors recognize that the layer of concepts and semantic relationships is positioned between the layer of text as an individual instance and the layer of a general context. Then, we must consider the relationship between researchers and the research domains of these concepts, as well as semantic relationships. Based on this hypothetical framework proposed in Kumazawa et al. (2017), we develop an analytical framework to assess the collaboration process in interdisciplinary research. This framework distinguishes clearly between a word referring to a particular object and a word referring to a general meaning. We call the former one a term and the latter one a concept. Even if we use the same word "groundwater", for example, we need distinguishing groundwater as a term from groundwater as a concept.

According to this hierarchical framework, we design a concrete research procedure. The record of discussion belongs to the sentence layer in Fig. 21.1. Based on this, we focus on the text analysis layer just below the sentence layer, so that we can understand the discussion situation. Using a text mining approach, we identify the researchers' concerns in a particular discussion context and represent the communication process focusing on keyword usage. We then describe the relationships between the fixed research issue and its basic direction and the researchers' domains based on ontology engineering. Finally, we examine the difference between the actual situations and semantic relationships by comparing the analysis results based on text mining with the analytical results based on an ontology engineering approach as an assessment process.

21.3 Collaboration Experiment of Research Development

21.3.1 Overview of the Experiment

We design an experiment on collaborative setting of research issue by researchers in different fields related to SS and environmental studies. We then implement the experiment for research issue setting in the style of a workshop and obtain a research issue and its basic research direction. The experiment is implemented at the Research Institute for Humanity and Nature (RIHN). The outline of the experiment is shown in Table 21.1.

21.3.2 Contents of the Discussion

The researchers involved in the experiment attempted to propose a research theme focusing on multifaceted values, and the discussion was steered toward an investigation of the relationship between the values of water and the flow volume, the

Table 21.1 Outline of the researchers involved in the experiment

Date	June 15th and September 14th in 2015 (approximately 2.5 hours)		
Site	Research Institute for Humanity and Nature		
Basic research issue	Research on water and water resource management		
Goal of work	Proposing the research theme and its framework		
Member of experiment group			
Water-Energy-Food Nexus Project researchers of RIHN		Other researchers	
ID	Specialty	ID	Specialty
	Affiliation		Affiliation
	Position		Position
1	Resource Governance	2	Soil Science
	Research Department		Research Department
	Project Researcher (Human-Environmental Security in Asia-Pacific Ring of Fire: Water-Energy-Food Nexus)		Project Researcher (Designing Local Frameworks for Integrated Water Resource Management)
4	Hydrology	3	Agricultural resource economics
	Research Department		Faculty of Agriculture Department of Agricultural Science, Kinki University
	Project Researcher (Human-Environmental Security in Asia-Pacific Ring of Fire: Water-Energy-Food Nexus)		Associate Professor

circulation velocity, and the quality of water. The following are the primary areas of focus: (1) the meanings of the values of water in the hydrological circulation process, (2) the scale of the hydrological circulation, and (3) how to distinguish field cases and select target cases. Table 21.2 shows the discussion issues and their order

Table 21.2 Discussion issues and their order

First discussion (June 5th)		Second Discussion (September 14th)	
Num	issue	Num	issue
1	With a focus on self-introduction of ID 3	**19**	Overview of the first discussion and introduction to the second discussion
2	With a focus on self-introduction of ID 2	**20**	Water circulation velocity and groundwater management in the policy context
3	With a focus on self-introduction of ID 4	**21**	Concept of flow and stock
4	With a focus on self-introduction of ID 1	**22**	Time scale of circulation velocity
5	Spring in Obama and participatory monitoring in Laguna de bay	**23**	Explanation about discussion composition
6	Valuation of spring and method of resource management	**24**	Water flowing through passing attributes
7	World-view of residents utilizing groundwater and value consciousness on hot spring	**25**	Case of iterative use in Echi River
8	Water resource management of Echi River and Yasu River	**26**	Different water values by circulation stages
9	Management case of groundwater as a regional resource	**27**	Originality of the idea of water circulation velocities and water governance
10	Value of water and water governance in the process of water circulation	**28**	Way of understanding the values related to water
11	Value of water changing depending on circulation velocity	**29**	Impact on local areas the water circulation changes in the watershed with accompanying climate change make
12	Handling circulation velocities and its case	**30**	Transforming pathway in action through selecting water values from the water attribute and feedback to the governance as a goal of the pathway
13	Assessment method of resource allocation	**31**	Making the research title through selecting keywords
14	Competition due to the difference of circulation velocity use and its adjustment		
15	Geographical condition and value of water		
16	Difference of water value depending on the difference of water properties		
17	Flow and stock on land water		
18	Dialogue to wrap up the first discussion		

Table 21.3 Fixed research issue and its basic research direction in the experiment

Fixed research issue	Main title	"Climate change and basin society"
	Sub title	"How will the value of water change in the circulation process?"
Basic research direction	Obtaining deep understanding of the value of water in the circulation process	
	Implementing water circulation simulation	

in two series of discussion experiments. Table 21.3 shows the fixed research issue and its basic research direction in the experiment as a result of the discussion.

The discussion of this experiment group included contents that were rather difficult to understand due to abstractness and complexity. For example, the keywords such as circulation and value are very abstract, whereas the targeting system is complex in terms of including multiple scales as a mechanism.

21.4 Representing the Communication Process in the Experiment

21.4.1 Identifying the Researchers' Concerns in the Discussion Context

For the purpose of representing the communication process in this collaboration experiment, we first identify the researchers' concerns in the discussion context. Second, we trace and discuss the communication process focusing on keyword usage. For the analytical approach, we focus on text mining. We then implement the morphological analysis through temporal text mining (TTM) software, which is a basic text mining method.

In Table 21.4, we show the top sixty terms used by each researcher, which correspond to about 10% of all terms used by the researchers in the discussion. The terms shown in Table 21.4 have been translated from the terms in Japanese. Therefore, on occasion different terms in Japanese are translated into the same terms in English. For example, value has several meanings from the different aspects of how much something is worth, being useful/important, beliefs and mathematics. On the other hand, Kachi in Japanese is translated into value in English, but it doesn't have the meaning of a mathematical quantity. In order to distinguish value translated into Kachi from value indicating a mathematical quantity we add the Japanese terms connected with underlines after the English terms. For example, in the case that Kachi is recorded as a term data we label this term value_Kachi.

Table 21.4 Terms in top sixty for each researcher

Rank	ID1		ID2		ID3		ID4	
1	value_Kachi	39	water	45	value_Kachi	59	value_Kachi	133
	water	21	groundwater	29	water	39	water	130
	resource	20	person people	21	person people	32	person people	82
	meaning	18	value_Kachi	20	stock	13	groundwater	70
	groundwater	17	dam	12	circulation	13	meaning	51
	problem	13	Echi River	11	measurement_Keisoku	12	attribute	41
	person people	12	climate change	10	culture cultural	12	hot spring	36
	local area	10	circulation velocity	9	local area	11	climate change	27
	forest_Mori	9	water circulation	9	upstream	10	circulation	26
	climate change	8	difference	7	watershed / basin	10	watershed / basin	26
	use	8	land improvement district improvement district	7	climate change	9	water circulation	25
	flow	8	local area	6	word	9	image	24
	change	8	use	6	image	9	change	23
	stock	8	rice field	6	multifaceted value	9	Echi River	23
	research	8	river	6	area	9	stock	22
	Yakushima Island	8	flow	5	groundwater	8	dam	21
	impact	7	change	5	flow	8	nature	19
	part	7	impact	5	downstream	8	river	18
	multifaceted	7	upstream	5	quantity	8	quantity	17
	flowing	7	local scale	5	situation	8	circulation velocity	17
	circulation velocity	6	culture cultural	5	economy economic	8	impact	17
	management	6	stock	4	management	7	multifaceted value	16
	national government	6	management	4	problem	7	difference	16
	interest_Kanshin	6	downstream	4	part	7	problem	15
	flow_Nagare	6	quantity	4	time point	7	place	15

(continued)

Table 21.4 (continued)

	history historical	6	the local area_Jimoto	4	calculation	7	flow	14
	society social	6	winter	4	discussion	7	spring (water from a spring)_Yusui	14
	dam	5	water quality	4	Philippines	7	the local area_Jimoto	14
	upstream	5	iterative use	4	meaning	6	use	13
30	relationship	5	Bali	4	way	6	possibility	13
	word	5	voice	4	domain	6	phenomena	13
	spring (water from a spring)_Yusui	5	problem	3	environment	6	alluvial fan	13
	movement	5	part	3	concept	6	interest_Kyoumi	13
	interest_Kyoumi	5	national government	3	water circulation	5	energy	12
	water circulation	4	relationship	3	change	5	perspective	12
	downstream	4	circulation	3	hot spring	5	area	11
	analysis	4	state	3	research	5	consciousness	11
	money	4	agriculture	3	Yakushima Island	5	local	11
	resource use	4	mountain	3	target	5	old days	11
	snow melting	4	agricultutal water	3	condition	5	culture cultural	11
	watershed / basin	4	lake	3	feedback	5	process	10
	situation	4	Kyoto	3	angle	5	recognition	10
	underground	4	soil	3	farming method	5	research	9
	human	4	caldera lake	3	breed	5	resource	9
	Obama City	4	explanation	3	objective function	5	carbonate spring	9
	development	4	project	3	pump	4	policy	9
	scientific	4	fertilization	3	flowing	4	Obama	9
	fishery	4	custom	3	forest_Shinrin	4	flow_Ryudo	9
	approach	4	Sabah River	3	year	4	CO2	9
	water source	4	multifaceted	2	conservation	4	holding pond	8
	economy economic	4	word	2	resource	4	simulation	8
	culture	4	spring (water	2	allocation	4	spring	8

(continued)

Table 21.4 (continued)

	cultural		from a spring)_Yusui				(water from a spring)_Wakimizu	
	difference	3	analysis	2	water resource	4	watershed water circulation	8
	circulation	3	money	2	judgment	4	basin_Bonchi	8
	image	3	resource use	2	land_Tochi	4	drainage water	8
	sea	3	image	2	work	4	Beppu	8
	power generation	3	sea	2	setting	4	assessment	8
	field site	3	power generation	2	objective	4	nexus	8
60	policy	3	old days	2	multiple	4	volcano	8
	handling	3	data	2	national wealth	4		
	piped water	3	Japan	2	Kona	4		
	sustained	3	watershed society	2	agricultural economy	4		
	discussion	4	cultural value	2	festival	4		
	amount of rainfall	3	consciousness	2	economic value	4		
	allocation	3	multifaceted value	2	constant	4		
	comprehensive	3	pump	2	coffee	4		
	nature	3	time point	2				
	Obama	3	ecosystem service	2				
	resident	3	children	2				
	Saijo	3	verification / confirmation_Kakunin	2				
	fishery_Suisan	3	holding pond	2				
	southeastern Asia	3	rain	2				
	scientist	3	period	2				
	connection	3	nutrient salt	2				
	Kumamoto	3	field_Hatake	2				
	virgin forest	3	north	2				
	government	3	ordinance	2				
	basin commission	3	questionnaire	2				
	national forest	3	competing	2				
	biodiversity	3	feeling	2				
	well	3	cattle	2				
	natural resource	3	water right	2				
	idea	3	treatment	2				
	electrically powered pump	3						
	finding	3						
	forestry	3						

(1) Usage of the terms related to the fixed research issue and its basic direction

We target "value_Kachi", "water", "climate change", "circulation", "water circulation", and "watershed/basin" as the terms related to the fixed research issue and its basic direction. Examining the researchers' usages in Table 21.4, we can find the following results.

First, all of the researchers use "value_Kachi" most often, use "water" frequently, and use "climate change" fairly often. Although "circulation" and "water circulation" are both used by all of the researchers, "circulation" is used more frequently by ID 3 and ID 4, whereas "water circulation" is used more frequently by ID 2 and ID 4. Lastly, "watershed/basin" ranked in the top sixty of the terms for just ID 1, ID 3, and ID 4.

(2) Analysis focusing on the top sixty terms

As shown in Table 21.4, many terms of rank lower than 30 are used only by one researcher as shown by red boxes. In the following, we examine the top sixty terms in terms of the number of the researchers who use them.

First, common terms in the top sixty terms used by all of the researchers are,"value_Kachi"," water", "groundwater", "problem", "person/people", "climate change", "flow", "change", "stock", "water circulation", "culture/cultural", "circulation" and "image", which are shown in green boxes in Table 21.4. These terms are keywords mainly for explicating the frameworks and mechanisms in the background of the research issue. Second, common terms in the top sixty terms used by three researchers (shown in blue boxes in Table 21.4) are "meaning", "circulation velocity", "use", "quantity", "management", "local area", "impact", "dam", "upstream", "part", "word", "spring (water from a spring)_Yusui", "downstream", "difference", "multifaceted value", "resource", and "research". These are basically the keywords following the first keyword group in terms of overall use, but "circulation velocity" in particular is the key operational concept proposed in the process of the discussion in this experiment. Third, common terms in the top sixty terms used by two researchers are shown in white boxes in Table 21.4. These terms represent what is shared and discussed mainly by the two researchers. Finally, terms in the top sixty terms used by a single researcher are shown in red boxes in Table 21.4. These terms represent the unique aspects of each researcher in this discussion group and context.

In this way, the target terms are categorized into several kinds of keywords, including terms shared by all of the researchers, terms shared by pairs of researchers, and terms unique to a particular researcher.

21.4.2 Communication Process Focusing on Keyword Usage

We consider the thirty most frequently used terms (accounts for about the top 5%) and discuss the communication process focusing on the usage of these keywords. Table 21.5 shows the frequency distribution of the top thirty terms[1] according to discussion issue.

[1] While the top thirty terms were intended to be selected for consideration, thirty one terms are listed in Table 21.5 because the last three terms are with the same usage frequency.

Table 21.5 Frequency distribution of the top thirty terms according to discussion issue

term \ issue number	1	2	3	4	5	6	7	8	9	10	11	12	13	14	15
value_Kachi						1	12		2	28	2	7	4	2	3
water		8	8		2	5	7	16	1	12	7	5	1	2	6
groundwater		1	13			2	3	16	1	4	10	7	1	10	6
person people	5	1	2	3	12	14	8	4		18		1		1	
meaning	4	1		2	1		3			5	1	4	2	4	2
climate change															
circulation		2	1				2		1	3	3	1			
stock															
flow															
attribute															
water circulation			1							1		1		3	
hot spring			11				13			1			3	2	
image				1	1							1		6	1
change				1	6							1			
dam		1						6				2		4	
discussion													1	2	1
watershed / basin									1						
problem	1			3	1	3	3	1			1		5		5
circulation velocity											4	15		5	
Echi River								7				4		8	1
resource				8	1	1	3		1	3			1	2	1
local area	5			3	1	3	1	2					1		
use				4			1	1		1	1	2	1	10	1
impact		1	1	3									1	2	2
culture cultural							2			6		1	5		2
quantity		1	1					1			9				
word	2									2					
multifaceted value															
upstream										13					
research			6	3		1					1	1	2		
river		1	5					2		2	2	1		3	1

(continued)

Table 21.5 (continued)

16	17	18	19	20	21	22	23	24	25	26	27	28	29	30	31	sum
10	8	1	3	1	13	2	6	8	2	23	31	22	23	10	38	**262**
3	2	1	2	1	15	3	1	15	11	15	23	12	16	11	30	**241**
	5			2	3	3			4	3	8	3	6	9	4	**124**
1	1	3	1		2			4	6	7	14	6	6	17	13	**150**
	7	2	3		5	2	1	2		3	3	5	5	3	11	**81**
													12	1	41	**54**
			4		6		2	2		1	3	1	3	1	17	**53**
	9		5	1	34								1		3	**53**
	8		7	1	26										4	**46**
1					2			19		1	4	17		2		**46**
		2	1							1	1	1	7		25	**44**
	4				1	1		2		1	2			3		**44**
	4		2		11	1	1	3		1	1		4		5	**43**
					5								21	1	6	**41**
	8					1			7	1	2	1	1	6	1	**41**
		4	15	2	2		5			1	1	1	1	1	3	**40**
										4	1	1	17	1	15	**40**
	1		1					1	1	1	2	2	3	4		**39**
1			1	1		1				1	2		2		3	**36**
	2								2		4	3			5	**36**
			1		7					2	2	1				**34**
				1						1	2		6		7	**33**
1	1		2		1									1	4	**32**
			3								1	1	6		11	**32**
					6			3		2	1	1	1		2	**32**
					10	1			2		2			1	3	**31**
	3		4		5						2	1			9	**28**
															28	**28**
								2	1	1	3		1	3	3	**27**
			2							4	2		2		3	**27**
	5					1						1			3	**27**

(1) Major transforming process of discussion issues focusing on keywords

Next, we consider the entire transformation process of the discussion issues by examining the change in number in terms used in particular trends of the discussion contents.

First, we focus on "water", "circulation", "culture/cultural", and "river" as keywords used through the entire process of this discussion experiment. These keywords form the basis of this discussion. As characteristic trends in the use of "circulation", the numbers of times these terms are used in the process increases from the starting point to the ending point for organizing the research issue in the second part of the discussion experiment. "River" is used in the first part of the discussion experiment, and both "culture/cultural" and "river" are characteristic in terms of the keywords used throughout the entire process, although these terms are used not so many times.

Second, we focus on "stock" and "flow" as keywords used only in the middle of this discussion. "Stock" and "flow" are used in the process from the end of the first discussion to the start of the second discussion. Taking the fixed research issue and its basic direction as a product into account, these keywords appear to have been fundamental concepts to understanding the mechanism of this research issue.

Third, we focus on "groundwater", "value_Kachi", "circulation velocity", "attribute", "climate change", "water circulation", "watershed/basin", and "multifaceted value" as the keywords used from mid-flow in this discussion. In issues 6 through 15, we found the perspective of "groundwater" became firmly fixed among the researchers. Moreover, in issues 7 through 10, we found the perspective of "value_Kachi" began to become firmly fixed among the researchers. In issues 10 through 12, i.e., starting from the point at which the perspective of "value_Kachi" became firmly fixed, "circulation velocity" began to be used, which is the key concept proposed in this research planning. As another key concept, "attribute", began to be used from the latter half of the discussion. "Climate change", "water circulation", "watershed/basin", and "multifaceted value" are used extensively as keywords for organizing the research issue in the concluding issue stage of the discussion.

(2) Analysis of issues for which particular keywords are used most frequently

The issues for which particular terms are used most frequently in each issue number are highlighted in green in Table 21.5. Based on the particular terms in the selected issues, we focus on and discuss the following terms in the following issues, of which numbers are shown in red boxes.

value_Kachi (7,10,16,23,26,27,28,29)
In issue 7, this keyword is used in the discussion of value consciousness of water, including groundwater and hot springs. However, after ID 4 stated that a difference in circulation velocity can generate a difference in water in issue 10, this keyword is used mainly in this context. This situation continues to the last issue.

circulation velocity (12)
Understanding was shared through discussions among researchers on the operation of water circulating velocity, the method of operation, and the value provided by water with some circulation velocity.

use (14)
A concrete method by which to use water with some circulation velocity, which ranges from surface water to deeper-lying groundwater, is shown in the discussion under issue 14.

flow (17,21)
In issue 17, ID 1 stated that it is necessary to use the concepts of flow and stock in the discussion, but the difference in these concepts between resource economics and groundwater hydrology was discussed among the researchers. Afterwards, in issue 21, the concepts of flow and stock are shared through commentary according to resource economics by ID 3.

attribute (24)
Water flows into the sea with not only attributes from nature, such as rain, but also cultural attributes. The idea that the values of water are represented by finding such attributes of water is shared through the discussion among the researchers.

climate change (31)
This keyword is used to indicate that the discussion of water circulation on a local scale can be connected to the discussion of climate change in the process of the final setting of the research issue.

In this way, we found that the basic framework for a research issue was set in the stage in which "value_Kachi" came to be used in relation to "circulation velocity" by means of text analysis and an examination of the discussion record. In addition, we also found that when concepts are proposed and shared, the terms related to the concepts are used many times in one issue, and the contents of this issue form the basis for the subsequent discussion.

21.5 Assessment of the Collaboration Process by Means of Ontology Engineering

How are the terms extracted in Sect. 21.4 linked semantically? We visualize such semantic linkages based on ontology engineering. We first describe the relationships between the fixed research issue and its basic direction and the researchers' domains based on ontology engineering. Second, we examine the difference between the actual situations and semantic relationships by comparing the analytical results based on text mining with the analytical results based on an ontology engineering approach as an assessment process.

21.5.1 Relationship between Researchers' Domains and the Fixed Research Issue and its Basic Direction

(1) Generating an ontology-based map reflected by individual researchers' interests and corresponding to the discussion context

We generate the conceptual map from the ontology used for the nexus system in the previous chapter by means of the Conceptual Map Creation Tool (map tool). The

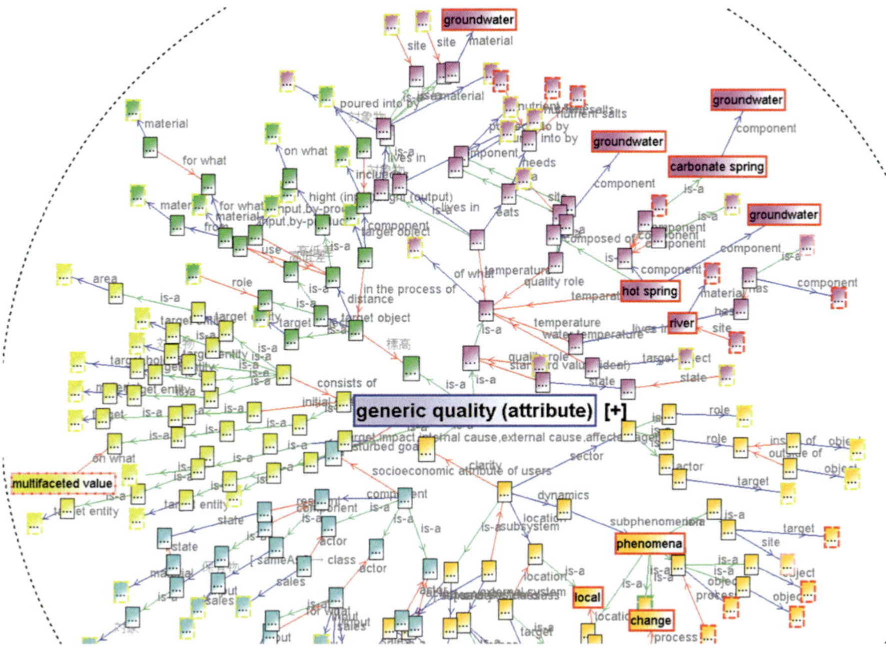

Fig. 21.2 Case of ID 4: Causal chain from an attribute to the terms related to the fixed research issue and its basic direction

words shown in this map represents concepts defined by the approach of ontology engineering. In order to generate the conceptual map, we set concepts represented by the terms that are used most frequently, and therefore characterizes each of the researchers, as a starting point. The terms set as concepts characterizing the researchers are "forest_Mori" (ID 1), "rice field" (ID 2)[2], "measurement_Keisoku" (ID 3), and "attribute" (ID 4). From these starting points, we explore the causal linkages to the concepts represented by the terms of "water", "climate change", "circulation", "water circulation", "watershed/basin", and "value_Kachi".

In this article, we target only the paths that include the top sixty terms of each researcher as concept nodes in the generated paths because these are the causal chains reflected by individual researchers' interests that correspond to the discussion context. This tool can show only the concept nodes of the top sixty terms of each researcher by changing the middle nodes other than the nodes of the top sixty terms of each researcher to empty node boxes. Using this function, we can determine the target paths. As an example, a part of the map for ID 4 is presented in Fig. 21.2. In Fig. 21.2 and 21.3 the colors of the node boxes correspond to the colors of the node boxes linking the concepts as a starting point directly. In addition, the

[2] In Table 21.4, we can find "land improvement district / improvement district" as a top term among the terms used only by ID 2, but this term corresponds not to a term showing a general concept but to a term for a concrete instance. Therefore, we select "rice field" in this case.

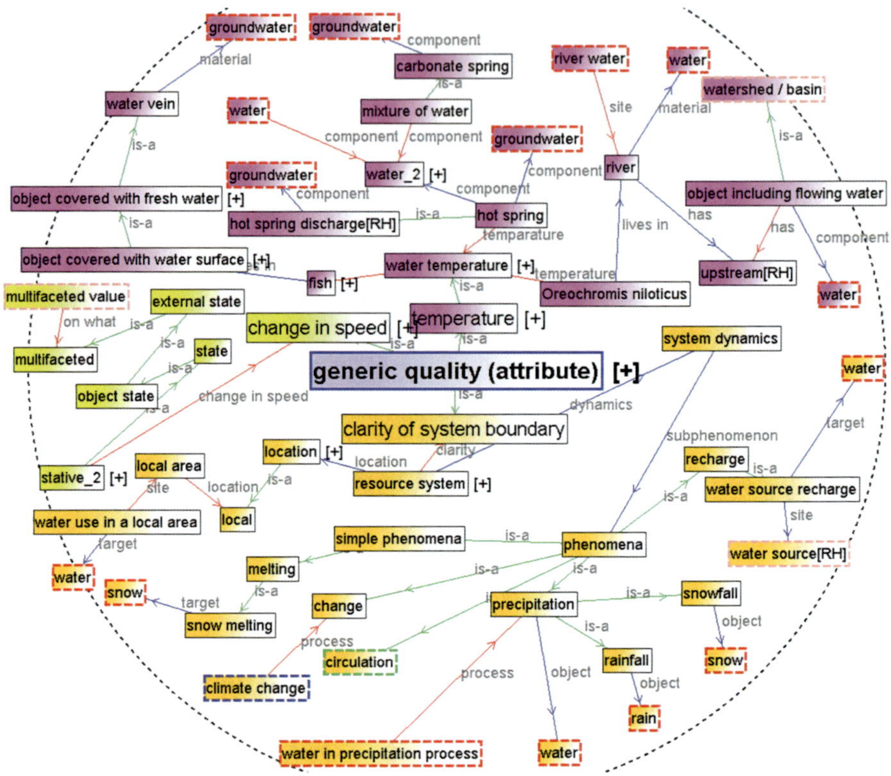

Fig. 21.3 Map showing all concepts in the paths through the concepts of the top sixty terms of each researcher. Case of ID 4: Causal chain from an attribute to the terms related to the fixed research issue and its basic direction

boxes consisting of dotted lines indicate concept nodes as ending points of exploration, while the solid boxes indicate the concept nodes of the top sixty terms of each researcher. We generated ontology-based maps for the other researchers in the same manner. The results and a discussion are presented in the following.

Table 21.6 lists the words through which the generated causal chains pass. As shown in the table, the top sixty terms used by ID 2 exhibited the greatest variety, whereas those used by ID 3 exhibited the least variety. The numbers of words used by ID 1 and ID 2 were small, but the results indicate that ID 1 and ID 2 provided many terms corresponding to the discussion context. On the other hand, ID 3 used only a few terms. This is because the starting point is "measurement_Keisoku", which is the concept represented by the term of general action, and several of the terms that are used are related to method and methodology. For this reason, tracing through concrete target entities is considered to be difficult. The map of ID 4 was also difficult to generate through the concepts represented by the terms of more concrete entities because the starting point, i.e., "attribute", is general and abstract.

Table 21.6 Words through which the generated causal chains pass (words in green boxes represent concepts shared by multiple researchers)

ID 1	ID 2	ID 3	ID 4
groundwater	groundwater	groundwater	groundwater
change	change		change
local area	local area		
resource use	resource use		
power generation	power generation		
	culture		culture
	river		river
	multifaceted value		multifaceted value
		hot spring	hot spring
resource	use	water resource	carbonate spring
national government	management	land	phenomena
society	problem		local
money	state		
snow melting	agriculture		
human	mountain		
fishery	lake		
piped water	soil		
amount of rainfall	caldera lake		
fishery	multifaceted		
scientist	watershed society		
basin commission	cultural value		
national forest	holding pond		
forestry	rain		
	nutrient salts		

These results indicate that, when numerous concepts represented by terms of a general action or property are shown in the map of a researcher, as in the cases of ID 3 and ID 4, they must be combined with concrete concepts as a starting point, as in the cases of ID 1 and ID 2.

In addition, the terms in green boxes represent the concepts shared by multiple researchers. This result represents the concepts to be shared from the perspective of semantic linkage, regardless of their frequencies by individual researchers.

(2) Domain analysis based on the paths through the concept nodes of the top sixty terms of each researcher

Next, we examine concepts represented by terms other than those of the top sixty terms of each researcher on the paths through the concept nodes of the top sixty terms of each researcher. Through this examination, we show the domains comprising the concepts used by each researcher, as well as the domains comprising the concepts linking concepts used by each researcher.

We show an example of a part of the map showing all of the concepts on the paths through the top sixty terms of each researcher in Fig. 21.3. This is the case of ID 4. Such maps were also generated for the other researchers in the same manner. We show the domains of the concepts on the paths through the top sixty terms of each researcher for individual researchers as follows.

ID 1
Concepts including excretion products of forests, solution by combination, industry (including agriculture and fisheries), buy, renewable, business, water use in human life, and natural and artificial are found as domains.

ID 2
Concepts including human, objects humans use (including forests, hot springs, and accommodations), human action and space for the action, systems and institutions, plants, sea, ecosystem, and numerous types of concepts related to nature and artifacts related to agriculture (including the relationship between an actor and another industry and excretion products) are found as domains.

ID 3
Concepts related to energy production, including power generation, which are ranked in the top sixty terms for ID 1 and ID 2 are found as domains. These concepts also include waste and emissions, mixture of water, industry (including agriculture), and coastal zone.

ID 4
Concepts including water temperature, fish, mixture of water, resource system, and particular phenomena including precipitation, melting, and recharge are found as domains. In addition, the generated map indicates that ID 4 has the perspective of a system boundary as the background where such phenomena and areas are shown.

21.5.2 Discussion

How are the shared concepts among researchers dealt with in the collaboration process? In this subsection, we discuss the semantic linkages shown in the previous subsection by comparing the communication process presented in Sect. 21.4. First, we compare Table 21.4 or Table 21.5, depending on the actual discussion, with Table 21.6, which is obtained as a result of semantically generated linkages.

The common keyword used by all of the researchers in both Tables 21.4 and 21.6 is "groundwater". On the other hand, "change" and "culture" are used by all of the researchers in Table 21.4, but "change" and "culture" are used by three and two researchers, respectively, in Table 21.6. The common keywords used by two researchers in both Tables 21.4 and 21.6 are "resource use" (ID 1, ID 2), "power generation" (ID 1, ID 2), and "river" and "hot springs" (ID 3, ID 4). On the other hand, while "local area" (ID 1, ID 2, ID 3) and "multifaceted value" (ID 2, ID 3, ID 4) are used by three researchers in Table 21.4, these terms are used by only two researchers in Table 21.6.

Based on the transition of the discussion shown in Table 21.5 and the list of Table 21.6, we can confirm that groundwater is a focal point of discussion both in a realistic sense and conceptually. Moreover, "change", "culture", "local area", "river", and "hot springs" are also used throughout the entire process of the discussion shown in Table 21.5. This result indicates that "change" and "culture" also form the conceptual basis for issue setting in terms of a quantitative difference in circulation velocity being able to generate a qualitative difference in value. On the other hand, this result also shows that "local area", "river", and "hot springs" represent explicitly shared spatial scale among researchers of this research planning. In addition, multifaceted value is an important concept for issue setting in terms of generating semantic linkage because multifaceted value is used extensively in the final issue of the actual discussion. In other words, the researchers discovered the importance of the concept of multifaceted value in the final stage of this discussion. Although" resource use" and "power generation" do not rank in the top thirty, we can confirm that these terms are important concepts for two researchers, who share these terms from the aspect of both actual discussion as well as semantic linkages.

As a second step, we compare the paths through the concept nodes of the top sixty terms of each researcher with the actual discussion. In the case of ID 2, we found concepts used by the other researchers in the actual discussion despite their infrequent use by ID 2. This indicates that ID 2 supported the sharing of concepts. In addition, in the case of ID 1, ID 2, and ID 3, we can find the concepts related to excretion and waste. Taking into account the actual discussion in which such a topic is not necessarily the focus, we can understand this topic as being still open to further discussion in the experiment group this time. On the other hand, in the case of ID 4, the target paths include the concept nodes on target entities and phenomena for several aspects related to water circulation, for example, river, water vein, hot spring, precipitation, snow melting and recharge. This result verifies that ID 4 provided the framework of the set issue.

21.6 Conclusion

In the present article, we discussed the assessment of the collaboration process in interdisciplinary research on the water-energy-food nexus by combining a text mining approach and an ontology engineering approach.

First, we overviewed a collaboration experiment designed for research development. Second, we identified researchers' concerns in the discussion context of this experiment and represented the communication process focusing on keyword usage.

Third, we showed the relationships between the researchers' domains and the fixed research issue and its basic direction by means of ontology engineering. Finally, we discussed how the shared concepts among researchers were dealt with in the collaboration process by combining the represented communication process as an assessment of collaboration.

In the future, we intend to enhance the assessment capacity by improving the ontology. Although, in the present article we constructed ad-hoc boundaries between issues after reading the discussion records, in the future, we intend to increase reproducibility by extracting points for issues by means of the topic model, which is a machine learning approach, and by creating boundaries between issues based on the extracted points for issues.

Acknowledgments The present research was supported by the Research Institute for Humanity and Nature (Project No. 14200097) and by KAKENHI (Grants 25340142 and 25280081). The authors would like to thank Dr. Kouji Kozaki, associate professor of Osaka University, for his technical support regarding the ontology-based mapping tool. We gratefully acknowledge all of the researchers who assisted with the workshop experiments.

References

Committee on Facilitating Interdisciplinary Research.; Committee on Science, Engineering, and Public Policy; Institute of Medicine.; Policy and Global Affairs.; National Academy of Sciences.; National Academy of Engineering (2004) Facilitating interdisciplinary research. The National Academies Press, Washington, DC.

Grin J, Rotmans J (eds) (2010) Transition to sustainable development—new directions in the study of long term transformative change. Routledge, New York

Holsapple CW, Joshi KD (2002) A collaborative approach to ontology design. Commun ACM 45(2):42–47

Kumazawa T, Hara K, Endo A, Taniguchi M (2017) Supporting collaboration in interdisciplinary research of water-energy-food nexus by means of ontology engineering. Journal of Hydrology: Regional Studies 11:31–43

Loorbach D (2007) Transition management: new mode of governance for sustainable development. International Books, Utrecht

Miller TR (2013) Constructing sustainability science: emerging perspectives and research trajectories. Sustain Sci 8(2):279–293

Repko AF, Szostak R (2016) Interdisciplinary research: process and theory. SAGE Publications, London

Stokols D, Hall KL, Moser RP, Feng A, Misra S, Taylor BK (2010) Cross-disciplinary team science initiatives: research, training, and translation. In: Frodeman R, Klein JT, Mitcham C, Holbrook JB (eds) The Oxford handbook of interdisciplinary. Oxford University Press, New York, pp 471–493

Chapter 22
Scenario-based Approach to Local Water-energy-food Nexus Issues with Experts and Stakeholders

Kenshi Baba, Naoki Masuhara, and Michinori Kimura

Abstract As the trade-offs within nexus issues of water-energy-food usually come along with scientific disputes, building a consensus among stakeholders with the scientific evidence is required to realize sustainable and resilient societies. In this chapter, we explore the capability of problem-solving for potential disputes that may occur among nexus issues at a local level from case studies in Beppu, Japan by using our scenario-based approach customized to integrate expert knowledge and local knowledge. Our scenario-based approach contains not only the elements of the existing scenario planning but also stakeholder analysis and the Delphi method which enables co-design and co-production of science and society. By integrating these methods, we intend to improve scientific evidence-based policy making processes. As we are now developing the Delphi method, we focused the first half of the process of stakeholder analysis and stakeholder meeting.

Keywords Co-design · Co-production · Local knowledge · Expert knowledge · Stakeholder analysis · Consensus building

---

K. Baba (✉)
Tokyo City University, Yokohama, Kanagawa, Japan
e-mail: kbaba@tcu.ac.jp

N. Masuhara
Research Institute of Humanity and Nature, Kyoto, Kyoto, Japan
e-mail: masuhara@chikyu.ac.jp

M. Kimura
Lake Biwa Environmental Research Institute, Otsu, Shiga, Japan
e-mail: kimura-m@lberi.jp

© Springer Nature Singapore Pte Ltd. 2018
A. Endo, T. Oh (eds.), *The Water-Energy-Food Nexus*, Global Environmental Studies, https://doi.org/10.1007/978-981-10-7383-0_22

22.1 Introduction

Recently, the expectation for the role of science in the policy process has been increasing. In the USA, the President's memorandum on scientific integrity for the heads of executive departments and agencies was issued in 2009, then Department of the Interior, NOAA (National Oceanic and Atmospheric Administration), and other agencies formulated some principles on scientific integrity. The UK and other European countries also formulated principles on scientific integrity. In Japan, after the Great East Japan Earthquake, JST (Japan Science and Technology Agency) formulated 10 principles on the role and responsibility of science and government in the policy process (Center for Research and Development Strategy, Japan Science and Technology Agency 2012). In this way, the importance of scientific evidence-based policymaking is growing.

Particularly in the policy process under scientific disputes, stakeholder involvement in the early stage of the process such as agenda setting is quite important. Recently the Future Earth Research Program, ICSU (The International Council for Science) proposed a framework of co-design and co-production (International Council for Science 2013). However, similar indications have been made before and repeated many times. This implies that difficulties in doing so exist.

As trade-offs within nexus issues of water-energy-food usually come along with scientific disputes, building a consensus among stakeholders using scientific evidence is required to realize sustainable and resilient societies. To this end, co-design and co-production of science and society is required and formulating a system of social decision making at not only a local scale but also multiple scales such as international and region are indispensable.

Scientific evidence (expert knowledge) such as for future projections gives quantitative forecasting results of continuous future from the present but it contains high uncertainty. Meanwhile the scenario planning envisions the qualitative discontinuous future from the present and thus gives policymakers insights of the structure of the future including uncertainties, sense of change and adaptability. Integrating the top-down sharing of scientific evidence (expert knowledge) of scientists and policymakers and the bottom-up sharing of local knowledge of stakeholders and the general public are effective to reduce the latent framing gap that can occur between relevant actors.

In this chapter, we explore the capability of problem-solving for potential disputes that may occur among nexus issues at a local level from case studies in Beppu, Japan by using our scenario-based approach customized to integrate expert knowledge and local knowledge (Fig. 22.1). Our scenario-based approach contains not only the elements of the existing scenario planning but also stakeholder analysis and the Delphi method which enables co-design and co-production of science and society. By integrating these methods, we intend to improve scientific evidence-based policy making processes.

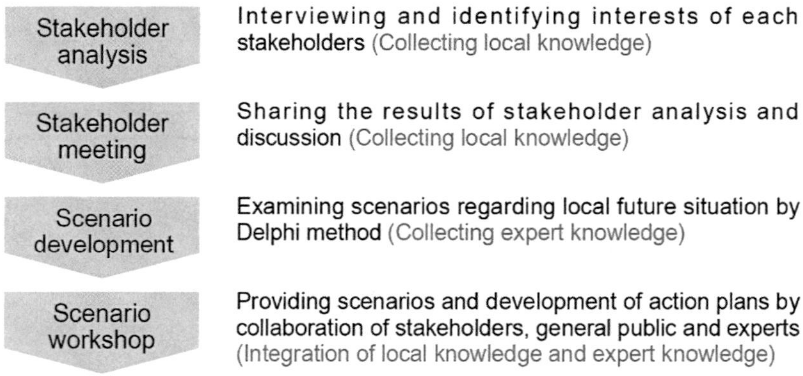

Fig. 22.1 Proposed scenario-based approach

22.2 Methodology

As mentioned above, our scenario-based approach consists of the following four steps; stakeholder analysis, stakeholder meeting, the Delphi method and scenario workshop. As we are now developing the Delphi method, we focused the first half of the process of stakeholder analysis and stakeholder meeting.

22.2.1 Stakeholder Analysis

The stakeholder analysis is founded on practices and research findings concerning negotiation and consensus building. It is implemented to classify, according to type, the stakeholders that should be included in agenda setting and to examine alternatives in the initial stages of policy formulation (Susskind and Thomas-Lamar 1999). It has been pointed out that it is extremely important that the study of policies start from the question of what should be recognized as the problem to be addressed by the policies, and that subsequent policies could vary greatly depending on how the study is designed (Stone 1997). This process is called "agenda setting," identifying recognition of the problem by stakeholders as a necessary condition for opening the "policy window" that makes policy formation possible (Kingdon 1995). This method attempts to make clear the issues in policy formation from a bottom-up approach, sorting out interests based on the point of view of the stakeholders. Developed chiefly at the Massachusetts Institute of Technology, this method has been applied to a variety of subjects worldwide (Baba K. a. 2015; Matsuura 2016).

The steps taken in stakeholder analysis begin with identifying and listing initial stakeholders. In setting up the questions, a common questionnaire is prepared through a literature review of subjects, with the literature including administrative documents related to the relevant policy. Also, initial stakeholder identification is prepared for through means such as these literature reviews and the pre-surveying of personnel responsible in key administrative authorities. After these preparations, each stakeholder is interviewed individually. Following completion of the survey, each stakeholder's interests are examined. This consists of identifying interests, searching for possibilities of mutual benefits, and identifying barriers to reaching consensus. It could be supported through techniques such as text mining as described in Chap. 19, but in this chapter, stakeholders' interests on a number of points are broken down from questionnaire items and collected together manually in matrix form while uniform criteria are employed among the persons conducting the analysis.

Interviews are conducted in semi-structured form. Stakeholders freely describe what comes to mind in connection with each question item provided; the interviewers keep their intervention to a minimum. In addition, stakeholders are added to the initial list using snowball sampling. The numbers of subjects are expanded through asking initial stakeholders (survey subjects) about possible subjects for the next round of interviews, ending when no information on new stakeholders is introduced, even when additional surveys are conducted. Since the stakeholders are classified according to their interests, the sample size N varies widely among individual categories, and it is conceivable that N = 1 in some cases. In this way, interests, attitudes and local knowledge of the stakeholders are collected during this step.

22.2.2 Stakeholder Meetings

To share the results, we held a stakeholder meeting in which the stakeholders who were the subjects of the survey and other stakeholders who are devotees of hot springs, the city councilor, as well as experts (in hot spring science and hot spring culture). The agenda included a presentation of the stakeholder analysis results by the authors, the sharing of expert knowledge by the experts, and then a workshop to dialogue with participants. The participants indicated diverse interests, attitude and local knowledge again and the results of this step were summarized by the authors as influence diagrams. This stage is the first opportunity for integration of expert knowledge and awareness of the issues based on local knowledge.

22.2.3 Scenario Development

Next, we prepared for scenario development. The authors do not mean scenario as quantitative used for quantitative models, but rather quantitative or narrative scenarios. One of the reasons why we develop such scenarios is to make scientific evidence easier to understand for local stakeholders and to give policymakers insights of the structure of the future including uncertainties, sense of change and adaptability. Using the results of stakeholder analysis, role-playing simulations have also had similar effect (Rumore et al. 2016) though from past experience it is difficult to collect the participants to conduct these simulations in Japan (Baba K. a. 2008). Scenario development and scenario workshop are more feasible because they have been conducted several times for wide-ranging subjects in Japan.

To initiate the step of scenario making, we hold an expert meeting consisted of hot spring science, geothermal engineering, hydrology, marine ecology, humanity and social science. The goal was to produce an initial story by adding expert knowledge to the influence diagram which described the structure of interests, attitudes and local knowledge of the stakeholders. With the results of the expert meeting, the authors produced 56 stories of hot spring and 10 stories of geothermal power based on natural environmental change and 8 stories of social change. Then, the stories were transferred to questionnaire for the Delphi method to get an evaluation from the experts in two rounds in terms of certainty and seriousness. After this survey, ultimately, the stories to constitute a scenario are selected and revised, and the authors then begin the second round of the Delphi method. This stage provides the second opportunity for integration of expert knowledge and local knowledge by verifying logical validity and probability of the collected expert knowledge, and produces narrative scenarios that can be easily understood by the stakeholders.

22.2.4 Scenario Workshop

Finally, held a scenario workshop to share the completed scenarios, get feedback and design action plans with the stakeholders, the general public, scientists and policymakers. This stage provides the final opportunity for verifying validity of the integration of expert knowledge and local knowledge, and for examining appropriate responses ahead of time for the future (we set a target of 2040 in this study). By doing so, the authors intended to find the likely “pathway of change” of natural and social environment does not simply follow a continuous pattern from the present but rather occurs discontinuously, allowing them to make a well-informed decision for the future.

22.3 The Study Site

22.3.1 Outline of Beppu City

Beppu city in Oita Prefecture has a population of 120,947 people making up 61,827 households. After reaching a peak in 1981, the city’s population has been on the decline. Meanwhile, the number of households has seen a marginal increase since the population began declining in 1981. There has been a progression of a declining birthrate and an aging population. The city is comprised of a gradual alluvial fan reaching from the mountains in the west to Beppu Bay in the east. Hot springs are located in the western mountains (Mt. Tsurumi) as well as the urban areas, and the temperature of these springs rises as one moves closer to the foothills of the mountains (Beppu city 2013, 2014a, b; 2012).

Oita Prefecture leads Japan in terms of the number of hot springs and the amount of hot water they produce. Within Oita Prefecture, Beppu produces more than any other location, as much as 87,032 L per minute from 2293 hot springs. The hot springs contain more than 10 different mineral substances, and the Beppu hot springs are unsurpassed in their variety compared to the rest of the country. With such high-quality and abundant geothermal resources, Beppu was developed into a hot spring town long ago. Beginning in the Meiji era, as transportation routes were expanded and techniques for drilling out the springs improved, Beppu became one of the most prominent hot spring (*Onsen*) regions in Japan. The *Yukemuri* (steam) landscape created by the hot springs has been designated as the "Important Cultural Landscape" by the Agency for Cultural Affairs (Beppu city 2014a; Oita prefecture 2014a; 2010).

Traditionally, homes in Beppu were not equipped with individual baths, as residents would take baths with their neighbors in communal bathhouses. Thus, the hot springs have also functioned as places of social interaction. People also utilized the steam from the hot springs to steam their food, using a technique called *jigoku mushi* ("hell steaming"). The springs have also been used in medical therapies, flower cultivation, and aquaculture. One of the distinguishing features of Beppu is the amount of research and development on hot springs conducted there. There are a number of research institutions within the city that work with the hot springs including national universities and prefectural government's institute. Through research efforts by Kyoto University, over 90 years of scientific knowledge has been accumulated on geothermal resources, and the immense amount of data collected on this unseen resource is considered rare even at the national level (Beppu city 2012).

22.3.2 Present Situation of Small-Scale Geothermal Power Development

Recently, some developers who are interested in the plentiful hot spring resources in Beppu have been exploring opportunities to introduce small-scale geothermal power. The unit dries off a medium such as pentane which has a low boiling point, using relatively low-temperature exhaust heat and hot water from hot springs, and drives a turbine with the steam to generate power. At present, several units are operating in Beppu and the number of units has been increasing as described in Chapter 15.4. Each unit has a capacity of 100–200 kW. A business model of one site, "*Goto-en*" geothermal power station whose capacity is approximately 100 kW is as follows (Fig. 22.2); (i) Four local and external small businesses have established SPCs (special-purpose company) who took a loan from the local *Shinkin* bank

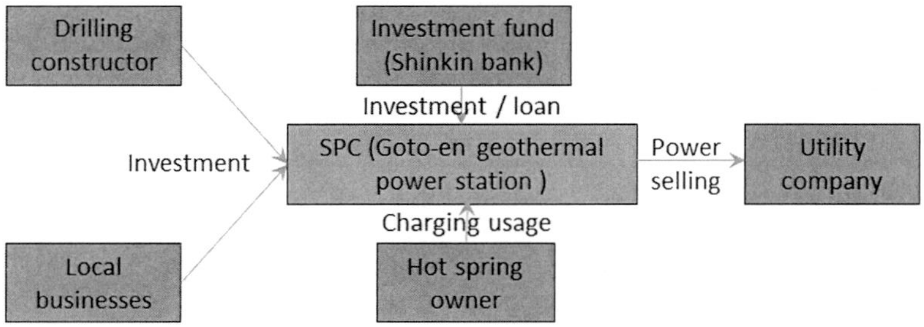

Fig. 22.2 Overview of business model of small-scale geothermal power in Beppu

(Japanese credit union for small businesses) and obtained a subsidy from the prefecture government for construction. (ii) The SPC continue to sell electricity to the utility company at the rate of 40 yen/kWh for 15 years, and the hot spring resource owner is financially rewarded for the term.

In terms of regulation, permission from the prefectural government is required in order to utilize the hot springs. In accordance with the Hot Spring Law and its enforcement regulations, Oita prefecture has provided the related prefectural ordinance and its enforcement regulations. The main applications required in relation to the use of the hot springs are as follows: (1) Application for permission to drill hot springs which are made for new drilling or for alternative drilling depending on the age of the drill site. (2) Application for permission to expand the drill sites or to install powered machinery which is to expand the bore diameter or the depth of the spring water supply pipelines, or to install powered units (pumps, etc.). The subcommittee of hot spring of the Oita prefecture environmental committee grants these permissions. With a part of the city designated as a special protection area, and most of the city designated as a protection area, the drilling of hot springs is subject to strong restrictions. New drilling is not permitted within "special protection areas" (but alternative drilling is allowed), and in "protection areas", drilling is not permitted in points 100 or 150 m from the existing well points. In other words, drilling is permitted in points 100 or 150 m away from the existing well points in "protection areas" (Oita prefecture 2014a; 2010; Beppu city 2012).

From a policy perspective, Oita prefecture aims to keep the highest rate of self-sufficiency in renewable energy in the country, and revised "Oita New Energy Vision" in 2011 and introduced even higher targets in 2014 (Oita prefecture 2014b). The prefecture has also established a subsidy to encourage the installation of new facilities as well as a fund. Meanwhile, Beppu city has proceeded with its own studies related to the introduction of renewable energy sources. In June 2014, the city released the "Beppu New Energy Feasibility Study Report" (Beppu city 2014a), and formulated "Beppu New Energy Vision" in 2015 (Beppu city 2015b). On the other hand, the city issued "Guideline of Prior Procedures Concerning the Introduction of New Energy," which was a precautionary measure for potential disputes caused by siting of renewable energy facilities (Beppu city 2015a). These two-pronged policy approach of promoting the introduction of renewable energy and regulating its development is now being brought into the municipality in relation to not only wind power generation, but also photovoltaic generation and geothermal generation.

22.4 The Results of Stakeholder Analysis in Beppu

22.4.1 Survey Outlines

The stakeholder analysis in Beppu was concerned with small-scale geothermal power, which has been operating and planned in the near future in Beppu. They are comparatively small-scale technologies that do not require new drilling and can generate electricity using existing wellheads in principle. They are thus expected to have a smaller environmental impact while they require a large volume of cooling water and a large amount of hot spring water for generation.

We conducted a stakeholder analysis during July and August 2014. We made interviews individually with the stakeholders of 38 organizations listed in Table 22.1. The main questions were as follows; (1) present situation of use of hot-spring, (2) interests on use of hot-spring and geothermal power, (3) future actions to use of hot-spring and geothermal power, (4) new stakeholders whose involvement will be needed (successive "snowball" sampling).

22.4.2 Analysis Results

Table 22.2 summarizes each of the stakeholder interests. The table indicates whether a particular stakeholder, shown on the horizontal line, holds an interest with regard to a particular issue, shown on the vertical line. In addition, it is somewhat relative whether an item is marked or unmarked with a circle, and thus an unmarked item does not signify that there was absolutely no interest in that particular issue.

Table 22.1 Interview subjects of stakeholders

Category of organizations	N of organizations	Category of organizations	N of organizations
City government	7	Commerce and industry	3
City council member	1	Sightseeing	1
Prefecture government	1	Consultant	1
Agriculture	6	Education	1
Fishery	1	Culture	2
Aquafarming	1	Inhabitants	2
Forestry	2	Civic organization	5
Food	4	Total	38

Table 22.2 List of stakeholder concerns in Beppu

Issues / Stakeholders	Concerns for Beppu City			Perceptions of hot springs/underground resources				Perceptions of small-scale geothermal power generation					Reconciliation between Stakeholders			Total
	1. Concerns for hot springs	2. Decline of tourism	3. Reduction of successors of hot spring culture	4. Usage unrelated to geothermal power (Economic values)	5. Usage unrelated to geothermal power (Non-economic values)	6. Usage for Geothermal power (Economic values)	7. Usage for geothermal power (Non-economic values)	8. Interest in small-scale geothermal power generation	9. Interest in gaining knowledge about small-scale geothermal power generation	10. Causal relationship between geothermal power and hot springs	11. Direct /Indirect profits	12. Related expense	13. Consideration of local resident perceptions	14. Concerns over speculators	15. Interest toward government	
Prefecture government				○		○	○	○	○	○			○	○	○	9
	○			○	○			○		○			○	○	○	8
																0
	○			○	○			○		○			○			6
	○			○	○			○		○			○	○	○	8
	○			○	○					○					○	5
City government	○					○		○	○	○			○	○	○	8
					○			○		○			○		○	5
	○				○			○		○					○	5
	○	○		○	○			○		○			○	○	○	9
Guesthouse/ Hotels		○	○	○	○			○	○	○		○	○		○	10
		○	○	○	○	○		○	○	○			○			9
	○		○	○				○		○			○		○	7
	○		○	○	○			○		○			○	○	○	9
		○		○		○		○		○			○		○	7
	○	○	○	○				○	○	○		○			○	9
Tourism Operator		○		○	○			○		○		○	○		○	8
Power producers/ Consultant	○			○		○		○	○	○	○		○	○		9
						○	○	○	○	○	○		○		○	8
			○	○		○		○	○	○	○	○	○		○	10
	○	○		○		○		○	○	○	○		○		○	10
						○	○	○		○			○		○	6
	○					○		○	○	○		○	○	○	○	9
Hot spring owners		○		○	○	○		○	○		○		○	○	○	10
	○	○	○	○	○		○	○	○	○			○	○	○	12
Hot spring operator	○			○	○			○		○		○	○		○	8
Local drilling operators		○		○		○		○		○					○	6
				○		○		○						○		4
Machinery manufacturer						○		○	○	○	○					5
Regional commercial cooperative	○			○	○	○		○		○		○			○	8
Fund	○	○	○	○	○	○	○	○	○	○	○	○	○	○	○	15
NPO	○		○		○			○		○			○		○	7
			○		○		○	○	○	○		○	○	○		9
Total	18	11	10	23	18	16	6	31	15	30	7	9	25	13	26	

(1) Overall issues structure

Looking at Table 22.2, the issues most stakeholders raised were "8. Interest in small-scale geothermal power generation", "10. Causal relationship between geothermal power and hot springs". "15. Interest toward government", "13. Consideration of local resident perceptions" and "4. Usage unrelated to geothermal power (Economic values)" are next to these. In addition, "1. Concerns for hot springs" and "5. Usage unrelated to geothermal power (Non-economic values)" were raised by relatively many stakeholders.

Basically, the concerns to small-scale geothermal power generation are increasing but there is no supporting evidence of consequences. As the scientific evidence itself is uncertain, it can be called "vulnerable concern". To change this situation, information provision from local governments and establishment of drilling standards are expected as one of the solutions by many stakeholders.

Many stakeholders recognized the geothermal resource as an important tourism resource rather than as a power generation resource which had not yet been a good source of profit. Also, many stakeholders recognized the geothermal resource as a common resource of the local community which led to giving consideration for neighborhoods in the sense that how wellheads are utilized in one's location may have an impact on others' locations.

This recognition resulted in concern that hot spring water, hitherto supplied based on trust between neighbors, may now cease to be supplied depending on the will of new owners. This is the concern that ownership of the hot springs falling into the hands of an outside party as the lack of successors would bring on serious impact on their customs of common resource.

The approaches to local revitalization and prevention of collapse of a local community depend on each stakeholders' query and understanding of knowledge of small-scale geothermal power generation, which cause the difference of views on causal relationship between geothermal power generation and hot springs.

(2) Preventive measures of potential disputes

While the interviews revealed certain differences in opinions, no obvious disputes of interest were observed. Even those stakeholders who expressed misgivings about small-scale geothermal power generation did not oppose its use by other business operators, nor did they take any particular action. Nevertheless, some stakeholders were considered likely to express an objection in the event that any type of adverse impact occurred, or toward any works requiring new drilling. Accordingly, while no clear disputes have occurred at present, this may be a possibility in the future.

In addition, experts indicated that although the present circumstances surrounding the hot springs and the geothermal resource are comparatively calm, protecting the flow of hot spring water at its source continues to be of the utmost importance. If the flow of hot water is jeopardized at its source, there will be a negative impact on the entire area downstream. The prospect of a distinct backlash against the uncontrolled development of the hot springs and geothermal resource in the upstream area indicates a continued risk of disputes of interest.

Accordingly, such risks should be avoided by providing in advance explanations about the possible impacts on the geothermal resource, and by building a consensus around solutions if any impacts did occur. To do so, it is important to integrate local knowledge which stakeholders have and expert knowledge, and stakeholders should be encouraged to participate in discussions aimed toward consensus building, and their level of understanding of the relevant issues should be raised to ensure that those discussions are of a substantial nature.

Stakeholders who have conflicting interests may be difficult to build consensus however they can take collaborative actions by sharing risk perception of depletion of geothermal resource and collapse of a local community that many stakeholders have. To realize it, "joint fact-finding" such as a collaborative monitoring, for example, of changes of the amount of hot spring water as well as the quality and temperature of the water, which a majority of stakeholders expressed strong concerns in the case of new drilling will be one of the effective ways.

22.5 The Stakeholder Meeting in Beppu

As mentioned earlier, we held a stakeholder meeting in which the stakeholders who were the subjects of the survey and other stakeholders who are devotees of hot springs and city councilor and so on as well as experts (respectively in hot spring science and hot spring culture) gathered to share the results of stakeholder analysis.

The meeting was held on August 7th, 2015 in a hot spring resort hotel in Beppu. The amount of the participants was thirty-two. As on the day at the same hotel, a NGO held a training session of their "Mister system of hot spring" just after our meeting, many devotees of hot springs from outside of Beppu also attended our meeting in addition to the stakeholders who were the subjects of the survey. The agenda included a presentation of the stakeholder analysis results by the authors and then a workshop to hold a dialogue among participants.

During the workshop, the participants were grouped into three (each group had about ten members) including the experts, and discussed two topics; i) favorites and challenges of hot spring in Beppu, and ii) how to utilize geothermal resources to realize sustainable Beppu.

The typical opinions for favorites of hot spring in Beppu were as follows; "various options to enjoy because of the plentiful amount of hot spring water as well as the variety of quality of hot spring water and wide range of how to take a bath", "take a bath casually" and "interaction among guests in bath". Also, the typical opinions for challenges of hot spring in Beppu were as follows; "wasteful use of hot spring resources though many people are concerned about the depletion", "fading of communal bathhouses due to difficulties of maintenance and management", "local residents tend to have a lack of understanding toward its own attractions or are overconfident in its own attractions".

Finally, the typical opinions for how to utilize geothermal resources to realize sustainable Beppu were as follows; "strengthening the regulation of utilize geothermal resource", "implementing the monitoring", "developing geothermal energy", "developing human resources for transmit the information for inbound

Fig. 22.3 Some aspects in the stakeholder meeting

tourist business", "unifying the tourist business of hot spring and geothermal power with learning from the examples of success and failure".

The authors summarized these opinions and restructured the elements of events and phenomena, and described them as influence diagram as pictured in the right of Fig. 22.3. Then we prepared for making scenarios with it.

22.6 Conclusion and Further Directions

In this chapter, we introduced our scenario-based approach which contains not only the elements of the existing scenario planning but also stakeholder analysis and Delphi method which enable co-design and co-production of science and society. We clarify any potential disputes that may occur among nexus issues at a local level from case studies in Beppu, Japan and explore the capability of problem-solving method by scenario-based approach. As we are now developing the Delphi method, we focused the first half of the process of stakeholder analysis and stakeholder meeting.

The main results demonstrate as follows. First, most stakeholders are interested in small-scale geothermal power generation basically but are not knowledgeable, and the expert knowledge on geothermal resource is still highly uncertain. Second, most stakeholders assume the geothermal resource as a sightseeing and management resource rather than power generation use, and acknowledge the value of not only economic but also non-economic. Third, explicit disputes are not observed at present because small geothermal power generation have not been utilized everywhere yet. However, the concerns over the depletion of geothermal resource are so serious that disputes might arise in future. Therefore finally, integrating expert knowledge and local knowledge is needed by sharing risk perception of depletion of geothermal resource and by reframing through joint fact-finding. Based on the above, a comprehensive arena of dialogue on geothermal resource is required to examine sustainable use of the resource.

Next step of our work is to prepare scenario design. We have already held an expert meeting consisted of hot spring science, geothermal engineering, hydrology, marine ecology, humanity and social science to produce an initial story by adding expert knowledge to the influence diagram which described the structure of interests, attitudes and local knowledge of the stakeholders. With the results of the expert meeting, the authors produced 74 stories. Then, the stories were transferred to questionnaire for application of the Delphi method to get evaluation of the experts in two rounds in terms of certainty and seriousness. After this survey, ultimately, the stories to constitute a scenario will be selected and revised, and the authors are now on the way to the second round of the Delphi method. Finally, we plan to hold a scenario workshop to share the completed scenarios, get feedbacks for them and to work out some action plans with the stakeholders, the general public, scientists and policymakers. With these steps, the authors intended to find the likely "pathway of change" of natural and social environment that not simply follow a continuous pattern from the present but rather occur discontinuously, in order to make an appropriate decision for the future.

References

Baba K, Matsuura M (2008) Examining Environemtal disputes solutions using negotiation Simulatoin – a case of wind power siting. Journal of Environmental Systems Study 36:149–158

Baba K, Tanaka M (2015) Challenges of implementing climate change adaptation policy for disaster risk reduction - implications from framing gap among stakeholders and the general public. J Disaster Res 10(3):404–419

Beppu city (2012) Cultural landscape, Yukemuri (steam) landscape conservation plan. http://www.city.beppu.oita.jp/education2/yukemuri_keikan/plan.html

Beppu city (2013) Statistics (FY2013). http://www.city.beppu.oita.jp/03gyosei/tokei/h25/01-tochi.html

Beppu city (2014a) Regional New Energy Feasibility Study Report. http://www.city.beppu.oita.jp/03gyosei/kankyou/alternative_energy/torikumi/pdf/hokokusho.pdf

Beppu city (2014b) Population in Beppu. http://www.city.beppu.oita.jp/03gyosei/jinko/index.html

Beppu city (2014c) Overview of Beppu. http://www.city.beppu.oita.jp/03gyosei/general/gaiyou/index.html

Beppu city (2015a) Summary of Prior Procedures Concerning the Introduction of New Energy in the Beppu City Region. http://www.city.beppu.oita.jp/03gyosei/kankyou/alternative_energy/jigyo/index.html

Beppu city (2015b) Beppu New Energy Vision. https://www.city.beppu.oita.jp/pdf/sangyou/environment/alternative_energy/torikumi/vision/new_energy_vision.pdf

Center for Research and Development Strategy, Japan Science and Technology Agency (2012) STRATEGIC PROPOSAL: toward the establishment of principles regarding the roles and responsibilities of science and government in policy making. CRD, JST

International Council for Science (2013) Future earth research for global sustainability draft initial design report. ICS

Kingdon WJ (1995) Agendas, alternatives, and public policies. Addison-Wesley Educational Publishers Inc, Boston

Matsuura M, Baba K (2016) Consensus building for long-term sustainability in the non-north American context: reflecting on a stakeholder process in Japan. Negot Confl Manag Res 9(3):256–268

Oita prefecture (2010) Precaution statement for utilization of hot spring. http://www.pref.oita.jp/site/onsen/onsen-tyuui.html

Oita prefecture (2014a) Hot springs data. http://www.pref.oita.jp/site/onsen/onsen-date.html

Oita prefecture (2014b) Revision of target of renewable energy introduction in Oita new energy vision. http://www.pref.oita.jp/soshiki/14200/sinene-kaitei.html

Rumore D, Schenk T, Susskind LE (2016) Role-play simulations for climate change adaptation education and engagement. Nat Clim Chang 6:745–750

Stone D (1997) Policy paradox: the art of political decision making. W. W. Norton & Company, New York

Susskind LE, Thomas-Larmar J (1999) Conducting a conflict assessment. In: Susskind LE, McKearnan S, Thomas-Larmar J (eds) Consensus building handbook. SAGE Publications Inc, Thousand Oaks, pp 99–136

Index

© Springer Nature Singapore Pte Ltd. 2018

A. Endo, T. Oh (eds.), *The Water-Energy-Food Nexus*, Global Environmental Studies, https://doi.org/10.1007/978-981-10-7383-0

FSC
www.fsc.org
MIX
Papier | Fördert
gute Waldnutzung
FSC® C083411

Kolibri 360 GmbH
Ferdinand-Jühlke-Straße 7
99095 Erfurt, Deutschland
produktsicherheit@kolibri360.de